GIS
for
SUSTAINABLE
DEVELOPMENT

GIS
for
SUSTAINABLE
DEVELOPMENT

edited by
Michele
Campagna

CRC Press
Taylor & Francis Group
Boca Raton London New York

CRC Press is an imprint of the
Taylor & Francis Group, an **informa** business
A TAYLOR & FRANCIS BOOK

Published in 2006 by
CRC Press
Taylor & Francis Group
6000 Broken Sound Parkway NW, Suite 300
Boca Raton, FL 33487-2742

First issued in paperback 2020

ISBN 13: 978-0-367-57804-6 (pbk)
ISBN 13: 978-0-8493-3051-3 (hbk)

**Visit the Taylor & Francis Web site at
http://www.taylorandfrancis.com**

**and the CRC Press Web site at
http://www.crcpress.com**

Library of Congress Cataloging-in-Publication Data

GIS for sustainable development / edited by Michele Campagna.
     p. cm.
  Includes bibliographical references and index.
  ISBN 0-8493-3051-3 (alk. paper)
  1. Sustainable development--Geographic information systems. I. Campagna, Michele.

HC79.E5G54 2005
338.9'27'0285--dc22
                                                       2005042098

Foreword

I was glad, but slightly puzzled, when Michele Campagna asked me to write the foreword for his book about GIS and sustainable development. In my planning experience I always welcomed and appreciated the arrival of the GIS cavalry, both in research and in the professional practice; I even found myself fostering its calling for in many occasions. Especially at the present time of spread diffusion and democratization of the computing power, and geographic data availability and access, planning professionals have the chance to experience new ways of exploiting geographic data management capabilities toward more creative analytical and design forms of planning. However, I am afraid that planning has perhaps more to take from GIScience than it has to give to it. Thus, I was puzzled — what should I have had to say about planning to introduce a book about GIS and sustainable development?

This happened before I read the table of contents first, and then the whole manuscript. Although it is not straightforward to accept a unique definition of planning — and perhaps of sustainable development either — nevertheless, reading this book I enjoyed discovering that it concerns sustainable development and planning as much as GIS. It concerns GIS but offers many useful insights for sustainable development planning practice. Definitely this is a book as much for the GISers as for the planners. I was quite relieved afterwards.

I think that there is not much more to say here about planning, but this book deals with crosscutting planning objectives and the way to tackle them. In the last century or so, planning evolution faced very different paradigms, spanning from the rational to the collaborative approach. In this evolution very different methods and techniques were proposed and applied, sometimes with consensus among practitioners and stakeholders, and success in the outcomes, other times not. It is perhaps now time for the planner to face the challenge to browse in this full box to find the right set of tools which best fit each individual local context, to design creative planning processes able to support democratic and informed decision-making, in this way aiding, as an expert, to foster the dialogue on the nature of the consistency of possible alternative courses of action with economic, social, and environmental concerns. Ample freedom is left to the reader to ethically interpret and address this challenge.

With this book the framework is set by the editor to discuss different calls for action proposed in Agenda 21. However, the focus on Agenda 21 is given instrumentally for the sake of clarity in the discussion, and most of the issues dealt with in the book may be applied to the many national and local programs and actions which, in one way or another, are consistent with a broader sustainable development framework. On the one hand, progress in GIScience is proposed to address specific problems such as socioeconomic and demographic analysis, environmental degradation, health care, or natural risk management. On the other hand, research results and experiences from practice are presented, which can be considered best practices in (geographic) information production, maintenance, analysis, and sharing. Moreover, several case studies are proposed which concern the collaboration of major groups in sustainable development planning and decision-making, such as institutional stakeholders, indigenous people, local communities, and citizens, undertaken in real settings to promote subsidiarity, transparency of administration, and public participation for democratic decision-making. In fact, in addressing many of the Agenda 21 objectives, the work itself conversely provides a contribution, although partially and at a conceptual level, to another specific call, namely capacity-building by carrying knowledge and knowhow. This book puts many problems on the table, illustrating in a sort of undeclared and implicit SWOT analysis, through documented case studies, strengths, weakness, opportunities, and treats of GIS application in the domain of sustainable development. This framework supplies many useful hints for the practitioner approaching the design of informational planning working spaces.

While one might be tempted to pay attention to selected chapters, as they concern a number of different particular GIS methods and applications addressing specific problems, I would suggest the reader to span throughout the whole book, as most of the chapters deal with the same overarching sustainable development issues with regard to the support GIS may offer for their solution, although from very different perspectives. As a matter of fact, topics such as data, technology, and knowledge integration, data sharing, and public participation, to mention only few, are dealt with through the different chapters in a diverse mixture of perspectives, giving as an overall result a much deeper insight — especially for the planner — than what may be achieved by reading certain selected chapters clearly related to particular issues or concerns. This is the major twofold value of this work, in that although avoiding a point-by-point answer to the call for sustainable development actions, on the one hand it aims at driving the GIS community toward a deeper awareness of sustainable development issues in setting research programs and in application design, while on the other hand it offers a wide spectrum of tools that professionals and practitioners may draw on after they understand how GIS can assist them in spatial planning, management, and decision-making to achieve sustainable development objectives.

This is a book for a broad readership. While most of the chapters will flow easily for the average reader, a few of them require some technical GIS background to be fully appreciated. Nevertheless, once Michele Campagna sets the framework in the first chapter suggesting crosscutting paths for reading, the reader will enjoy discovering the further facets of GIS application for sustainable development thanks to the diverse perspectives offered by the contributors in each chapter.

Thus, I would like to conclude this foreword suggesting, as an added value, considering this book not so much a conclusive work, but rather as a starting point to trigger further discussion, which may eventually lead to defining a structured research agenda for GIS use in sustainable development processes.

Giancarlo Deplano
Professor of Urban Planning
Università degli Studi di Cagliari
Cagliari, Italy

Editor

Michele Campagna is lecturer in urban and regional planning in the Department of Land Engineering (DIT), Universitá Degli Studi di Cagliari, Italy, where he teaches planning and GIS. His research focuses on GIS applications in urban, regional, and environmental planning, and on planning support systems.

Contributors

Seraphim Alvanides
School of Geography, Politics and
  Sociology
University of Newcastle
Newcastle upon Tyne, United Kingdom
s.alvanides@newcastle.-ac.uk

Mette Arleth
Department of Planning and Development
Aalborg University
Aalborg, Denmark
marleth@land.aau.dk

Dimitris Ballas
Department of Geography
University of Sheffield
Sheffield, United Kingdom
d.ballas@sheffield.ac.uk

José I. Barredo
European Commission — DG Joint
  Research Centre
Institute for Environment and
  Sustainability (IES) — Land
  Management Unit
Ispra, Italy
jose.barredo@jrc.it

Anthony Beck
Geography Department
Durham University
Durham, United Kingdom
a.r.beck@durham.ac.uk

Stefania Bertazzon
Department of Geography
University of Calgary
Calgary, Alberta, Canada
bertazzs@ucalgary.ca

Sandrine Billeau
Department of Geography
University of Geneva
Geneva, Switzerland
sandrine.billeau@cueh.unige.ch

Bernadette Bowen Thomson
Safer Cardiff
Cardiff, United Kingdom
safer.cardiff@virgin.net

Bénédicte Bucher
Laboratoire COGIT
Institut Géographique National
Saint Mandé, France
benedicte.bucher@ign.fr

Michele Campagna
Dipartimento di Ingegneria del
  Territorio, Sezione Urbanistica
Università degli Studi di Cagliari
Cagliari, Italy
campagna@unica.it

Vania A. Ceccato
Divison of Urban Studies
Royal Institute of Technology
Stockholm, Sweden
vania@infra.kth.se

Luisella Ciancarella
Ente per le Nuove Tecnologie l'Energia
  e l'Ambiente Unità Tecnico Scientifica
  Protezione e Sviluppo dell'Ambiente e
  del Territorio
Bologna, Italy
cianca@bologna.enea.it

Piergiorgio Cipriano
CSI-Piemonte
Torino, Italy
piergiorgio.cipriano@csi.it

Jonathan Corcoran
GIS Research Group
School of Computing
University of Glamorgan
Pontypridd, United Kingdom
jcorcora@glam.ac.uk

Giuseppe Cremona
Ente per le Nuove Tecnologie l'Energia
    e l'Ambiente Unità Tecnico Scientifica
    Protezione e Sviluppo dell'Ambiente e
    del Territorio
Bologna, Italy
giuseppe.cremona@bologna.enea.it

Konstantinos Daras
School of Geography, Politics and
    Sociology
University of Newcastle
Newcastle upon Tyne, United Kingdom
k.k.daras@newcastle.ac.uk

Andrea De Montis
Dipartimento di Ingegneria del
    Territorio, Sezione Costruzioni e
    Infrastrutture
Università degli Studi di Sassari
Sassari, Italy
andreadm@uniss.it

Gilles Desthieux
GIS Laboratory
Swiss Federal Institute of Technology
Lausanne, Switzerland
gilles.desthieux@epfl.ch

Alexandra Fonseca
Centro para a Exploração e Gestãode
    Informação Geográfica
Instituto Geográfico Português
Lisbon, Portugal
afonseca@igeo.pt

Sébastien Gadal
Université de Marne-la-Vallée
Master AIGEME
Marne-la-Vallée, France
sebastien.gadal@wanadoo.fr

Marina Gavrilova
Department of Computer Science
University of Calgary
Calgary, Alberta, Canada
marina@cpsc.ucalgary.ca

Andrea Giacomelli
CH2MHILL s.r.l.
Milano, Italy
pibinko@tiscali.it

Phil Gibson
School of Geography
University of Leeds
Leeds, United Kingdom
P.D.Gibson@leeds.ac.uk

Cristina Gouveia
Centro para a Exploração e Gestãode
    Informação Geográfica
Instituto Geográfico Português
Lisbon, Portugal
cgouveia@alum.mit.edu

Laura Harjo
Cherokee Nation GeoData Center
Tahlequah, Oklahoma
lharjo@alumni.usc.edu

Florent Joerin
Centre for Research in Regional
  Planning and Development
Laval University
Quebec City, Québec, Canada
Florent.Joerin@esad.ulaval.ca

Marjo Kasanko
European Commission — DG Joint
  Research Centre
Institute for Environment and
  Sustainability — Land Management
  Unit
Ispra, Italy

Alenka Krek
Department of Geoinformation
Salzburg Research
  Forschungsgesellschaft m.b.H.
alenka.krek@salzburgresearch.at

Carlo Lavalle
European Commission — DG Joint
  Research Centre
Institute for Environment and
  Sustainability — Land Management
  Unit
Ispra, Italy

Alexandr Napryushkin
Cybernetic Center of TPU
Computer Engineering Department
Tomsk Polytechnic University
Tomsk, Russia
nadryuskinaa@yandex.ru

Aurore Nembrini
University Centre of Human Ecology
  and Environmental Sciences
University of Geneva
Geneva, Switzerland

Walter Oostdam
City of s-Hertogenbosh
s-Hertogenbosch, The Netherlands
waoo@s-hertogenbosch.nl

Krištof Oštir
Institute of Anthropological and Spatial
  Studies
Scientific Research Centre of the
  Slovenian Academy of Sciences and of
  Arts
Ljubljana, Slovenia
kristof@zrc-sazu.si

Tomaž Podobnikar
Institute of Anthropological and Spatial
  Studies
Scientific Research Centre of the
  Slovenian Academy of Sciences and of
  Arts
Ljubljana, Slovenia
tomaz@zrc-sazu.si

Aimée C. Quinn
Richard J Daley Library
University of Illinois at Chicago
Chicago, Illinois
aquinn@uic.edu

Laxmi Ramasubramanian
Department of Urban Affairs and
  Planning
Hunter College of the City University of
  New York
New York, New York
laxmi@hunter.cuny.edu

Tarek Rashed
Department of Geography
University of Oklahoma
Norman, Oklahoma
rashed@ou.edu

Claus Rinner
Department of Geography
University of Toronto
Toronto, Canada
rinner@geog.utoronto.ca

Linda See
School of Geography
University of Leeds
Leeds, United Kingdom
L.M.See@leeds.ac.uk

Assaad Seif
Directorate General of Antiquities
National Museum
Beirut, Lebanon
assaadseif@culture.gov.lb

Robin S. Smith
Department of Town and Regional
  Planning
University of Sheffield
Sheffield, United Kingdom
digital_participation@yahoo.co.uk

Susanne Steiner
Institute of Surveying, Remote Sensing
  and Land Information
BOKU University of Natural Resources
  and Applied Life Sciences
Vienna, Austria
susanne_steiner@gmx.at

Eugenia Vertinskaya
Cybernetic Center of TPU
Computer Engineering Department
Tomsk Polytechnic University
Tomsk, Russia
napryuskinaa@yandex.ru

Klemen Zakšek
Scientific Research Centre of the
  Slovenian Academy of Sciences and of
  Arts
Institute of Anthropological and Spatial
  Studies
Ljubljana, Slovenia
kzaksek@zrc-sazu.si

Alexander Zipf
Department for Geoinformatics and
  Surveying
University of Applied Sciences
  of FHMainz
Mainz, Germany
zipf@geoinform.fh-mainz.de

Acknowledgments

In the second half of the 1990s, the GIS academic community has grown considerably in Europe. Many research conferences, workshops, summer schools, or other GI-related meetings were held, contributing to the creation of a multidisciplinary network of researchers sharing the common interest for GIScience, with the active participation of young researchers collaborating and sharing their achievements. Thus I would like to acknowledge the work carried out by the following organizations: the European Science Foundation, for promoting the European Research Conferences on GIS; the Association of European Geographic Information Laboratories in Europe (AGILE), for organizing the annual conferences; the Centre for Spatially Integrated Social Sciences funded by the National Science Foundation, for the CSISS summer workshops; the Vespucci Initiative Founders, for the Vespucci summer schools; the eduGI.net, for the first summer school in GIScience; and the UNIGIS, for the international summer schools in GIS. All these initiatives contributed to stimulate not only scientific interest and research results exchanges, but also overall networking by early-career scientists. A special thank you goes to those individuals within or collaborating with these organizations for contributing to the success of these events.

It is within this framework that I was tempted by the challenge to have this established yet informal network of scientists, researchers, and GI practitioners discuss opportunities for GIS application in a cross-cutting field of utmost importance for our society such as sustainable development planning and decision-making by integrating our diverse perspectives in the present work. Most of the invited contributors gave immediate positive responses to the first call for expression of interest. They come mainly from European and North American academia, but also from the public and private sectors.

I am very grateful to the 44 contributors from Austria, Canada, Denmark, France, Germany, Italy, Lebanon, Portugal, Russia, Slovenia, Switzerland, The Netherlands, United Kingdom, and the United States for taking the time and making the effort to write the chapters presenting their research results in light of the common topic of sustainable development, and for their valuable collaboration to the peer review.

Thanks to Max Craglia, Andrea De Montis, Giancarlo Deplano, Werner Kuhn, Ian Masser, Jonathan Raper, and two anonymous referees for their encouraging comments to the early project proposal and their advice and suggestions, which were essential for the editorial work.

Finally, I wish to thank Randi Cohen, Taisuke Soda, Yulanda Croasdale, Theresa del Forn, and Amy Rodriguez from Taylor & Francis for their kind support to the editorial project.

Michele Campagna
Cagliari, Italy

Contents

Introduction

PART I
General Issues for GI Use in Planning Sustainable Development

PART II
GIS Research Perspectives for Sustainable Development Planning

PART III-A
Learning from Practice: GIS as a Tool in Planning Sustainable Development
Urban Dynamics

PART III-D
Learning from Practice: GIS as a Tool
in Planning Sustainable Development
Public Participation

PART III-E
Learning from Practice: GIS as a Tool
in Planning Sustainable Development
SDI and Public Administration

Introduction

1 GIS for Sustainable Development

Michele Campagna

CONTENTS

1.1   INTRODUCTION

Sustainable development is the term commonly and broadly used to describe a complex range of objectives, activities, and mankind behaviors with respect to the environment which should be consistent with the aims of meeting "the needs and aspirations of the present without compromising the ability of future generations to meet their own" [1]. This concept implies that both technological and social settings should be organized so that human activities would not overload the capacity of the biosphere to absorb their impacts [1]. This, which may be agreed upon as a general definition, is yet a rather vague concept to define sustainable development for operational purposes. To this end, this introductory chapter starts with a brief note on the history of sustainable development, outlining some milestones that eventually led to international consensus and a widely agreed-upon adoption of common principles and plans of action to pursue sustainable development. Along a half-a-century path, the role of the United Nations (UN) has been fundamental in promoting international awareness among nations and the wider public. As an outcome, in 1992 nearly 180 countries convened in Rio de Janeiro at the Earth Summit and agreed on the principles of the Rio Declaration and on the programs of its plan of action, Agenda 21 [2]. The success of the initiative was reaffirmed ten years later in Johannesburg, where successful results and proposals of new ways were presented. Indeed many other international organizations, governments, and individuals contributed much to define, promote, and achieve sustainable development objectives; nevertheless the widespread consensus on a such comprehensive plan of action as Agenda 21, makes it a fertile reference framework deserving special attention here to discuss the application of GIS to sustainable development planning, decision making, and management.

1.2 THE WAY TOWARD SUSTAINABLE DEVELOPMENT

With the Industrial Revolution, human activities started to produce new impacts on natural resources. Factories were built, producing new sources of pollution on air, water, and soil; many towns and cities started to grow, generating social and human health problems. Progress in science and technology continued to grow until after the Second World War, when it reached unprecedented rates, raising enthusiastic optimism on development. In many industrialized countries, the achievement of better life conditions, economic growth, increased production and distribution of goods, infrastructures, and housing generated an ideal trust in development, eventually changing radically the relationships between man and the environment. The outcomes would become evident soon. The unlimited growth of most developed countries would compromise seriously the terrestrial ecosystem, destroying limited natural resources, causing dangerous conditions for human health, and augmenting poverty in less developed countries, which were unable to contrast the exploitation of resources carried on for the sake of development at their expenses. Soon the awareness arose of the need for new sustainable development models.

Rachel Carson's *Silent Spring,* first published in 1962 [3], reporting the negative impacts on human health and animal species caused by the use of pesticides in modern agricultural production processes, is widely acknowledged as an embryonic alarm call from which the debate on the environmental issues has arisen and evolved until the present day. In 1972, under the aegis of the Club of Rome — an organization of economists and scientists — the "Meadows" proposed their catastrophic vision with their *Limits to Growth* [4], a report of a model-based forecast according to which trends in demographic growth, increase in production and consumption, and widespread pollution diffusion would have led in a few decades to the collapse of the terrestrial ecosystem. These widely known early works in the history of the socio-cultural debate on environmental issues are only two examples of the enormous work promoted in the last fifty years or so by the establishment in many countries and carried on with the support of the international scientific community. The evolution of the environmental issues debate and of the initiatives of the many organizations spread internationally, which led to the definition of the principles of sustainable development and the way toward their practical implementation, is rich and is characterized by the important role played by the United Nations. In 1972 the UN promoted the Conference on Human Environment held in Stockholm, where those issues which would have later become the principles of sustainable development were discussed. The view of man as "creature and molder of its environment" is proposed in the Declaration of the UN Conference on Human Environment [5]. The document acknowledges the ability of man, enhanced to an unprecedented scale by the progress in science and technology, to transform his surroundings. This ability can be used wisely to bring improvements in the quality of life of the people all around the world, or conversely, if used wrongly can produce incalculable harms to human beings and the environment. Seven proclamations are given in the document, and twenty-six statements are proposed as guiding principles for sustainable development [5]. The UN continued to build on the outcomes of the Stockholm conference, forming in 1983 the World Commission on Environment and Development, chaired by Gro Harlem Brundtland. The commission worked for three years, eventually producing a report on social, economic, and environmental issues [1], which brought the idea of sustainable development into the international view in 1987. The results presented in the report titled *Our Common Future,* also known as the *Brundtland Report,* were discussed at the UN General Assembly, and in 1989 the UN made the formal decision to convene the UN Conference on Environment and Development, which was held in Rio de Janeiro in 1992. The Summit agreed on the Rio Declaration establishing twenty-seven general principles. Moreover, the action plan, Agenda 21, was issued, and it was recommended that all countries adopt national strategies and promote local practices according to sustainable development principles and programs. After ten years, a second World Summit on Sustainable Development was held in Johannesburg (Rio +10), being one of the most important international meeting ever held on economic, environmental, and social decision-making, and focusing on promoting further actions to put Agenda 21 into practice.

This brief yet oversimplified discussion gives just an outline of the history of sustainable development. The reader is invited to consider reading the original documents mentioned above for a thorough definition of the principles of sustainable development — which are out of the scope of this discussion — whereas in the following section the contents of Agenda 21 are reviewed critically with the aim of discussing the implications for the GIS application as support to spatial planning practice according to sustainable development principles.

1.3 AGENDA 21: PUTTING SUSTAINABLE DEVELOPMENT PRINCIPLES INTO PRACTICE

Since its adoption in 1992, Agenda 21 plans of action have been implemented at different rates worldwide. Meanwhile, the UN promoted the monitoring of the initiatives at the national level, and after 10 years of implementation a thorough report was presented highlighting the monitoring results with respect to the Agenda 21 calls at the national, regional, and global level [6].

Agenda 21 is divided into 40 chapters. While some of the chapters are concerned with specific objectives — or objects — such as promoting sustainable human settlements development and sustainable agriculture and rural development, or protecting human health conditions, some others deal with the ways or processes, in both societal and operative terms, that should be adopted to pursue these objectives, such as integrating environment and development in decision-making, promoting collaborative decision-making processes, fostering participation, and promoting public awareness and training. Both *objects* and *processes* are derived from and consistent with the principles agreed to within the Rio declaration. These broad categories are proposed in Agenda 21 in the different chapters within four parts taking into account, respectively, the social and economic dimensions, the conservation and management of resources for development, strengthening the role of major groups, and means of implementation.

While Agenda 21 is directed primarily to national governments, the role of local public administration is acknowledged by the action plan as fundamental, since the problems addressed have their roots at the local level (Agenda 21, Chapter 28). It is also often at the local level that national and regional environmental policies and planning processes are implemented. Locally, moreover, the level of governance is closest to people. Thus public administrations face the challenge to act locally to achieve the objectives of Agenda 21. In order to achieve sustainability, subsidiarity is seen as a means to integrate, both vertically and horizontally, global and local development frameworks. Besides, public participation is seen as a means to achieve democratic decision-making through transparency of public administration and citizens' involvement in sustainable development processes. According to the plan of action, a comprehensive group of actors is proposed to be involved with local authorities in sustainable development decision-making, such as local (indigenous) communities, nongovernmental organizations, workers and trade unions, business and industries, farmers, and the scientific and technological community. Moreover, the role of women, children, and youth should be fostered. The promotion of awareness and education with regard to sustainable development issues, and development of know-how and skills is a prerequisite for socially inclusive collaborative decision-making. To this end the scientific and technological community faces the challenge to develop methods and tools for supporting sustainable development practices.

Information (Agenda 21, Chapter 40) plays a major role in planning and decision-making for pursuing the objectives of sustainable development. This is a common prerequisite for all the Agenda 21 plans of action at all levels. Geographic Information Systems are proposed generically in Agenda 21, Chapter 40.9, as one of the tools to be used to produce, maintain, analyze, and disseminate environmental data. However, as it is discussed in the reminder of this chapter, GIS offers a wide range of reliable tools to support sustainable development-led activities, such as problem setting and solving, planning, decision-making, and management. Thus, further insights are required to fully understand opportunities for this application field and to promote geospatial technologies application as a valuable support to sustainable development processes.

1.4 GIS FOR SUSTAINABLE DEVELOPMENT

Economic, social, and environmental processes are inherently spatial. They can hardly be fully understood without taking into account their spatial dimensions. The relationship between man and the environment cannot be represented without a reference to a special location, because the environment is described by the topological relationships among physical objects (e.g., the soil or the air composition in a given space-time location, the solar radiation on a given piece of land), and human activities produce impacts on the environment spatially.

As introduced in the previous section, Agenda 21 focuses both on special objectives — the objects — and on the ways to be followed to pursue these objectives — the processes. The object is related to solving spatial problems, while the process implies sharing knowledge for collaborative, transparent, and participatory decision-making. Both serve to achieve the higher objectives set by the principles of sustainable development.

GIScience has been proven to offer theories, methods, and applications to effectively support the following categories of tasks, which together find wide space for application in the implementation of Agenda 21 to fulfill the principle of sustainable development:

- Producing and maintaining geographic information (by definition)
- Supporting distributed access to (environmental) information (i.e., spatial data infrastructures)
- Solving spatial problems (i.e., spatial analysis and environmental modeling)
- Supporting collaborative decision-making (i.e., group spatial decision-making)
- Supporting public participation (i.e., public participation GIS)

In planning, decision-making, and management GIS may be considered just one among the most advanced tools available to deal with complex problems — the spatial problems — in a balanced mediation of economic, environmental, and social objectives. It is an essential tool though, which, when properly used, may offer effective support to spatial planning and decision-making, because the geographical component of the problem at hand is determinant when dealing with sustainable development. Thus, geospatial technologies should be a driving engine in the technical, but also socio-organizational, implementation of knowledge-based open and integrated platforms for informed analysis, collaborative problem solving, planning, and decision-making.

According to this general premise, this book presents recent research results and case studies which offer a diverse perspective of the problem at hand, taking into account methodological and technical — but also organizational and societal — issues related to the use of GIS to solve complex problems faced by practitioners in planning and implementing sustainable development objectives.

The aim is to deal with a wide range of topics related to how GIS application may contribute to improve vertical and horizontal collaboration in decision-making among all the actors involved in sustainable development processes at all institutional levels (national, regional, and local). The growth in spatial data availability and the developments in GIScience allow us to carry on "informational planning" processes (analysis, design, evaluation, decision, management, and communication). In fact, whatever the planning paradigm adopted, a knowledge-based approach is required to carry on sustainable development processes.

The book is structured in three parts.

The first part sets the framework for GI-based collaborative spatial e-planning processes. Cyber planning is defined in Chapter 2, while societal, technical, and

organizational issues are proposed in the following chapters, giving particular attention to digital public participation, interfaces, data accessibility, and economic value
of geographic information.

The second part of the book presents GIScience methods and techniques, which
can be used to solve particular problems — objects — commonly addressed in
sustainable development planning and decision-making. A number of topics are
proposed, such as the GIS integration with simulation and microsimulation models,
spatial multimedia, online computer-based collaborative tools, spatiotemporal databases, remote sensing (RS) data collection, geodemographics, (multivariate) spatial
analysis, and zone design techniques and tools to solve problems such as environmental modeling, socioeconomic system analysis and planning, health care planning
and management, urban settlement monitoring, community safety, risk prevention,
and hazard mitigation. Practitioners in sustainable development processes commonly
address all these problems, and the methods proposed here offer techniques and
tools which can be used in integrated sustainable planning support systems.

In the third part, the book presents GIS applications and case studies from
research and real practice projects. The chapters are grouped by topic according to
the following categories:

- Urban dynamics
- Natural and cultural heritage
- Society, health, and environment
- Public participation
- SDI and public administration

For each category, several examples are given of application methodologies and case
studies.

In the remainder of this chapter, the overall perspective is described in detail,
discussing the opportunities GIScience theories, methods, techniques, and tools offer
to support the work of practitioners and of all the actors involved in sustainable
development planning, decision-making, and management processes.

1.5   REQUISITES FOR GI-BASED COLLABORATIVE
SUSTAINABLE DEVELOPMENT PLANNING SUPPORT

Sustainable development is a multi-actor process that involves all levels of society
globally and locally. The process is inherently collaborative and participatory in its
own nature. Senior government decision makers at the international to the local
level, organizations, entrepreneurs, interest groups, social minority advocates, and
citizens are involved; individuals, groups, and organizations should have equal access
to information for decision making.

The first part of this book pays attention to these characteristics of sustainable
development decision making, spatial planning, and management processes, dealing
with the issues of public participation, in terms of theoretical and methodological
premises, and of accessibility to (GI) data, in cognitive, technical, and economic
terms.

Information Communication Technology (ICT) has granted freedom from distance and from the cost of digital data reproduction, virtually giving ubiquitous access to information at no (or low) cost. Nevertheless, developments in information communication infrastructures do not warrant *per se* the absorption by society of the newly available ICT. Societal, cultural, cognitive, organizational, and economic issues, among others, have to be seriously taken into account when implementing new processes on innovative technology platforms. While one may acknowledge that the Internet affects people's everyday activities, the research is still ongoing about how members of a community adopt technology and telecommunications and use them to enhance their capabilities to perform a given task. On these premises Chapters 2 to 6 set a framework of basic assumptions for the implementation of e-platforms to support governance, spatial planning, decision-making, and management.

In Chapter 2, Andrea De Montis introduces the concepts of cyber planning. The pervasive diffusion of ICT is deeply affecting all sectors of society, generating cultural mutations. In many sectors the Internet has become an everyday tool to access information and communicate, fostering changes in the traditional way of working, and offering new possibilities of economic development. New professions were born, and other professions have substantially changed, while some others are changing with less radical differences. In spatial planning, technology adoption has been partially exploited, with differences depending on the planning processes and on the different local contexts in the different countries. While the planning professionals have enjoyed the support of ICT and GIS in many way [7–10], nevertheless we are perhaps still far from a sound mutation of the planning theories and paradigms, and thus from the core professional practice. Andrea De Montis argues, in line with an ongoing theoretical debate and on the bases of recent planning research results, that ICT can favorably support the implementation of collaborative information-hungry planning processes, such as those proposed by Agenda 21, to achieve sustainable development objectives. He envisages, moreover, that cyber planning instances are emerging in practice as a sort of digital evolution of planning in the Information Era, which might eventually lead to more substantial changes in the way of making plans. Chapters 9, 10, 26, and 27 propose methodologies and present case studies of digital planning experiences.

The inspiring principles of sustainable development require planning and decision-making processes to be participatory. The collaboration at the global and the local levels, and between major groups, such as institutional stakeholders, interest groups, local communities, and citizens, demands complex forms of participation. The implementation of e-government and e-governance processes, triggered by the availability of ICT, has fostered a new interest rising about democracy, transparency of administration, and public participation. To the latter, particular attention has been paid by the GIS and planning community in the last decade as many scholars, denying the idea of GIS as an elitist tool in the hand of power, on the stream of Pickles' *Ground Truth* [11], demonstrated that ICT and GIS together may help to support public participation, empowering marginalized communities and citizens [for an example, see 12]. Along with this debate, the renowned Arnstein's ladder of participation [13] has been reinterpreted to adapt the current digital e-government practices [14–16].

In Chapter 3, Robin S. Smith, on the basis of representative examples taken from U.K. local authorities' practices, discusses present achievements and opportunities for public participation processes and settings in Internet-based environments. The analysis of practical experience developed so far opens the way to theoretical argumentation dealing with the definition of public participation, the way actors are chosen to get involved, and the methods that can be implemented. He advises, "participation is not a unique or shared construct, and failure to recognize different views can lead to unsatisfactory outcomes for all." Attention is paid in this chapter to the differences participation implies in actual and virtual environments, and how traditional and digital participatory methods should be integrated. Participation processes are complex, and special care should be devoted to their design as well as to the analysis of current practices to elicit critical success factors. Finally, as he explains, the use of GI in digital participatory settings inherits the same characteristics from generic participation processes, yet presents new issues to which GIScience research should dedicate further analysis in the long run.

The theoretical issues dealt with by Smith in Chapter 3 are considered with different perspectives in Chapters 4, 5, 9, and 10, which discuss, in turn, information perception and access problems, and propose methods and tools which can support public participation; case studies are proposed in the third part of this book in the section on public participation.

Prerequisite to build collaborative (spatial) decision-making processes is information being available to all the actors involved. Thus, it must exist, it must be accessible, but also it must be comprehensible to those who use it. Information production, sharing, and integration involve high cost. Thus, the information cycle of life, from production to exploitation has to be cost effective. As the costs are sensible, new business models should be developed. The last three chapters of Part 1 address these problems in turn.

In Chapter 4, Bénédicte Bucher discusses the problem of data accessibility. Complete information for decision-making is often gathered from multiple sources; hence the role of metadata in its retrieval and exploitation is outlined, and reference to current interoperability standards is given. She illustrates then the problems data producers such as national mapping agencies face in implementing reliable data catalogs and tools to assist the users in discovering data sources to find suitable data to solve spatial problems they face. Current results and further research questions are proposed, aiming at improving the user interface for data discovering and retrieval and geolibraries exploitation.

When data are available, the new challenge is to use them to produce suitable information for decision making. Geographic information is characterized by representation models, which are not always intuitive or easily readable for the lay user. Sustainable development decision-making involves a variety of actors with different backgrounds and sometimes even different cultural underpinnings, and it may be sometimes difficult for them to agree even on basic geographic constructs such as boundaries [17]. Thus, when building GI-based web applications for spatial planning and decision-making, or territorial governance, special attention should be paid to geographic information modeling and representation and to interfaces design.

Mette Arleth discusses in Chapter 5 the issue of laypeople understanding geographic information. With the widespread diffusion of GI-based web applications by pubic administration and citizens, even in countries such as Denmark where public administration services on the web have reached a high level of development and diffusion among citizens, GI-based applications for spatial planning and governance are not accepted by different categories of users at the same pace. She argues that these difficulties are related to the fact that some geographic information representation models are not equally well understood by different categories of users. Hence, more research should be devoted to investigate user understanding of GI and GI-based web applications usability. Several methods relevant for such investigation are proposed, with reference to selected case studies.

Finally, *conditio sine qua non* for GI production, sharing, and exploitation in spatial decision-making, as required by sustainable development processes, is its economic feasibility. Information production, maintenance, and even its distribution, although digital technology helps a lot especially with reference to the latter, are high-cost activities. Thus, appropriate business models should be developed and implemented to make the information life cycle cost effective. The definition of a suitable economic value for geographic information is fundamental to design reliable GI products pricing policies. Alenka Krek in Chapter 6 argues classical economic principles may cause difficulties, giving an economic value to geoinformation, and proposes value-based pricing as a means to develop sustainable pricing policies.

To summarize, the digital media offer favorable opportunities to improve efficiency and efficacy in (spatial) decision-making processes. Nevertheless, the implementation of ICT and GIS settings to support collaborative spatial planning and decision-making processes presents several challenges. To view it with Latour [18], from a general perspective the black box (technology) is ready; hence the current challenge is to build on it effective collaborative processes. New hypotheses should be tested, and analysis of current practices should be carried on, in order to draw general guidelines for implementation. To this end, the first part of this book deals with several fundamental issues to which further attention should be paid in the planning and GIScience research agenda, so that practitioners may adopt more effective tools to fully exploit the opportunities GIS offers in implementing sustainable development planning and decision-making processes.

1.6 SOLVING SUSTAINABLE DEVELOPMENT PROBLEMS WITH GIS

Sustainable development planning, decision-making, and management are comprehensive processes which deal with multiple-dimension problems aiming at achieving balanced economic development, environmental protection, and social equity and welfare. The use of (geographic) information to support (spatial) decision-making requires availability of data, and tools to analyze data to be integrated in complex information systems.

In the second part of this book eleven chapters are presented which are concerned with the application of GIScience core methods and tools to support specific tasks commonly found in spatial planning and decision-making. Methods and techniques for data production, data modeling, system integration, and advanced spatial analysis are proposed in this part of the book aiming at presenting recent research results in hot topics within this general theme. Methods and tools presented here should be liable for being integrated in broader information systems to support spatial planning processes.

Data production is often by far the most costly part of information system development. Within GIS, besides traditional expensive field surveying techniques, remote-sensing data have been proven to supply, at a relatively lower cost for the end users, the possibility to collect and maintain large datasets in terms of spatial footprints and time series. Environment can thus be monitored constantly, and changes detected promptly. Last-generation satellite sensors have improved significantly and offer very high-quality data in terms of spatial and spectral resolutions. Geographic information a few years ago collected once every so many years can now be recorded with daily temporal resolution. Once raw data are collected, techniques should be available for data processing in order to supply useful information for decision-making. This is still an open broad research field, and improved techniques for semiautomatic data processing and thematic information production are proposed unceasingly by scientific research.

In Chapter 7, Alexandr Napryushkin and Eugenia Vertinskaya present an advanced method for thematic mapping based on remote sensing data processing. A number of complex stages including an adaptive classification procedure allow classifying raw data in an efficient way. This technique is illustrated with reference to multispectral imagery from RESURS-01 satellite for mapping ecosystem themes, and tested in the Tomsk area in Siberia.

Improved availability of spatial data allows recording geographies in close time intervals. To analyze changes in physical environment, data models and analysis tools able to take into account the time dimension of data are required. Conventional GIS by now have shown difficulties in managing efficiently the dynamics of geography. More sophisticated data models and analysis tools are required to integrate the time dimension together with geographic objects' geometry and attributes.

In Chapter 8 Alexander Zipf presents an object-oriented model developed to manage temporal dimension of 3D geographic objects. The framework proposed by Zipf offers a sophisticated way to model time, hence contributing to extend database models for dynamic information systems. A case study is described, showing the application of the framework to a 3D historic city model for an urban information system, yet the model can be adapted to other domains. Finally, opportunities are proposed for further research developments in this field.

Besides geometry and thematic attributes of objects, environment can be effectively described and analyzed with the support of multimedia data. Multimedia data such as text, images, videos and sounds, have the advantage that they are immediately communicative even for the lay user. Multimedia data are usually handled with different technologies than spatial data. However, as thoroughly explained by Alexandra Fonseca and Cristina Gouveia in Chapter 9, multimedia information systems can be integrated with geographic databases. As a result, spatial multimedia information systems are successfully implemented to support spatial planning and decision-making. The communicative capability of such systems makes them particularly effective in supporting collaborative and participatory planning processes. As a matter of fact, multimedia formats are particularly efficient in integrating the expert knowledge of the professionals with the common experiential knowledge of the laypeople that may be involved in collaborative planning, particularly in participatory processes. Different examples are proposed by Fonseca and Gouveia, showing different online implementation settings for spatial multimedia information systems, respectively, to disseminate environmental data, to support public involvement in environmental impact assessment, and for concerned citizens to collect environmental data. This discussion is integrated by the examples on the use of spatial multimedia in participatory planning given in the case study presented in Chapter 27, which address the same issues from a different angle.

The possibility for the actors involved in the planning process to supply their own information to enrich the dialogue with their local knowledge and advocate their interests is another interesting aspect to which research in computer support for collaborative planning has paid attention. Groupware is broadly used as an umbrella term to indicate computer tools to support collaborative work distributed in time and space. This characteristic is particularly useful when stakeholders have difficulties in arranging real-time face-to-face meetings. In Chapter 10 Claus Rinner argues how the integration of groupware with GIS would effectively support spatial planning and decision making. He discusses opportunities for the spatial extension of argumentation theory and proposes a conceptual model for implementing what he calls *argumentation maps.* Existing applications are analyzed in the chapter, and further developments for computer support to collaborative planning are outlined.

Another possible GIS capabilities extension is the integration with operational models to simulate and forecast environmental and social processes, with the aim of evaluating possible consequences of given courses of action. In the following Chapters 11 and 12, Andrea Giacomelli and Dimitris Ballas discuss, respectively, the GIS model integration from a general perspective and the implementation of spatial microsimulation models within GIS environments.

Giacomelli opens Chapter 11 with a general discussion of the capabilities GIS offer to support spatial planning and decision-making, particularly when the integration of environmental, economic, and social issues should be implemented in a common analytic framework. Then he outlines further advantages given by the integration of GIS with simulation models, which extend GIS capabilities for dealing with dynamic processes. The discussion addresses both technical and societal issues with regard to the GIS–models integration and gives general yet enlightening insights into the use of decision support systems in sustainable development planning and management, with reference to their main components — data, tools, people, models — and their integration.

After Giacomelli's framework discussion, in Chapter 12 Ballas goes straight to the core of the topic, proposing the integration of GIS with microsimulation models for the evaluation of socioeconomic and spatial effects of major developments. According to a system approach which considers a socioeconomic system as composed of a number of interacting subsystems, spatial microsimulation modeling offers a potentially powerful framework for the analysis of urban and regional systems, as the integration of microsimulation models based on individual's behaviors within GIS allows investigation of the relationships between socioeconomic processes, spatial planning policies, and environmental settings.

Another class of GIS applications based on the analysis of small area statistics is geodemographics. Initially developed for research and planning support, these applications spread in the business and retailing field as analytical tools to supply operational, tactical, and strategic functions of an organization in locating business and services in the short, medium, and long term. Although geodemographics may give enlightening insights in understanding rich patterns of urban social structures, they can be fruitfully applied also in sustainable development processes to effectively tailor public services planning and policy design according to the local societal needs.

In Chapter 13 Linda See and Phil Gibson give a thorough overview of geodemographic underlying methods and application tools. Then they analyze recent experiences of geodemographics application to sustainable development processes and discuss further potential opportunities. They conclude the chapter not only arguing geodemographics may supply effective support to policy making and planning, but also suggesting its use to set up sustainable development sensitization and participation programs. Once again, it is shown how a creative application of a single GIS tool may address sustainable development planning issues in a multiplicity of ways.

Chapters 7 to 13 present several GIS methods, which may help to solve information production, management, and analysis for sustainable development planning and decision-making. The remaining chapters of the second part, in turn, address specific sustainable development objectives with regard to health, safety, and risk mitigation. Human health deserved a special chapter in Agenda 21, which proposes several program areas to meet primary health care needs, control the spread of diseases, and reduce health risks from pollution and hazards.

In Chapter 14 Stefania Bertazzon and Marina Gavrilova discuss the application of advanced multivariate spatial analysis to address the need for efficient models in health care management. They argue that spatial regression analysis based on alternative distance functions of non-Euclidean metrics constitutes an efficient tool for human health research and its integration with environmental processes.

Spatial aggregation of area units is a common problem in several policy fields and may be an efficient tool to help solve different sustainable development planning and management problems. In Chapter 15 Kostantinos Daras and Seraphim Alvanides discuss the application of zone design methods to support public health management policies which better fit patients' needs. The new tool for zone design proposed by the authors takes into account objective functions as well as constraints of the zones' shape, offering reliable tools for effective health care management.

Besides human health, according to sustainable development principles, the well-being of all urban dwellers must be improved so that they can contribute to economic and social development. Urban areas often face high rates of violence and crime occurrences, which underline symptoms of social disorder. Vania A. Ceccato discusses this important issue for social welfare in Chapter 16. On the basis of the recent literature, she argues crime events tend to be related to particular socioeconomic conditions and functional and physical settings within the urban environment.

To fight crime events and promote community safety and social well-being, GIS and spatial statistics, she argues, should be used in planning and policy design in defining measures to reduce crime events and risk. Ceccato illustrates in detail several techniques for crime analysis and outlines advantages and limitations of the different methods. Then she suggests potential applications of spatial pattern detection and explanation techniques to strategically support safety planning. Later, in Chapter 22, Corcoran and Bowen Thomson present a real-world case study of community safety policy making.

Human health protection also involves risk prevention and mitigation. Hazards originated by natural and human causes can create serious dangers to populations especially when people live in concentrated areas. Vulnerability assessment and hazard mitigation is of utmost importance in settlement planning and management.

Tarek Rashed sensibly addresses this important issue in Chapter 17 with a thorough explanation about the concept of vulnerability and hazard, and about possible approaches for its mitigation, before turning to a detailed discussion about the GIS and remote sensing methods in vulnerability analysis, which may be helpful in sustainable development planning. Rashed outlines the opportunities GIS and RS offer for vulnerability assessment for planning hazard mitigation and proposes a GIS/RS-based methodology for understanding spatial and temporal vulnerability patterns in urban areas.

In line with a common perspective in this book, once again attention is paid to the opportunities geospatial technology offers in supporting an integrated knowledge-based holistic approach to sustainable development planning.

1.7 GIS FOR SUSTAINABLE DEVELOPMENT IN PRACTICE

The history of scientific and technological innovation is studded by adaptive evolutionary processes of putting scientific research findings into practice. Discovering a new theory, method, or technology is not always a warranty of its recognition and acceptance by users. When this is the case, it often requires a lot of effort by the end user to put *science into action* [18] by an adaptive process through which the user might modify the initial product into a new one. Differences among the two might be significant. Economic, cultural, societal, and institutional factors influence the innovation adoption in given contexts at different rates.

The GIScience and geoinformation industry in the last decades produced a plethora of theories, methods, and tools to solve spatial problems. However their diffusion and adoption in professional practice varies in the many application fields. Data availability, funding supply for implementation, and training facilities determined a wide range of GIS diffusion patterns in a variety of contexts. It is interesting to note how general patterns of GIS innovation and diffusion have recently been subject to a general turn toward giving broad access and diffusion to small pieces of (geo-) information, to use Frank and Raubal's words [19], rather then continuing to search for advanced spatial analysis tools. That is to say that putting GIS into practice is not just a matter of quality data and efficient tools availability. The application user requirements often drive technology adoption toward unexpected trails in an adaptive process, which sometimes may offer as many interesting hints as research findings. This is particularly true in planning, which is an application field strongly related to economic, institutional, and sociocultural settings, where disciplinary theories, methods, and tools vary sometimes substantially in different local contexts. To this end, the third part of this book presents a series of application research and real practice case studies aiming at showing GIS theories and methods application to different tasks of spatial planning and decision-making with reference to sustainable development objectives and processes.

Chapters 18 to 29 are grouped in thematic sections regarding different sustainable development issues, namely concerning urban settlement dynamics analysis and forecasting, natural and cultural heritage preservation, wise use of energy, water resource and community safety management, public participation, and information system management to support governance and decision-making.

Sébastien Gadal opens the first section with Chapter 18. As introduced in Chapter 7 by Napryushkin and Vertinskaya, remote sensing data processing is a reliable source of information for detecting by thematic mapping anthropic impacts on the environment. Gadal analyzes the case study of the urban dynamics in Maghreb, Morocco. The Morocco Atlantic Metropolitan Area is subject to considerable growth forces. The lack of quality data availability to monitor ongoing urban sprawl causes serious problems with regard to settlement expansion control. To address this problem, which is especially common in developing countries where there is a lack of quality socioeconomic and geographic data, Gadal presents a methodology for multiscale geographic dataset generation by integrating different satellite data sources and processing techniques. The result is the availability of updated datasets, at different time steps, at a cost relatively lower than those required by traditional surveying campaigns, to monitor the urban dynamics and support settlement development control.

When quality geographic datasets are available on top of that, it is possible to implement, as discussed earlier in Chapters 11 and 12, sophisticated simulation and forecasting models to assist decision-makers in planning processes.

José Barredo, Carlo Lavalle, and Marjo Kasanko present the successful results of the MOLAND research project. Barredo et al. discuss the application of the MOLAND methodology to the case study of Udine, Italy. The methodology, based on a multiscale modeling framework that integrates several submodel components representing environmental, social, and economic subsystems, allowed them to implement long-term development forecasts for settlement development in an area characterized by residential discontinuous urban fabric and industrial land uses.

The second section deals with the preservation of natural and cultural heritage.

In Chapter 20, Susanne Steiner presents a case study of rural landscape analysis in the transnational border between Austria and Hungary. Different historic vicissitudes of the region in the last century deeply affected the agricultural landscape differently on the two sides of the border. Thus, current development pressure requires seriously taking into account such differences while planning national policies on the two parts of the same region. To this end, a detailed explanation is given on data acquisition and database modeling to implement an information system able to support an understanding of rural landscape evolution in time as a basis for further development planning, sustainably exploiting local resources, and safeguarding the regional historical identity.

On the basis of research and practice experiences in archaeology, in Chapter 21 Anthony Beck and Assaad Seif discuss the role of geoinformation technology in the discipline. As argued by the authors, the latest theoretical developments in archaeological data collection and interpretation require computerization with high potential for improving analytical frameworks. Geospatial technologies offer reliable tools to integrate diverse data in a coherent whole, which, according to the authors, is the current challenge in archaeology.

Three examples are given in the third section that address specific societal and environmental issues.

In Chapter 22 Jonathan Corcoran and Bernie Bowen Thomson present a case study of community safety policy-making in the United Kingdom. They explain a GIS-based methodology to support crime analysis, giving particular attention to institutional problems of collaboration among the different actors involved in the process. This is an interesting real-practice example for the discussion of data sharing and integration process with its institutional, organizational, and technical problems. The efforts to achieve a balance between the application of advanced spatial analysis tools and data integration requirements in a collaborative process guided the application of the methodological approach, and they outline the satisfactory compromise solution implemented and the reliability of results.

Giuseppe Cremona and Luisella Ciancarella present a strategic plan for water resource management. The authors illustrate the methodology implemented according to an environmental suitability approach in water facilities planning, aiming at achieving a balance between the demand for the water supply and natural protection in such environmentally sensitive context as the Mediterranean Eolian Islands.

In Chapter 24 Tomaž Podobnikar, Krištof Oštir, and Klemen Zakšek present a case study of solar radiation modeling for the evaluation of solar energy resources in Slovenia, whose exploitation can reduce the need for other energy sources, which generally produce much higher negative impacts on human health and the environment.

A crosscutting theme found in Chapters 22 to 24 is the data availability, which deeply influences the application of methodological approach and the reliability of the analysis results.

The theme of geospatial technologies as empowering tools for communities and citizens, discussed from the theoretical and methodological points of view, respectively, in Chapter 3 and in Chapters 9 and 10, is further investigated in its practical implementation by Chapters 25 to 27.

The case studies presented offer three different perspectives, which together give a multifaceted and thorough insight into the application of GIS and spatial multimedia in collaborative and participatory sustainable development planning processes.

In Chapter 25 Laura Harjo discusses how GIS is used to empower the local community in a sensible context such as a tribal nation in the United States. Thanks to the implementation of a dedicated information data center, decision-making processes are supported in a number of application fields such as health care, cultural heritage, land and property use, negotiations, and others. Of special interest in the experience presented by Harjo is the integrated approach, according to which geographic information is analyzed to support multiple decision-making processes within the community. This is an excellent example of how the tools can be used in an integrated manner to pursue sustainable development strategies. Moreover it is interesting to note how the experiences presented by Harjo show the opportunities for GIS to represent local identity and the citizens' common knowledge.

A different perspective is offered by Aurore Nembrini, Sandrine Bileau, Gilles Desthieux, and Florent Joerin, who in turn pay attention to the use of geoinformation technologies in participatory urban planning. Nembrini et al. focus on a traditional planning process in which citizens — and this is the original experience of the case study of Geneva — are called for involvement in the plan-diagnosis phase. The authors describe a methodology to elicit citizens' views and aspirations about future urban development. The methodology proposed in this chapter is used to perform analysis on spatial indicators, which represent citizens' concerns. Furthermore GIS representations are used instead of traditional means to support the dialogue among professionals and citizens. Nembrini et al. conclude the chapter by discussing the role GIS played in the process, and to what extent citizens accepted the use of geoinformation technology support.

As a third example of participatory planning practices, Laxmi Ramasubramanian and Aimée C. Quinn in Chapter 27 present a successful experience of GIS and geovisualization tools integration in a web system to support a broader participatory planning experience in Chicago. A spatial multimedia web application was used, together with traditional face-to-face methods, to promote concerned citizens' involvement and help them to discuss future scenarios for their neighborhood development. This paradigmatic example thoroughly shows both societal and technical aspects of digital support to participatory planning in the detailed description of this best practice.

The last section presents two experiences, in Italy and The Netherlands, respectively, concerning organizational and technical underpinnings in developing geographic information systems.

In the last decade, growing attention has been given to the development of spatial data infrastructure at the global, national, and local levels in order to integrate geographic information sources produced by the public sector (i.e., government bodies, national mapping agencies, local administrations) according to interoperability standards for its exploitation in decision-making, and to promote its reuse by the private sector as an opportunity for economic development.

Piergiorgio Cipriano presents the latest developments of the spatial data infrastructure implemented in the Piemonte Region in Italy. A joint consortium of regional and local administrations undertook its development according to European and international standards for data, metadata, and technology platforms. The results achieved so far represent a best practice in this field.

The last section is concluded by Walter Oostdam with Chapter 29, which gives a detailed description of an urban geographic information system developed by the municipality of the City of 's-Hertogenbosch in The Netherlands. This case study offers several interesting insights into both the organizational and technical problems afforded in information system development. Unlike major national organizations and research centers, smaller organizations such as local administrations often barely meet the demand for funding, infrastructure, and skills required for enterprise GIS implementation. Oostdam explains in detail how technological implementation has been accompanied by multistepped adaptive processes of organizational changes. He explains also how pilot projects were implemented to promote the awareness of the advantages offered by GIS at a managerial and operational level.

1.8 CONCLUSIVE SUMMARY

Sustainable development is a primary objective and an urgent problem to be addressed by our society. Geospatial technologies offer reliable tools to support analysis, problem solving, planning, decision-making, and management of the processes required to pursue this common objective. GIScience theories, methods, techniques, and tools are presented and discussed here in the role they play in setting up sustainable processes to achieve sustainable development objectives.

The GIScience community has a role to play in further investigating opportunities to solve sustainable development problems by means of sustainable development processes and in promoting awareness of all the actors involved about the potential of geospatial technologies.

This book aims at addressing this challenge by presenting a diverse set of contributions, which together propose a comprehensive perspective of the issues and possible solutions for implementing knowledge-based sustainable development support systems. To this end data production, maintenance, and access issues are discussed together with their economic, technical, and organizational feasibility. Moreover, tools for inclusive collaborative planning are presented, which may constitute reliable platforms for participatory analysis and problem solving and decision-making. These tools, although promising from a technical perspective, need to be implemented carefully, taking into account the impact new technologies may have on diverse groups of users.

The overall discussion aims at dealing in an integrated way with the many issues outlined in this chapter; nevertheless, further efforts are required to develop a more focused research agenda for GIS application for sustainable development. To this end, it is the hope of the editor that this work, by paying attention to a broad range of problems and suggesting possible solutions, may contribute in raising the awareness of researchers, developers, and end users of the opportunities for the application of GIS for sustainable development.

REFERENCES

1. Brundtland, G., Ed., *Our Common Future: The World Commission on Environment and Development*, Oxford University Press, Oxford, U.K. 1987.
2. United Nations, Agenda 21, Earth Summit, Rio de Janeiro, 1992, http://www.un.org/esa/sustdev/documents/agenda21/index.htm, last visited on January 13, 2005.
3. Carson, R. *Silent Spring*, 2002 edition, Houghton Mifflin Company, Boston, 2002; originally published in 1962.

4. Meadows, D.H., Meadows, D.L., Randers, J. and Behrens, W., The Limits to Growth, A Report to The Club of Rome, 1972, short version available at http://www.clubofrome. org/docs/limits.rtf, last visited on January 13, 2005.
5. United Nations, Report of the United Nations Conference on the Human Environment, Stockholm, 1972, http://www.unep.org/Documents/Default.asp?DocumentID=97& ArticleID=1503, last visited on January 13, 2005.
6. United Nations, National Implementation of Agenda 21: A Report, Department of Economic and Social Affairs Division for Sustainable Development National Information Analysis Unit, United Nations, New York, August 2002.
7. Harris, B., Beyond GIS: computer and the planning professional, *J. Am. Plann. Assoc.,* 55(1), 85–90, 1989.
8. Hinnes, J.E. and Simpson, D.M., Implementing GIS for planning: lessons from the history of technological innovation, *J. Am. Plann. Assoc.,* 59(2), 230–236, 1993.
9. Budic, Z.D., Effectiveness of GIS in local planning, *J. Am. Plann. Assoc.,* 60(2), 244–263, 1994.
10. Heikkila, E.J., GIS is dead; Long live GIS! *J. Am. Plann. Assoc.,* 64(3), 350–360, 1998.
11. Pickles, J., *Ground Truth: the Social Implication of GIS*, Guilford Press, London, 1995.
12. Craig, W., Harris, T. and Weiner, D., *Community Participation and Geographical Information Systems*, Taylor and Francis, London, 2002.
13. Arnstein, S., A ladder of community participation, *J. Am. Inst. Planners*, 8, 216–224, 1969.
14. Weidemann, I. and Femers, S., Public participation in waste management decision making, *J. Hazard. Mater.,* 33, 355–368, 1993.
15. Kingston, R., Web Based GIS for Public Decision Making in the UK, paper for the Project Varenius Specialist Meeting on Empowerment, Marginalization and Public Participation GIS, Santa Barbara, Cal., 1998, http://www.ncgia.ucsb.edu/varenius/ ppgis/papers/kingston/kingston.html.
16. Carver, S., Participation and Geographical Information: a position paper for the ESF-NSF Workshop on Access to Geographic Information and Participatory Approaches Using Geographic Information, Spoleto, 2001, http://www.shef.ac.uk/~scgisa/spoleto/ workshop.htm.
17. Turk, A., Tribal Boundaries of Australian indigenous peoples, in *Geographical Domain and GIS*, Winter, S. (ed), Institute for Geoinformation, Vienna University of Technology, Vienna, 2000, pp. 117–118.
18. Latour, B., *Science in Action, How to Follow Scientists and Engineers through Society,* Harvard University Press, Cambridge, Mass., 1987.
19. Frank, A. and Raubal, M., GIS education today: From GI science to GI engineering, *URISA J.,* 13(2), 5–10, 2001.

Part I

General Issues for GI Use in Planning Sustainable Development

2 The Rise of Cyber Planning: Some Theoretical Insights

Andrea De Montis

CONTENTS

2.1 INTRODUCTION

According to recent estimates [1], the number of personal computers in the world currently amounts to around 600 millions units and, by 2010, is expected to reach 1 billion. This means that on average, almost one out of six persons on the planet is forecasted to have a personal computer and, most likely, to be able to connect to worldwide networks. Studies on real complex networks [2, p. 10] reveal that in 2003 the number of World Wide Web pages linked by the sole search engine AltaVista equaled 203,549,046, while the number of connections among them was 2,130,000,000. With respect to regional distribution of information technology, according to a recent UN report [3, p. 4], the so-called digital divide is shrinking: the number of personal computers per 100 inhabitants in 1992 in the developed countries was 27 times more than in the developing countries, while in 2002 it was only 11 times more. Moreover, the number of Internet users per 100 inhabitants in 1992 in the developed countries was 41 times more than in the developing countries, while in 2002 only 8 times more.

Even though twenty years have passed since Gibson's *Neuromancer* was published in 1984 [4], by looking at the reported figures it is possible to acknowledge the power of the previsions envisioned in that famous novel, which introduced the term "cyberspace" into our current ways of speaking and thinking. What perhaps Gibson was not able to foresee was the exact size of this particular space and its immediate reflections onto societies, economies, and cultures: the rise of the information and network-based society will keep on producing even sharper changes in lifestyle, and thus in the patterns citizens think, work, organize, communicate, speak, buy, invest and plan their own future.

While the spread of the digital culture involves mutations and may display its effects in more visible and touchable ways in other sectors of our societies, in planning it is possible to detect the rise of a new kind of player, the cyber planner, who has developed his or her skills, apart from traditional issues, in new branches of knowledge, such as information technology, geo-informatics, communications technology, software engineering, and network and distributed computing. This professional is confronted with the need to communicate, involve, and stimulate groups of other practitioners and citizens in order to sustain a social-consensus-based and collaborative style of planning. One of the milestone principles of sustainable development can be found in the empowerment and auto-determination of local societies, which should be made able to master their own plans and programs for future development. According to this perspective, this new figure of practitioner, the digital info-planner, may be believed to be the suitable professional, as far as he is able to bring the required endowment of transparency, trustworthiness, and responsibility into the procedures of analysis and production of structured information supporting the activities of planning.

In this chapter, the author aims at providing insights on the rise of cyber planning by examining the diffusion of digital informative culture across all the sectors of our society and by suggesting relevant relations among the strategies toward sustainability, distributed computing, and digital planning. The arguments are presented as follows. In the next section, the concept of cyberspace is first presented from a theoretical point of view and then applied to the mutations of some leading sectors of society. In the third section, cyber planning is introduced and described as a new style of practice. In the fourth section, concluding remarks of the chapter are drawn, by viewing the concepts of cyberspace and cyber planning with sustainability-driven processes and emphasizing the key role informational endowments may play for decision-making, planning, and management in a perspective of sustainability.

2.2 CYBERSPACE, VIRTUALIZATION, UBIQUITY: A GENERAL THEORY AND SOME APPLICATIONS TO PRACTICE

In a thought-provoking article, Batty [5, p. 1] stated that "by 2050, everything around us will be some form of computer," referring in the end to the evidence that everything, and the city as well, may soon become computable. According to Batty [5, p. 3], the main point, which induces a very real revolution and leads to a novel kind of space and metric, relies on the convergence between those computers and telecommunications. Starting from this statement, a possible definition in complex terms of cyberspace should apply not only to the ways information, models, geographical displacement are stored in their digital format into an electronic domain but also, and especially, to the patterns in which they are transmitted along clusters of networked hard disks. Other scholars refer to cyberspace, invoking "any types of virtual space generated from a collection of electronic data that exist within the Internet" [6, p. 2]. Thus, a precise definition of cyberspace has to be given in connection with the discourse on remote exchanges of data in the network of the networks.

2.2.1 CYBERSPACE: STARTING FROM INFORMATION AND TELECOMMUNICATIONS TECHNOLOGY (ICT)

Information and communication technology (ICT) can be interpreted as the current system of thought and associated tools that make an individual able to manage information, meant as data structured into an informative framework. This system allows one to construct, gather, edit, and transfer information from a transmitter to a receiver device. A particular ICT has been the hallmark of every historical era. Thus, information and communication technology can be considered not only as the cultural product of a certain community, but also as a crucial factor in the behavior and thoughts of that society.

McLuhan [7] believes that an affinity can often be found between the content of the information and the medium used to transfer it from a transmitting to a receiving system. The sentence "the medium is the message" is the starting point of the McLuhan hypothesis and provides an instrument for the interpretation of the relationship between media and society. According to McLuhan, the medium can be considered as an extension of human possibilities, a tool for widening the field of action, either in material or in cultural terms. The innovative process of technological advance is principally responsible for the changes in the medium throughout the last millennium and, above all, in the last century.

McLuhan's thoughts seem to be relevant, as they focus on the relationships between the medium and the cultural infrastructure of a society. Every time there is a change of the nature of the extent of the medium, it is associated with a disturbance in the categories of perceived reality and in the individual's relationship with space.

In the contemporary era, telecommunications represents the current innovation. Definable as a medium in the McLuhanian sense, this instrument is believed to finally remove the obstacle of the physical distance. Telecommunications allows the contemporaneous transmission of information to a theoretically unlimited number of destinations. Thus the crucial cultural repercussions of telecommunications are that it eliminates space or, more simply, eliminates the category space in Euclidean terms. In this sense, the "message" embodied in telecommunications can be interpreted as the system of social, cultural, and productive opportunities stemming from the enlargement of the number of users and from its "real-time" aspect. The sensorial sphere of the individual widens and, theoretically, can become ubiquitous. Virtual reality technology is an example of the artificial extension of human capacities. Through this instrument an individual becomes able to perceive sensation, such as the sense of touch or smell, about realities located in remote places or, sometimes, in unreal environments.

Currently societies are being affected by a huge diffusion of information technology, whose products are becoming accessible to everybody and are likely to become necessary components of daily life. These strategic innovations can be seen in digitalization and miniaturization. The bit and the microprocessor are nowadays really the masters of current culture and design. These objects, when linked to the development of distributed computing, yield what is known as the Internet work environment. One common hypothesis is that the Internet can be considered as the medium, which allows the digital revolution to explode, following the same pattern as the Industrial Revolution in the eighteenth century. The latter caused the exponential increase of industrial production and, above all, of goods. The former permits a similar increase in information transmission. According to studies about the social mutations caused by technological change [8], the contemporary era is going through a painful transition to a new interpretative paradigm of reality, a "techno-communicative transition" from a sociocultural system dominated by communicative technology to a sociocultural system dominated by another communicative technology. Currently, humankind is experimenting with a techno-communicative transition from a system dominated by the analog and spatial communication technology of the Industrial Era to the digital and cyber spatial technology of the Informatics Era.

2.2.2 VIRTUAL VERSUS ACTUAL

Two phenomena can be considered the immediate consequences of the aforementioned current changes being related to a process of undermining the status of reality and, hence, becoming crucial keys to understand the revolutionary concept of cyberspace: deterritorialization and virtualization.

With respect to the deterritorialization, telecommunications allows reaching through the Net places located even quite far away in a very short period of time; even if the time of the so-called death of distance has not come so far, nevertheless a deep mutation affecting the concept of geographical space might result in the beginning of social uprooting and the progressive waning of the sense of belonging to a certain place. Hence, telecommunications can result in the absence of identity.

On the other hand, the virtualization can be interpreted as an activity connected to an enlargement of human actions and their perception of remote objects. As Steven Spielberg has foreseen, soon it will be possible to have neuronal and psychic contact with anyone on the planet. In this sense, the tele-transmission of sensorial experiences is the final objective, which has not yet been achieved by virtualization. According to Lévy, the cultural impact of new information technologies can be studied under the umbrella concepts of virtualization and of collective intelligence [9–11]. Lévy defines virtualization as a change of identity, a displacement of the ontological center of gravity of the case-study object. In his view, the virtualization of any entity whatsoever consists of discovering the general idea beyond it and of the redefinition of the starting reality as an answer to a precise question. In this way, virtualization makes the established differences fluid, increases the degrees of freedom, and turns the empty creative space into dynamic moving power [9].

Digital advances allow a virtualization of the concept of geographical displacement, until the sense of "*hic et nunc*" is dissolved, as is a feeling of cultural identity with a precise place. In the case of the transmission of information through the Internet, a text, an image, or a form are virtually present because they are available in whatever personal computer is connected to the Net; no location or address need be indicated. Telecommunications leads to situations where digital communities can meet and express their opinions together. Deterritorialization, in the sense of the contemporaneous presence in many places, can be seen as one of the characteristics of virtualization. Without the sense of geographical location, collective intelligence is able to evolve. It can be defined as a ubiquitously distributed, ever-present, real-time coordinated intelligence that leads to an effective mobilization of abilities [10]. It is now accepted among sociologists and communication philosophers that current telecommunications technology is able to generate a true digital culture. Interconnection seems to be the principal task of cyberculture, a new paradigm for the digital communities. The culture of cyberspace aims at a civilization of the generalized tele-presence [11].

In the remainder of this section, cyberspace is described with respect to the changes it determines on a variety of social and economic domains.

2.2.3 Cyberspace and Economy: Disintermediation and Destructuralization

The rise of cyber spatial patterns into entrepreneurship, finance, and commerce keeps on producing structural mutations that often bring benefits to clients by means of the progressive abolition of the intermediaries.

A study on the virtual enterprise in Italy detects a positive movement of large northwestern firms toward the introduction of ICTs into their management systems [12]. According to this research, Italian medium and small northeastern firms, while considered the engine of development for the entire country, risk being trapped in their current scarcity of digital infrastructures. The most relevant changes affect the relations within the production and delivery systems, while a collaborative attitude involves the firm and its external partners, which are considered not only as simple deliverers of services but also as contributors to the efficiency of the system in its whole. ICTs may be introduced along different patterns; they can support activities such as research of alternative delivery channels, customer relationship management, supply chain management, and enterprise management. The highest level of penetration of the ICTs corresponds to a reengineering process toward a new map of production and service/goods delivery processes. Nevertheless, it is worth considering the mutations that ICTs, and the embedded concept of cyberspace, provoke in the strategies of customer relationship management. One of the key concepts of digital commerce, the abolition of the intermediaries, is led by the possibilities opened by the use of the Internet as a common marketplace. Nowadays its users, the customers, have the opportunity to directly access digital catalogs and archives of goods and services, compare them, and judge the convenience of each purchase.

According to many scholars [13], the spread of cyberspace into commerce will bring extreme consequences to the already studied gap between economics of ideas and of objects [14–16]. Economics of ideas, information economics, are going to separate from the economics of goods, since the vector is fading into a less physical and tangible support. It is also possible to recognize the influence of cyberspace-inspired concepts on the calibration of novel econometric models related to the link between economies "located" in digital spaces and in physical places for urban domains [17]. The introduction of web-based patterns for presentation of the information about commercial products is predicted to abate in a few years the current system of consolidated comparative advantages due to imperfect information throughout the markets. With respect to selling strategies, soon it will be possible to solve the dilemma between depth and wideness. By means of strategies aiming at the digital affiliation of the customers, they are now becoming digital navigators and self-instructed miners of commodities. In this way, the traditional compensatory relationship between depth and wideness will be overcome. There soon will be a deconstruction process of the traditional roles and professions linked to commerce; somehow commercial information delivery services are likely to become more profitable than selling activities themselves.

On the side of finance, the development of computerized trading has led to an often-anonymous market environment. Deterritorialization acts as a potential cause of elimination of any difficulties connected to physical distance. These aspects parallel the rise of what we know as globalization of financial markets. Wider possibilities to directly access financial markets can open unimagined options for investors to browse into the catalog of products and choose the most suitable one for their own needs. Also in this case, the intermediaries, formerly the financial promoters, are going to be replaced by personal consultants, who will be in charge of guiding and suggesting appropriate paths to the investors. Deterritorialization also fosters the birth of parallel systems, such as Island, an electronic communication network (ECN) that hosts a number of electronic terminals connected to online unofficial, although actual, marketplaces. The ECNs display a series of advantages: they grant low transaction costs, and they allow buying and selling for a longer period of time each day, since the open time period is longer than in the official markets. Recently an ECN, like Island, applied to be recognized as an official stock exchange. In addition, other ECNs are willing to list themselves at the stock exchange [18].

2.2.4 CYBERSPACE AND NEW JOB DESCRIPTIONS

The first immediate, and perhaps also most quoted, consequence of telecommunications can be considered teleworking: every place, even home, when connected to a central organizing body, may become a workplace. Deterritorialization might result in the beneficial creation of an unexpected number of new jobs. Despite the hopes for this generalized 24-hour-work world, after more than two decades it is possible to state that, especially at a directive level, strategies should be set during face-to-face meetings when physical space, emotions, touch, and smell still do matter. What is recognized as a radicalization of teleworking, meant as the link between telecommunication and job strategies, is the widespread rebound effects of electronic remote control on almost every production process over the shape and role of traditional professionals. According to Rifkin [19], workers currently live in a post-market era ruled by digital technologies. The introduction of the ICTs implies a sharp reduction of the employees, since higher levels of productivity may be reached, encouraging, however, a deep transformation of the skills and education required of the incoming labor force. Knowledge workers represent the actors of the Third Industrial Revolution, since they are required to master the high-tech information. Among these professionals, web architects occupy a particular niche, which will widen its embrace. They are expected to acquire a high credit for the design of large-scale web sites, their maintenance and future development [20].

2.2.5 THE VIRTUALIZATION OF THE GOVERNMENT: TOWARD A DIGITAL AGORA?

How does cyberspace reflect upon the strategies of government reform? According to Lévy, the invention of new forms of political and social systems seems to be one of the main duties of contemporary humanity [11]. He stresses the opportunities offered by communication technology in the fields of political participation and representation. While in the past, one of the main obstacles to direct democracy was that it was impossible for a large number of people to collect in a single place, nowadays, a number of personal computer terminals could be used as diffuse interfaces between citizens and political bodies. There could be a revolution in political style, because of the innovative utopia created by dispersed decision-making. Cyberspace, according to Lévy, is to become the place where problems are explored and pluralistic discussion will focus on complex questions, where collective decisions and evaluations will be adjusted to the needs of interested communities [11].

Political institutions, however, seem to react slowly to these suggestions, since the changes in the ways of receiving and processing information imply a painful abandonment of the old political procedures and the start of a new era. The environment of this democratic decision-making would "take place" in a digital arena dispersed among many terminals participating in the political debate. In this way the problem of finding the meeting place for a great number of people can be overcome. Some signs of this mutation are already visible in many digital civic activities, such as social networks and online forums. But the way ahead is directed to scenarios where the simultaneous digital expression of the political ideas of each citizen will acquire an importance, which will be impossible to ignore. Real-time democracy needs new forums, new agoras, new places for socializing and government that help people and groups to recognize each other, meet each other, negotiate, and draw up contracts [11].

Through the Internet, each citizen could virtually participate in government processes. The current form of digital dialogue between governmental bodies and citizens takes place inside the civic networks. In these cases, the virtual agora means speeding up administrative processes and simplifying control procedures, since data can be transmitted to a virtually infinite number of users.

2.2.6 Cyberspace, Architecture, and Planning

Among the scholars who have conceptualized the influence of cyberspace on architecture, Maldonado refers to dematerialization [21] as a parallel counterpart of virtualization. According to his thought, just as, in "microphysics," the studies on subatomic processes have revealed the existence of antimatter, in "macrophysics," theorists try to suggest the development of similar paths toward the dematerialization. While he is skeptical about the rise of worlds populated by ectoplasms, Maldonado stresses the new role of virtual modeling. According to him and to Eco [22], semeiotics should receipt the changes of the nature of the vectors that bring the iconic meanings. Cyberspace seems to be acting either on the introduction of even less material digital models, as a means of design and support to knowledge and control, or on the use of lighter materials for building. During the Renaissance age, architects had a relative advantage over the other artisans, since they were able to previsualize the future products of their craft. The development of computing performances has enhanced their role of previewers, opening novel opportunities to redirect cyberspace in terms of the aid to design both a single building and a group of buildings within an urban fabric. A sort of obsession for space representation has characterized architectural curricula, while it has been considered a potential source of physical determinism by planning theorists. The contamination of planning with other disciplines, such as sociology, anthropology, economics, and statistics, brought as an immediate consequence a part of the evident beneficial effects for the foundation of a multilayered complex field, a clear, although transient, indifference of planners for physical space [23]. Langendorf recognized an appreciable development of the visualization methods, due to the higher performances allowed by current network-based information technologies. Three ages can be individuated [24]: during the 1980s, the birth of computer graphics and 2D digital representation with analogical use of movies, pictures, and audio documents; during the 1990s, the research of integration among different information systems to link spatial with other related multimedia information; and during the current age, the 2000s, the experimentation of further integration of systems, such as multisensorial systems, multimedia data sets, hypertexts, and geographic information systems, that enable the design of informational landscape, digital libraries, and electronic laboratories. In the information landscape, visualization of cyberspace can be interpreted as the creation of informational domains where knowledge is linked in a continuous virtual context, which opens new and unexpected scenarios for aiding the design. The evolution of the visualization techniques, inspired to cyber spatial modes, has followed a path along with representation and interpretation of information in a heuristic pattern able to support actions for planning and design [25–29].

2.3 CYBERSPACE AND PLANNING: COULD IT BE THE END OF GEOGRAPHICAL LOCATION?

How does cyberspace relate with planning?

It could be advanced that digital technologies contribute to a sort of attempt to change the nature of geographical space by mining its own physical distance-based properties. Deterritorialization might cause a transition from a cities-based to cyber cities-based world and society. Again, the absence of the sense of belonging to a specific location might imply also that cultural identity, based on geographical location, may be in danger of extinction. Thus, the focus of planning has changed; planners are now confronted with the task of managing cyberspaces. On the other side of the coin, planning itself has deeply changed: traditional blueprint professionals, used to drawing by means of pencils and afterwards to discussing their master plans with citizens and stakeholders, are currently engaged in a transition to soon become cyber planners, always connected to their digital draft plans, which most of the time will be considered in progress and will be distributed and accessible by 24-hour-living communities.

The disciplinary paradigms of urban and regional planning do not seem to be adequate to provide correct analysis and to deal with complex changes affected cyberspace, in its wider sense. Graham and Marvin confirm this crisis in the interpretative framework [30–31]. They complain that urban planning researchers and scholars are not very interested in the relationship between the digital field of telecommunications and the stony hardware of the city: "Urban analysts and policy makers still see cities through analytical lenses which actually have less and less to do with the real dynamics of telecommunication-based urban development" [30, p. 48].

Batty agrees with them: "Understanding of the impacts of information technology on cities is still woefully inadequate" [32, p. 250]. The specialist literature itself shows the signs of a sort of scientific inertia, since the attempts to classify do not go beyond the metaphorical transposition between the dual virtual/actual fields and avoid describing the real changes induced by digital telecommunication into the city. Graham and Marvin [30] and Couclelis [33] after them quote more than twenty different terms coined ad hoc for illustrating the revolutionary nature of cyber cities.

However, the dichotomy of urban places/electronic spaces seems to leave the directions of future research open. The key to the problem is the correct interpretation of the related material and immaterial flows between city and hyper city. These are characterized by synergy and not only by simple duplication of social fields of study.

The unspoken background of the above problem is the need to establish new paradigms for urban and regional planning. In this transition process, planners have to adapt to the demands of new spatial settlements and infrastructure, listening to both the displaced and the digital communities. Digitalization encourages changes in the types of planning tools through the introduction of digital formats and the need to negotiate digital draft procedures. The imperative seems really to be to discover the new sense of location displayed by the "collective intelligence."

Nevertheless, planning still seems to be connected with geographic systems of real displacements, even if telecommunications allows people to work without moving, to vote without going to the ballot box, or to watch movies without entering a cinema. This global interconnection, through virtual presence, means an expansion of opportunities and also of the need to move, act, travel and picture.

The rise of the Internet mode of exchanging information truly opens aspatial ways of relating with others. Even without the indication of addresses and locations, the Internet is configured as a "place" where it is possible to meet people, to work, and to live an associative life. In this respect, William Mitchell describes the place Internet,

[Internet] subverts, displaces, and radically redefines our perceived conceptions of gathering place, community, and urban life.... The Net negates geometry.... The Net is ambient — nowhere in particular but everywhere at once. You do not go to it; you log in from wherever you physically happen to be [34, p. 8].

Simultaneous contact admits the existence of a third dimension, the "real time," beyond space and time. It is easy to understand how the system of geographic spaces implies different relationships among its points, with respect to the relationships linking the points of the virtual spaces. These fields have different topologies.

Planners are engaged in interpreting the evolution affecting the topology of urban environments, while bearing in mind that there are important interactions, sometimes invisible, from electronic spaces. Telecommunications modify the sense of living and the related architectural design. They modify regional relationships and the planning processes connected to them. Virtual locations dominate real situations, as in the case of telecommanded houses or of telesecured offices.

In this "digital era" [35], professionals have to think about their working instruments and disciplinary paradigms. Their subjects are going to change and be complemented with elements coming from different subjects, such as geomatics, geographic information science, remote sensing, and fractal and cellular modeling. This era seems to be characterized by the use of network cooperation between remote professionals and scientists. According to Howkins [36], who describes the transition to a new style for planning, the old style planner talked about physical zoning, the balance of employment, housing, and open space and traffic flows. In contrast, Howkins stresses how the new style planner, which might be termed the cyber planner, has to consider the configuration of electronic systems and local area networks (LAN) and the provision of bandwidth to each urban area. The town planner dealt with the stock and flows of vehicles. Today's public authorities have to face the stock and flows of information [36, p. 427]. Furthermore, according to Machart, "Telecommunication is becoming a new component in urban and regional development planning. [The] desire is to use telecommunication as a structuring element in cities and regions and to incorporate telecommunications in economic and social development" [37].

The actual challenge is to interpret how the suggestions of high-tech solutions for communications can be used to design new relationships and cultural geographic spaces. According to Mitchell, the physical integration of electronic devices will characterize future planning and design practice: "... architects and urban designers must gracefully integrate the emerging activity patterns created by pervasive digital telecommunication into the urban forms and textures inherited from the past" [38, p. 35].

2.4 CONCLUSION: CONFRONTING CYBERSPACE AND CYBER PLANNING TO SUSTAINABILITY

The variety of declinations of cyberspace introduced above can be thought to constitute an ideal basis to translate into current practice some of the most important and often-abused concepts inspired to sustainable development.

The solemn declarations formulated at the end of the well-known conferences held in Rio de Janeiro in 1992 and in Johannesburg ten years later (Rio +10) seem to agree on this topic: achieving full access to information in order to strengthen the deliberative capacity embedded in groups of as many citizens as possible [39, p. 102]. This is believed to be the basis for increasing the level of empowerment of local societies and stimulating self-driven patterns of decision-making and planning. Furthermore, according to Agenda 21, the subsequent operative document, one of the most important tasks in a process toward sustainability should be "improving the use of data and information at all stages of planning and management" [40].

Information can be made entirely open and accessible either by disseminating it to remote communities and groups or by bringing those societies to it. In the last hypothesis, cyberspace might play a leading role, by inducing innovative channels for digital information distribution and exchange, by individuating and constructing common, sharable, and thus transparent datasets, and by opening an era of collective and interactive processes developed by local societies on self-built scenarios. The institution of a common and always-accessible informational endowment can be considered a fertile humus for encouraging the diffusion of behaviors inspired to Local Agenda 21 protocols, with respect to trustful, transparent, consensus-built, and self-reliant planning. In this perspective, tools for managing, enhancing, and distributing (spatial) information are particularly welcome: web-based maps, GIS, images, movies, other multimedia, checklists, networks, forums, and newsgroups are the necessary bricks to conceive innovative digital planning environments. The supply of these tools is already well grounded on a wealth of software and GI-based applications available online; on the other side, though, the social demand might not meet this level of diffusion. A widespread and acceptable level of social trustfulness for digital processes and tools is still lacking; this constitutes one of the most difficult barriers to a current practice of cyber planning. After creating a common ground for the culture of bottom-up self-planning, and sustainability, society should produce its efforts for reducing the large digital gap that still divides information-rich domains in cyberspace from the corresponding information-poor excluded communities in the geographical space.

While the broadness of this mission cannot be deferred only to a single kind of institution, public sector bodies seem to be directly charged with the commitment of introducing local communities to the potentials of cyberspace and planning, by displaying, and often also explaining, the revolutionary meanings of activities, such as online retrieval, manipulation, editing, and interactive upload of each one's own informative experience to a common spatial database [33]. In many cases, municipal web sites show an important effort for the diffusion of the culture of digital geography and information and thus of cyber and shared planning [41].

These can be considered the necessary steps toward the construction of what might now be termed "informational digital heritage," the personal endowment communities actually leave to their future generations.

REFERENCES

1. Un miliardo di persone userà il Pc entro il 2010, Vunet, August, 3, 2004, http://www.vnunet.it/detalle.asp?ids=/Notizie/E-business/Mercati/20040803006&from=hemeroteca&pagina=1.

2. Newman, M.E.J., The structure and function of complex networks, *SIAM Review,* 45, 167, 2003.
3. United Nations, Economic and Social Council, *Second Annual Report of the Information and Communication Technologies Task Force,* New York, 2004.
4. Gibson, W., *Neuromancer,* 1st ed., ACE Books, Penguin Putnam, New York, 1984.
5. Batty, M., The computable city, *Online Plann. J.,* http://www.casa.ucl.ac.uk/planning/articles21/city.htm.
6. Shiode, N., An outlook for urban planning in cyberspace, *Online Plann. J.,* http://www.casa.ucl.ac.uk/planning/articles21/urban.htm.
7. McLuhan, M. *Understanding Media. The Extensions of Man,* Reprint ed., MIT University Press, Cambridge, Mass., 1994; originally published in 1964, http://www.ifi.uio.no/~gisle/overload/mcluhan/umtoc.html.
8. Berardi, F., *Mutazione e cyberpunk. Immaginario e tecnologia negli scenari di fine millennio,* Costa & Nolan, Genova, 1994.
9. Lévy, P., *Qu'est-ce que le virtuel?* Éditions La Découverte, Paris, 1995.
10. Lévy, P., *L'intelligence collective. Pour une anthropologie du cyberspace,* Éditions La Découverte, Paris, 1994.
11. Lévy, P., *Cyberculture. Rapport au Conseil de l'Europe,* Éditions Odile Jacob, Paris, 1997.
12. Capitani, G. and Di Maria, E., Le nuove tecnologie dell'informazione e della telecomunicazione come fattore strategico di sviluppo locale, in *Distretti industriali e tecnologie in rete: progettare la convergenza,* Micelli, S. and Di Maria, E., Eds., Franco Angeli, Milano, 2000, 41.
13. Evans, P. and Wurster, T.S., Strategy and the new economics of information, *Harvard Bus. Rev.,* Sept.-Oct., 10, 1997.
14. Romer, P., Endogenous technical change, *J. Political Econ.,* 98(5), S71, 1990.
15. Romer, P., Idea gaps and object gaps in economic development, *J. Monetary Econ.,* 32, 543, 1993.
16. Negroponte, N., *Being Digital,* Alfred A. Knopf, New York, 1995ı
17. Shibusawa, I., Cyberspace and physical space in an urban economy, *Pap. Reg. Sci.,* 79(3), 253, 2000.
18. Magrini, M., *La ricchezza digitale. Internet, le nuove frontiere dell'economia e della finanza,* Il Sole 24 Ore, Milano, 1999.
19. Rifkin, J., *The End of Work—The Decline of the Global Labor Force and the Dawn of the Post-Market Era,* Putnam & Sons, New York, 1996.
20. Russo, P. and Sissa, G., *Il governo elettronico,* Apogeo, Milano, 2000.
21. Maldonado, T., *Reale e virtuale,* Feltrinelli, Milano, 1992.
22. Eco, U., *Sugli specchi e altri saggi,* Bompiani, Milano, 1985.
23. Langendorf, R., Visualization of architectures and cities, *Urbanistica,* 113, 159, 1999.
24. Langendorf, R., Computer-aided visualization: From applications to information environments and the implications for planning and urban design, in *Proceedings of the 7th International Computers in Urban Planning and Urban Management Conference (CUPUM),* July 18–20, 2001, University of Hawaii at Manoa, Honolulu, 2001.
25. Engeli, M., Ed., *Bits and Spaces. Architecture and Computing for Physical, Virtual, Hybrid Realms, 33 Projects by Architecture and CAAD,* ETH Zurich, Birkhäuser, Basilea, 2001.
26. Engeli, M., The digital territory, in *Bits and Spaces. Architecture and Computing for Physical, Virtual, Hybrid Realms, 33 Projects by Architecture and CAAD,* ETH Zurich, Engeli, M., Ed., Birkhäuser, Basilea, 2001, 83.

27. Engeli, M. and Miskiewicz-Bugajski, M., Information landscape and dreamscape, in *Bits and Spaces. Architecture and Computing for Physical, Virtual, Hybrid Realms, 33 Projects by Architecture and CAAD,* ETH Zurich, Engeli, M., Ed., Birkhäuser, Basilea, 2001, p. 75.

28. Sibenaler, P., Visdome, in *Bits and Spaces. Architecture and Computing for Physical, Virtual, Hybrid Realms, 33 Projects by Architecture and CAAD,* ETH Zurich, Engeli, M., Ed., Birkhäuser, Basilea, 2001, p. 157.

29. Cooper, M. and Small, D., Visible language workshop, in *Information Architects,* Wurman, R.S., Ed., Graphis, Zurich, 1996, p. 202.

30. Graham, S. and Marvin, S., *Telecommunications and the City. Electronic Spaces, Urban Places,* Routledge, London, 1997.

31. Graham, S. and Marvin, S., Planning cybercities: integrating telecomunications into urban planning, *Town Plann. Rev.,* 70(1), 89, 1999.

32. Batty, M., Invisible cities, *Environ. Plann. B Plann. Design,* 17, 127, 1990.

33. Couclelis, H., The construction of the digital city, *Environ. Plann. B Plann. Design,* 31(1), 5, 2004.

34. Mitchell, W.J., *City of Bits, Space, Time and the Infobahn.* MIT University Press, Cambridge, Mass., 1995.

35. Batty, M., Evaluation in the digital age, in *Evaluation in Planning,* Lichfield, N. et al., Eds., Kluwer Academics Publishers, Dordrecht, 1998.

36. Howkins, J., Putting wires in their social place, in *Wired Cities: Shaping the Future of Communications,* Dutton, W., Blumler, J. and Kraemer, K., Eds., Macmillan, New York, 1987.

37. Machart, J., Roubaix Euroteleport, *Technopolis Int.,* 3, 1994.

38. Mitchell, W.J., The era of the E-topia: the right reaction to the digital revolution can produce lean and green cities, *Architectural Rec.,* 3, 1999.

39. United Nations, *Report of the World Summit on Sustainable Development,* Johannesburg, South Africa, 26 August–4 September 2002, Δ/conf.199/20, http://www.johannesburgsummit.org./html/documents/aboument.html.

40. United Nations Division for Sustainable Development, *Agenda 21–Chapter 8,* http://www.un.org/esa/sustdev/documents/agenda21/english/agenda21chapter8.htm.

41. Campagna, M. and Deplano, G., Evaluating geographic information provision within public administration websites, *Environ. Plann. B Plann. Design,* 31(1), 21, 2004.

3 Theories of Digital Participation

Robin S. Smith

CONTENTS

3.1 INTRODUCTION

Although "sustainable development" lacks a universally accepted definition, it can be seen as a policy area that attempts to draw together, compare, and resolve economic, social, and environmental issues as a principle or "working ethic." The inclusion of the social factor not only adds an important dimension to economic/environmental problems but also identifies the need for local actors' support, particularly through the policy area of Local Agenda 21. It is public participation that primarily draws together citizens and decision-makers in this context so that information can be obtained, understandings increased, and solutions reached. However, in as much as definitions of sustainable development can vary, "public participation" is equally difficult to discuss, and the simplistic way that many in research and practice view it needs to be challenged. Participation is not a unique or shared construct, and failure to recognize differing views can lead to unsatisfactory outcomes for all. With this comes a need to understand the ways actors choose to become involved in public participation and the methods they use, from the perspective of both a participant and those that wish to consult.

One particular context involves Internet-based activities, which provide interesting avenues for research and opportunities for fuller forms of participation. Research by Smith has investigated the role of information and communication technologies (ICTs) for public participation in U.K. local authorities, or "digital participation," not only exploring the hyperbole of the information revolution but also offering greater insight into the nature of public participation in the digital age [1]. The general findings from this research also provide a means to explore the wider role that geographic information (GI) and associated technologies can provide for public participation, particularly where there has been increased interest in sharing GI across the Internet through spatial data infrastructures (SDIs) and the application of participatory approaches using GI (PAUGI) that were examined in a recent transatlantic research agenda [2].

This chapter can only offer an introduction to public participation, highlighting some of its main features and issues that require further work. Firstly, a framework is discussed that demonstrates the various "components" that come together to describe people's understanding of participatory activities, mapping out actors' notions of participation, the issues under consideration, who is involved, views of the outcomes of activities and the methods employed. A discussion of the literature in this field, mainly drawn from political and planning theories, helps the reader to understand the range of participation's complexity and how activities can differ in online environments. Secondly, as space is often a connecting point for many sustainable development problems, the emergent role of PAUGI is also considered alongside the theme of "access" to activities in a very broad sense. Before turning to the theoretical discussion it is useful to outline some examples of "digital participation."

3.2 DIGITAL PARTICIPATION

In the spring of 1999 a survey was made of around 300 U.K. local government websites (.gov.uk) to find examples of public participation online. The survey looked at the broad themes of environment, planning, governance, and community, with Local Agenda 21 being one of the key areas for exploration. Content, in terms of the amount of information online, was found to vary greatly, and methods of communication ranged from a local authority's switchboard number on its homepage to online chatrooms and bulletin boards.

To "classify" the participatory nature of the websites, these two features of communication and content were analyzed using Arnstein's ladder of citizen participation [3]. This frequently cited model provides an initial means to contrast instances where the public have a limited say (toward the bottom "rungs" of the ladder) to those occasions where they are given full control, toward the top. The "most" participatory examples in this government-driven/"top-down" setting were placed in the middle rungs of "consultation" and "partnership," accounting for 13% of those websites surveyed, but with examples from all levels of local government in Great Britain. Below this was a group of websites that provided information (29%), but whose content was limited, or where no evidence of active participatory activities could be found. Equally common were websites that tried to replicate the organizational structure of the authority (28%), typified by "a–z of services" that frequently acted as online telephone directories of service departments or officers. Less participatory still were those websites that appeared to advertise their areas for economic development or tourism purposes (16%). Often graphically intensive, these would have been rated as "good" websites in other surveys, but they did not provide information for a potential participant to become involved. The last two categories included those with very limited content (6%) and those that could not be accessed after several attempts (8%), potentially offering the greatest barrier to digital participation.

To understand the forms of participation taking place in the leading examples, it was important to look behind the "digital façades" of the websites (and associated methods) and to explore the social context of the technology, through interviews with officers in several authorities. Three leading cases were then chosen for in-depth case-study analysis that, importantly, included interviews with citizens who had participated online.

The first case occurred at a local policy level through Rushcliffe Borough Council's interim local plan consultation exercise, where residents were asked to respond to a housing allocation from Nottinghamshire County Council and central government. A leaflet was sent to every household, a dedicated website and an e-mail address were established, and public meetings were held throughout the area. The exercise generated a great deal of public interest compared to previous activities, and there were just fewer than forty e-mails sent as formal responses. Interest also led to residents groups generating a number of petitions and a "standard letter" that residents were asked to add comments to, sign, and send to the authority.

The second case was at a strategic policy level, through the City of Edinburgh Council's pilot "community plan" consultation exercise that related to the "sustainability" of the city. The local authority and its community plan partners (local businesses, voluntary groups, other public sector actors, etc.) had developed a draft document for wider consultation with selected representative groups in the city and the general public. This was made available in print but also online, through a dedicated website and e-mail address. The local authority wanted a wide range of opinions and held several meetings with targeted representative groups. They introduced a telephone call center to conduct a survey with their "citizens' panel" (a demographically representative group of around 1,000 residents) and provide a service that allowed some participants to telephone their responses. Compared to Rushcliffe, overall response rates from the general public were lower, and only twelve e-mails were received.

The third case was perhaps the best example of digital participation at this time. The London Borough of Lewisham was involved in the pan-European Dialogue Project that used sixty selected members from their citizens' panel to inform elected member decision-making through facilitated chatroom discussions, e-mail, bulletin boards, and in-person meetings in the council chamber. This group included people who had never used computers before and participants received training, their own computer, free Internet access, and support, for both technical and project-related activities. Usage was high, and the project received additional funding from the authority to extend it for several months.

From the survey and the cases it should be noted that all these are examples of public participation. Some may appear more successful than others, but this assumes that response rates, in a broad sense, are the only outcome by which success can be measured, which can be misleading. From this empirical starting point it is possible to identify five main components that those inside and outside organizations will use to construct their understanding of "public participation": notions of participation, issues, audience, outcomes, and methods. The discussion draws on these cases and reflects upon the possible conditions involving GI.

3.3 NOTIONS OF PARTICIPATION

As notions of participation explain why people participate, and much about the nature of participation itself, they take a dominant role in the literature. Daniels et al. divide this material into four categories (theoretical perspectives, strategic objectives, inventories or explanations of techniques, and evaluations of agency implementation [4]), but few have looked specifically at the contribution and role of methods, particularly for digital participation or PAUGI. Such categories also demonstrate that "participation" has both theoretical and practical components that can be readily examined and evaluated. Similarly, several roles for participation have been identified where participation can, for example, "further democratic values, ... educate the public [and] enable social or personal change" [5]. Although these competing goals are relevant to the top-down focus of the cases, less theoretical comment is offered about grassroots activities, and there is limited opportunity to discuss them here. An issue then arises about participation and "power relations," where citizens' say can be limited by decision-makers [6].

Holden sees participation in terms of "deciding on ideas" and "choosing among options" [7], with Pateman suggesting that if a process is used to gain the acceptance of ideas, other than the citizens' own, then this is only "pseudo-participation" [8]. For Pateman, "genuine consultation" must occur before agenda-setting, and if final decisions are made by those outside the "rank-and-file," then this is merely "partial participation." For all the cases mentioned above, the public was choosing among options, because the authorities controlled most of the activity by initiating consultations, selecting certain methods, and supplying particular information. The authorities also expected responses to be formed in certain ways that were formal, structured, and often written (except for the call center). This influenced (or actively selected) which citizens would participate and how they could contribute, simultaneously impacting on their ability to access the process. As such, the examples cited exhibit partial participation but variation occurs between the cases. It could be argued that Edinburgh's approach was too strict and that there was no opportunity for potential participants to contribute to the draft consultation document as partners in the community planning process. In contrast, although Rushcliffe wanted to hear residents' concerns about a narrow issue, these participants had more opportunity to express varied opinions, and in Lewisham, genuine discussion was promoted and supported by the facilitator.

Such variation can be theorized through Arnstein's ladder, but it should be noted that this model has a number of flaws. Least of all is its structure: that by being a ladder it is a continuum; that one is forced to ascend it; and that (once climbed) one reaches the pinnacle of "citizen power," something that may not be appropriate in all settings involving governments. Secondly, although it provides a useful starting point, Arnstein's ladder was developed for a specific context of the U.S. civil rights movement in the late 1960s, and some have started to question relying on it to describe participatory activities [9], with other theories offering useful avenues that explore some of the concepts built into the ladder.

One such framework is put forward by Christiano, who outlines three different positions of citizens and their wills: the direct, constructive, and epistemic conceptions [10]. In the direct conception, "one's participation in making laws is a direct expression of one's will" — representing exactly what a participant wants. In the constructive conception, "one's participation is an attempt to define what one wills" — by participating, citizens are trying to understand the activities they engage in. In the epistemic conception, "one's participation is an attempt to discover what one wills with regard to political society" — participation is a means to understand political problems but participants may not know what they desire.

These three conceptions relate directly to choices that individuals make. In some instances, citizens participated because they wanted a policy to be shaped in a particular way, a form of the direct conception. The constructive and epistemic conceptions can be seen where interviewees, both inside and outside organizations, saw participation as a more exploratory activity, closely associated with a right and/or duty to participate, something that Holden relates to citizen theory [7]. Actors also saw participation as a means to impress some of their principles and beliefs upon a general process and not particular ideas in response to a plan or policy, reflected in the Not In My Back Yard-ism (NIMBY-ism) associated with neighboring developments in Rushcliffe or having transportation issues emphasized in Edinburgh's community plan. Interestingly, both have clear spatial characteristics as parcel-based and predominantly linear features.

However, actors were quick to recognize the thoughts and actions of others, particularly where the mass of people come together to make decisions about policies that affect their lives, or the "popular sovereignty" discussed by Christiano [10]. Citizens stated they wanted the "common opinion" to be listened to, and officers felt they needed to take account of participants' "representative" nature. A "common opinion" introduces the "incompatibility problem," where an individual's ability to express ideas is impacted upon when many other participants are involved. Similarly, a "representative" set of participants is ambiguous (discussed below). People may also become involved for other reasons, such as learning to use a computer or socializing, highlighting the important role of methods to attract participants. This applies equally to modern digital technology and to traditional methods, where some would view a public meeting as a form of theater or an opportunity to socialize. One set of ideas that begins to look at a more social participatory context is discussed by Holden through instrumental, developmental, communal, and philosophical arguments [7].

Under instrumental arguments, participation is not an end in itself but is instrumental in achieving another objective, as people protect their own interests by participating in the decisions that affect them. This rests well with the idea of NIMBY-ism, and Holden feels that this benefits from a nondilution effect, where protected interests mean more to the individual.

Through developmental arguments, participation is seen as valuable in itself and is thought to be greater than a means to an end. Participation has an educational function, leading to "political efficacy," which Birch thinks provides confidence with

decisions, adding to participants' sense of control and increasing their overall participation [11]. Similarly, Christiano feels that participation teaches the individual about
"the nature and importance of the community and of their place within it," partly solving
the incompatibility problem [10]. However, there are both physical communities and
"communities of interest" that can be heavily dispersed, showing a variation in spatial
distribution and potentially impacting on the role methods can have.

Communal arguments expand this theme, and Holden notes their benefits to the
state, as public participation increases a decision's legitimacy and provides a "political obligation" for participation. However, this could potentially "force" people to
become involved who do not necessarily want to be, not recognize those who choose
to abstain, or incorrectly label them as "contented."

Philosophical arguments, by contrast, are the hardest to obtain and "relate to
basic theoretical issues and contend that only in participatory democracy can they
be resolved," perhaps tied to the epistemic conception. Participation is thought to
fill "the vacuum" between individuals and their governments by having all involved
in what Holden sees as a "proper participatory democracy" [7]. Participation is seen
as the "fundamental nexus" between the community and the individual, between the
individual and the state, and between individual autonomy. This nexus provides a
theoretical means to question some of the possible advantages that ICTs and GI(S)
brings to public participation, connecting citizens who wish to debate with each
other and the authority and drawing together the necessary information to make
informed decisions. Although interconnectivity is influential, such a view requires
a sense of "universal enlightenment" and assumes that individuals are equally capable of contributing, an issue that can be related to our understanding of "access" in
a participatory context.

Access can be related to equality and openness, and "participation" and "access"
can be seen to exist on the same spectrum of interaction between citizens and
government. When governments only offer information, then participants' access is
low, but once public opinion is sought, a two-way process of access begins, helping
to generate more robust democracies. This can be seen in the variation between the
surveyed websites, where some offered greater access to information, sought more
public opinion, and were seen as more participatory. In addition, "access" can have
other meanings, but it is not possible to expand on their "notions," although the
broad concept reemerges several times throughout this chapter.

This discussion demonstrates the ways in which participation can be viewed
from a theoretical perspective, but it also shows why people can have varying views
or notions about what "participation" may involve. However, it is important to
recognize that such ideas only present part of what can occur in participatory settings,
and that the remaining components also have significant roles to play.

3.4 ISSUES

"Issues" relate to the topics or concerns that actors may have. Pacione suggests that
there is a competition between creating "effective administration" through central-
izing activities and attempting to obtain "maximum accountability ... [that requires]
greater decentralization" [5]. Participation can operate at different scales, through

different levels of government, and within varying departments, for example. In part, this competition may help governments to outline which issues are suitable for public participation. This can be related to the problem Edinburgh officers had when trying to find a representative public and a practicable strategic activity within a set time. In contrast, Rushcliffe's local exercise occurred because of a county council's strategic policy, and unitary authorities have issues that often span strategic and local contexts, as found in Lewisham. This variation was partly the basis for case selection, but specific matters should also be of concern.

Illeris's work in Danish land-use planning showed that most citizens were effective when commenting on very local issues, with some capable of viewing a "greater context" [12]. The quantity of responses in the cases would seem to reinforce this point, where the strategic community plan had less response than the local housing allocation. Participants in Lewisham, in contrast, demonstrated Illeris's "greater context" and developed a degree of political professionalism. During a discussion on library closures, panel members made tactical rather than general suggestions, indicating their ability to deal with issues that did not necessarily immediately apply to them in either a physical, social, or emotional sense. As such, certain issues will appeal to different audiences, impacting on the numbers participating.

Once a consultation process is in place, an audience may have their own issues and use it as a means to voice their general concerns, relating well to Illeris's suggestion that other "political problems" will surface during consultation [12]. For example, in Rushcliffe some citizens presented issues that should have been expressed during the previous County Council's structure plan consultation exercises. Officers expected this but noted that such submissions could not be formally accepted. Interviewees also suggested that responses could have been greater if some residents understood that, unlike the local plan, the structure plan's allocation was not fixed. Because citizens see local authorities as one entity, they frequently do not recognize separate issues, consultation exercises, departments, or services, particularly where services are delivered by more than one authority. It is important that those involved in initiating exercises not only expect this but also make potential participants aware of the restricted nature of an exercise, explaining how a participant's contributions will be dealt with if they do not meet the focus of the exercise.

In contrast, Holden sees "bottom-up" (or grassroots) activities that include more "individual specialization," where citizens have their own questions to ask of experts and have "special interests" [7]. Participants are more likely to group and become organized around an issue in the first instance, either for a specific location or an issue, as a community in space or one of interest. Specifically, the spatial element also appears at varying levels and issues where:

- The development of a building is normally very "local"
- The regeneration of city centers can impact heavily on "neighborhoods"
- "District-wide" policies occur for activities such as housing allocations or community plans (noting a difference in the authorities that may implement them)
- "Regional" issues that occur for matters such as the site of a new national park or the location of an oil refinery

This last level may accrue enough interest to become a national or even international issue, and all relate to sustainable development. The research agenda on PAUGI offers examples of these issues, particularly relating to "jumping-scale," where issues in disparate communities gain more importance as they begin to recognize shared common problems [2], perhaps initially identified through electronic fora. Such circumstances raise questions about who the audience for an issue should be and who they are in practice.

3.5 AUDIENCE

A discussion of the "audience" of a participatory exercise draws on two themes. The first relates to the nature of those actors from outside and inside organizations who are involved in activities. The second is the interaction between elected representatives and the people they represent that impacts on democracy.

"External audiences" vary, and a single "public" does not exist. An authority's desired external audience may include responses from groups who would not normally participate, shown by the desire to recruit younger participants to citizen panels or offering to translate participatory materials into community languages. Such intentions draw on ideas of equality of access and similar notions of participation. The actual external audience of participants who respond can be contrasted with the "representative" (geo-) demographic of participants an authority wants. A difference also emerges between how an authority will want to create such groups, for example: "young people," "the elderly," "residents," "ethnic minorities," "the socially excluded," "middle-class homeowners," or even "those with Internet access" (local government officers), and how citizens choose to view themselves. In part, this relates to the analysis and implementation of participatory activities and how organizations analyze citizens' contributions, but it also demonstrates the complexity of social entities as an external audience.

Participants in land-use planning have been classified as major elites (e.g., other local authorities), minor elites (e.g., community councils), and individual members of the public [13]. A desire to consult with ready-made community representatives relates well to these elites (as found in Edinburgh), but there is possibly a need to include another category between the individual citizen and the minor elites. This takes the form of organized, but possibly unexpected, reactive and rapidly created groups of citizens, such as the Rushcliffe residents' associations. Similarly, in previous accounts of participation in planning contexts, there has been another "elite" that Thomas characterizes as educated, middle-class, middle-aged, and predominantly male [6]. Although the demographics in the cases were not complete, younger people did not seem to be participating, online or otherwise, and the majority of participants interviewed appeared to reflect the findings of Thomas, but a question remains about their dominance. The Lewisham case, by comparison, showed that it was possible to bring together people from a variety of backgrounds to participate in online activity, including those with no experience of computers.

In contrast, internal audiences can include officers from other service departments, those involved in improving service (such as the United Kingdom's Best

Value policy, which draws on public participation to democratize and continuously improve public services), politicians, or possibly other public sector organizations (as found in Edinburgh's plan partners). Internal audiences may help to select the form of the exercise, either by providing guides to participation (found in Edinburgh and Lewisham), producing consultation methods and materials (as in Rushcliffe), or recognizing the need to employ a facilitator from outside of the organization (as in Lewisham) to minimize the influence of the organization on participants' issues. Internal audiences may also influence budgets for exercises or not see the need for a consultation, as O'Doherty found in a survey of senior planning officers [14]. Recognizing the role of various actors inside and outside organizations in shaping exercises is an emerging research area, including work by Tait that utilizes an actor-network theory approach that makes planning documents an equal "voice" alongside the officers, politicians, and participants [15].

Public participation in local government is often imbedded in the democratic relationships between politicians and citizens, both practically and theoretically. This in itself can construct what participation means and should be viewed alongside the discussion of notions of participation. Holden discusses "conventional" and the "radical" forms of representative democracy that help to explain some facets of the nature of an external audience [7]. Under the conventional system, citizens believe their representatives are more knowledgeable than themselves and that their representative is a "trustee" of public opinion. In contrast, representatives in the radical perspective are delegates, "conveying the policy decisions of their constituents."

The cases from this research seem to relate closer to the radical perspective, but a number of the citizens interviewed defined participation as supporting elected members' decision-making, seeming to demonstrate a belief that politicians are better equipped to deal with certain issues. Arblaster suggests that, in order to achieve a successful participatory democratic system, there needs to be wide, free, and open discussion that is accessible and where representatives have a readiness to listen [16]. Some citizens felt that the authority was not listening to contributions, but, remarkably, they simultaneously wanted to remain part of ongoing processes. As such, some actors saw certain facets of participation as welcome but hidden within the overall activity, whereas others rejected the same facets or did not recognize them. Internal and external audiences can, therefore, change their views of an exercise as it travels through varying stages of the process, including periods long after exercises have finished.

In another relationship, Birch notes that participation can be seen as the "self-determination" of "amateurs" or "codetermination" that utilizes the aid of professionals [17]. Birch prefers the latter, because complicated information or expert knowledge will then be available to all those involved, perhaps required in many PAUGI settings, given the complexity of GI(S). An example of codetermination was found in Lewisham, where a policy officer was invited into a chatroom to answer citizens' questions. Like all the cases, this was somewhat incomplete, because final decisions were taken by elected members so that those who did not participate were given a voice, reflecting Arnstein's "consultation" rung and generating particular outcomes.

3.6 OUTCOMES

There is little theoretical discussion relating to the outcomes of participatory exercises or what they mean. In part, this can be related to the traditional linear view of consultation exercises, where the publication of response rates relates to the decline of an organization's interest in an activity. Outcomes can be actual or perceived, and the difference between the two can influence the types of activity likely to take place. Actual outcomes include the ability to gain democratic legitimacy for a solution or the completion of "successful" activities. Although identifiable, they often have less influence on the nature of participation than perceived ones. This is compounded by the idea that different actors' objectives/notions can vary greatly and that they may not be clearly stated [13]. An authority may desire a certain level of participation from the public, want useful contributions to inform their decision-making, see some contributions as less relevant, inform the public about certain issues, and complete an activity "appropriately," using the correct methods and gaining a "representative" voice from their public. The authority may also have more negative views, such as seeing consultation as unnecessary or believing that the public will not have understood the importance of the issue and failed to respond. Similarly, citizens' outcomes may include the adoption of their ideas in policy, that they will have performed their duty, that they will have learned something about their environment, a policy, or a technology, and/or that they will have socialized with their neighbors and friends. Their more negative views may include that the authority is not listening. These ideas present some of the concerns or risks that may occur in participatory settings and, in many ways, the questions that actors will tacitly ask themselves.

This can lead to different understandings of what forms of participation are being offered by the authority or desired by the potential participants. Given varying notions, the same outcome will not necessarily be treated by different actors in similar ways. This emerges through the documents that are created as an outcome of an activity, often produced by officers or consultants for elected members. There is possibly a need to consider better ways of communicating "findings" that preserve some of the nuances of participants' responses and relate more to qualitative rather than quantitative analysis, influenced by the methods used.

3.7 METHODS

The term "methods" relates to the ways in which people engage in participatory activity. This could include leaflets, meetings, exhibitions, proposed policy documents, questionnaires, and letters. It also relates to the media that can be used to communicate this information from "traditional" accepted communication modes, such as the postal service and the telephone, to more complicated "digital" methods involving ICTs, such as e-mail, websites, and chatrooms. Such digital methods are closely related to GI technologies because of the increasing use of the Internet to share and analyze spatial information (through spatial data infrastructures), and these methods will continue to rely upon communication similar to the methods discussed in the cases.

Pacione suggests that participation "consists of many different approaches" and refers to the methods that can be employed (37 in all under 6 headings) [18]. He also suggests that a "technique" (method) must fulfill the "ideas" (notions and issues) of both experts and citizens (audience), and that a different strategy will be appropriate for each set of goals (outcomes). However, Shucksmith et al. recognize that "the same method may be used in a participatory or manipulative manner" [13]. Additionally, digital and traditional methods cannot be seen as separate, and certain methods may apply to certain audiences. For example, in Rushcliffe and Edinburgh traditional leaflets and press adverts were used to guide participants to online facilities, taking particular audiences from one method to another. Alty and Darke note that "any programme of public participation must include a range of techniques and approaches if it is to be more than tokenist" [19]. An activity that only involved digital methods would have only represented the views of the "digital haves." It is not likely that a tool can be produced with participative ideology in mind, unless all actors have input into its design. Even then, users will use and abuse that technology for their own purposes or desired outcomes, consciously or otherwise, following ideas of the social construction of technology [20]. If an exercise involves mass participation, then a variety of methods will be needed to engage a variety of groups. This varied between cases. Where Rushcliffe's residents used both local authority and community-led methods; Edinburgh's "representative voices" required the use of several methods; and Lewisham's hand-picked participants were used to test new Internet technologies for participation.

Two caveats need to be applied to discuss this further. Firstly, it is assumed that it is possible to examine methods and determine philosophical or theoretical positions, a "theory-identifier" view. Secondly, it is also assumed that it is possible to identify philosophical or theoretical underpinnings/notions of public participation through empirical research and relate these to certain appropriate methods of engagement, a "theory-driven" view. From this research, two examples support an idea of the theory-identifier view: the local authority website survey classification through Arnstein's ladder and Lewisham's classification of their participatory activities (based on a similar model by Burns et al. [21]). These examples show that it is possible to establish a framework that examines methods of participation, although with the criticisms noted above. By comparison, a theory-driven view demonstrates that it is possible to select particular methods that are based on some notion of citizen engagement. For example, officers' guides recommend using particular methods for certain situations, related to particular notions of participation and specific or broad audiences.

However, when examining a method, these two views do not necessarily reach agreement. When a theory-driven method is chosen, certain groups may identify with it, but others may see different characteristics that can be related to several theory-identifier possibilities. Misinterpretation of the method's role can lead people to believe they will have a greater say in decision-making than is being offered and, in part, explains why some citizens felt that they were not being listened to online, even in instances where they had a great deal of support and training. Those dealing with participation, therefore, need to choose the methods they employ with a great deal of care. It may be necessary for practitioners not only to understand and be open about the type of public participation taking place but also the part that the methods can play. Clarity is important, and expectations require management.

Shucksmith believed that participation in land-use planning would also create "new institutions or new approaches to old institutions" [13]. If digital methods are new approaches to old institutions, then the application of technology has offered a limited number of new participants the opportunity to become engaged, although the situation is very much in its infancy. For those who had Internet access, digital methods did offer several notable features. O'Doherty highlights communication problems between planners and citizens, especially in terms of the language used during meetings and the "style of the documentation produced" [22], a potential barrier to access and participation in both digital and traditional contexts. E-mails were seen as more informal in terms of their writing style and overall appearance. Interviewees felt that some formality had been given to e-mails through printing and adding them to files of letters. However, planning officers were concerned that Planning Inspectors (who judge the fairness of consultation exercises in the United Kingdom) would see e-mails as inappropriate contributions, because they were anonymous and lacked information about location. Concerns of anonymity also emerged in Edinburgh, where a participant with a Scandinavian name responded through a website-based e-mail. Although they could have been a resident, it raised questions about the validity of contributions from people outside of the city. Officers wondered if the Internet had opened up their consultation to an unexpected, legitimate, and geographically-separate audience, given the international importance of the city center as a World Heritage Site and the broad issues of the community plan.

Digital methods' effectiveness as a communication tool and aspects of "access" also require examination. In traditional settings, "few individuals could represent themselves ... purely because such meetings were held during normal working hours" [23]. Digital methods offer the possibility for large numbers to participate and particular features for some. One participant in Lewisham with disabilities found it difficult to travel to meetings and said that chatrooms had allowed participation from home and the ability to have breaks without feeling embarrassed. Technology also offered greater interaction between internal and external audiences, supporting e-government's idea of "twenty-four–seven" (24/7) access to services, particularly for caregivers of children and the elderly, who benefited from being able to contribute at more convenient times from home. Kling offers a useful general discussion of Internet access in "technical" and "social" senses, in terms of the means and ability to use the technology [24]. However, participation requires an additional view of "political access" to obtain information, understand it, and respond appropriately to help the leverage of funds [25].

There is, however, a sense of scale that needs to be applied, as geographically dispersed individuals interact with space in different ways. For example, the "friendships" established between participants in Lewisham overcame geographical separation, but the other cases had less interparticipant involvement. Those who responded by e-mail in Rushcliffe and Edinburgh were, arguably, in more "socially impoverished" disparate settings, with less cohesive groups. When contrasted with public meetings based in certain locales, dispersion may impact on consensus building (a notion of participation) and the ability to build relationships of trust between all actors involved.

As noted above, the notion of being "representative" impacts on methods through three examples. Firstly, O'Doherty's survey also notes that 17% of officers gauged the intensity of public feeling solely through the numbers of responses, ignoring their quality [14]. Secondly, some officers noted that there was a need to restrict responses for analytical purposes, through mechanisms such as questionnaires. Thirdly, in discussions about a hypothetical large response, some officers suggested that they would have to classify or take a sample of responses. A question then arises about whether a citizen would welcome this approach, having contributed only to find themselves in a "lottery" to have their voice heard, particularly when their issues were more than a "vote" to be categorized (such as a referendum). The methods of participation being offered are related to the analysis of ideas, and being "representative" is often seen more in terms of statistical validity rather than democratic ideals, especially in terms of the audience being sought. This is a concept that requires further investigation and relates to the ways activities are reported as outcomes (noted above).

3.8 CONCLUSION

As an important facet of sustainable development, this brief discussion of public participation has tried to highlight current practice, theoretical understandings, and how the findings of the Internet-based activities of digital participation apply equally to any method, including activities that involve spatial information. More specifically, the survey of U.K. local government websites showed that there were some leading examples of digital participation at all levels of U.K. local government, but with a great deal of variation. Evaluating participation solely by examining such digital methods or response rates is problematic because it takes no account of the understandings of those involved. As such, the research investigated actors' views of participation through three case studies, but it should be noted that there is also a need to explore more grassroots activities. Participation is not a unique or shared construct, and the five components of notions, issues, audience, outcomes, and methods offer one approach to investigate meanings further.

Notions of participation are the fundamental democratic ideals and philosophies that actors express when trying to describe or articulate participation in practice, described through political and planning theories. "Power" is often the focus of our understandings of such activities, but it is only one "notion" amongst several components, and other areas could be seen as important by those involved. When notions are examined, they often have relatively practical implications relating to "accessing" the process, who has a voice, and the extent to which activities match their desired outcomes. This can be seen where the cases exhibited what has been seen as partial, or less than optimal, forms of participatory democracy. Variation between these power relations and access can be described through Arnstein's ladder, but other literature may prove more useful. Christiano's "conceptions" show some of the reasons why citizens wanted to participate, and possibly why local authorities wanted to initiate exercises [10]. The direct conception is perhaps too simplistic, and people are more likely to participate in an exploratory behavior to find out what they want or to express some particular principle, such as NIMBY-ism. Similarly, more societal activities, related to popular sovereignty and the "incompatibility problem," are reflected in citizens' desires for a common voice to be listened to or for governments trying to find a "representative voice." Holden's "arguments" offer a means to examine some of these more group-based principles [7]. People participate because it helps to protect their interests, so that participation is not an end in itself but "instrumental." In contrast, it can be "developmental," reflecting participation's ability to educate, either by learning about a new technology, policy-making, or the nature of participation. Participation can also increase legitimacy and provide an obligation to participate, through "communal" arguments, and it can be seen as the "fundamental nexus" between actors under philosophical arguments. ICTs, perhaps, offer this nexus a vehicle, but with variations in "access" competing with the notions of equality and openness. Both "participation" and "access" exist on a spectrum of interaction, where limited access (for both citizens and decision-makers) relates to limited participation, and increased access is felt to be more favorable.

"Issues" help actors to understand what is suitable for participation or worthy of concern. They occur at different scales and levels of government decision-making, where activities at one level of government can impact on the activities of another, in terms of response rates and defining the issue requiring consultation. Participants are not equally capable of dealing with issues and, as Illeris suggests, people will tend to have a better understanding of the issues that impact on them immediately [12], often in a spatial sense. However, he also notes the ability of some to demonstrate understandings of a greater context that was indicated by the political professionalism of some participants in Lewisham. In another instance, citizens will also have their own concerns of problems that require the action of their governments, often departing from an authority's required response because "government" is seen as one entity, and contributions have not been made at the "correct" time. As such, those who consult need to make potential participants aware of the restricted nature of an exercise and how an issue that departs from this would be dealt with. Often these special interests have a spatial element at varying scales, from the local to the global, with the research agenda on PAUGI noting a need to explore the accrued importance that a shared issue has for disparate communities and the role of ICTs and GI(S).

The "audience" of an activity relates to two themes: those actors that are internal and external to organizations and the impact that relationships between elected members and citizens have on democracy. As much as understandings of participation vary, so does the "external audience" of the public, with differences occurring between the desired "representative" voice that an authority wants, the measured (geo-) demographic of participants and how internal and external audiences choose to view "the public." Those who are concerned with an external audience's elites need to recognize the groups created from grassroots activities that mobilize and influence public responses and that the "typical" participant online is often not the young but a replication of those who have always had a strong voice. However, technology has possibly offered others the opportunity to contribute with an equal voice, and certainly in Lewisham, those from a variety of backgrounds were equally capable of participating online. Internal audiences, also, vary from officers in other departments to other public sector bodies and politicians, influencing activities through guides to participation or activities' budgets. The political relationship of the audiences characterizes the second theme. Many participants chose to see politicians as more knowledgeable than themselves, rather than vessels of public opinion, and participation as Holden's "conventional" representative democracy [7]. Some participants felt that they were not being listened to but wanted to continue to contribute, showing that not all elements of an exercise are welcomed by those involved and that views of participation can change throughout and long after exercises. To counter these difficulties, it may prove useful to draw on professional expertise in a number of digital participatory contexts, especially those where the more complicated GI technologies are to be used. However, as in most top-down situations, it is elected representatives that are required to give a voice to those who did not participate and to make final decisions.

Outcomes of activities are rarely examined as a component, in part because participatory processes are often seen as linear, ending with the publication of analyzed responses. Both actual and perceived outcomes can influence the activities that take place and actors' views of participation. Perceptions possibly have more influence, because they may lead some citizens to become nonparticipants. Desired outcomes can vary for those involved and can be seen as positive or negative if they meet, or fail to meet, ideals of participatory democracy. Such differing perceptions may lead to different understandings of the forms of participation taking place. There is a need to communicate the intended outcomes of an activity with greater clarity and analyze responses in ways that are easily understood by all internal and external audiences, helping to foster relationships of trust that will help future activities.

Methods are the artifacts through which participation is enacted and provide a focus for study. They can be "digital," by involving ICTs, which in some senses can include GIS in online and offline environments, or "traditional," often typified by activities that do not draw on computers. There is also a relationship between traditional and digital methods, with the former often acting as a link or advert for online activities, engaging and notifying particular audiences. No one method can be developed for participation. Participants will use and abuse technology to suit their own ends, and if the notion of mass participation is sought, then several methods will be needed, with different methods capable of offering access to (or manipulating) different audiences. Confusion surrounding the purpose of a method can be further understood by contrasting "theory-identifier" and "theory-driven" views and by realizing that a method chosen from a particular notion contains several entities that can be related to various other notions. Practitioners should, therefore, not only understand and be open about their notions of participation but also about how they want methods to be used. Perhaps the greatest contribution of the digital methods has been increased "access" for particular groups, offering an example of the 24/7 access to services that e-government aims to provide and inclusion for some with disabilities. In participatory settings, ideas of Internet access need to include "political access" that reflects abilities to make use of information to good effect. However, those who respond online may not experience the social networking found in public meetings, creating implications for building trust between actors and developing a consensus. The idea of being "representative" also needs further examination, particularly in terms of a desire to quantify responses, restrict input through questionnaires, or classify and/or sample responses, and the impacts such "lotteries" have on future participation.

This research has primarily dealt with digital participation in the context of U.K. local government. Further work is needed in exploring the notions of participation that exist in other contexts and the relationships between notions of participation and those of access. There is also a need to explore grassroots activity in greater depth. Finally, participation can relate to varying geographies, and the recent research agenda on PAUGI [2], as well as access to geographic information [26], should provide a useful guide to more in-depth longitudinal studies that do not rely on the flawed short-term projects that much current activity relies on, both in research and in practice.

REFERENCES

1. Smith, R.S., Public Participation in the Digital Age: a focus on British local government, Ph.D. thesis, University of Sheffield, Sheffield, UK, 2001.
2. Smith, R.S., Participatory Approaches Using Geographic Information (PAUGI): towards a Trans-Atlantic Research Agenda, presented at 5th AGILE conference on GI Science, Palma de Mallorca (Spain), April 25–27, 2002.
3. Arnstein, S. R., A ladder of citizen participation, *J. Am. Inst. Plann.*, 35 (4), 216–224, 1969.
4. Daniels, S.E., Lawrence, R.L., and Alig, R.J., Decision-making and ecosystem-based management: applying the Vroom-Yetton model to public participation strategy, *Environ. Impact Assess. Rev.*, 16 (1), 13–30, 1996.
5. Pacione, M., Public participation in neighbourhood change, *Appl. Geogr.*, 8 (3), 229, 1988 [after Alterman, 1982].
6. Thomas, H., Public participation in planning, in *British Planning Policy in Transition: Planning in the 1990s*, Tewdwr-Jones, M., Ed., UCL Press, London, 1996, pp. 168–188.
7. Holden, B., *Understanding Liberal Democracy*, Harvester Wheatsheaf, London, 1993.
8. Pateman, C., *Participation and Democratic Theory*, Cambridge University Press, Cambridge, UK, 1970.
9. Sharp, E. and Connelly, S., Theorising participation: pulling down the ladder, in *Planning in the UK: Agendas for the New Millennium*, Rydin, Y. and Thornley, A., Eds., Ashgate, Aldershot, UK, 2002, pp. 33–63.
10. Christiano, T., *The Rule of the Many: Fundamental Issues in Democratic Theory*, Westview Press, Oxford, UK, 1996.
11. Birch, A.H., *The Concepts and Theories of Modern Democracy*, Routledge, London, 1993.
12. Illeris, S., Public participation in Denmark: experience with the county 'regional plans,' *Town Plann. Rev.*, 54 (4), 425–436, 1983.
13. Shucksmith, D.M., Rowan-Robinson, J., Reid, C.T. and Loyd, M.G., Community Councils as a medium for public participation: a case study in Grampian Region, *J. Rural Stud.*, 1 (4), 307–319, 1985 [after Boaden et al., 1980].
14. O'Doherty R., Using contingent valuation to enhance public participation in local planning, *Reg. Stud.*, 30 (7), 667–678, 1996.

15. Tait, M.A.A., Room for manoeuvre? An actor-network study of central-local relations in development plan making, *Plann. Theor. Pract.*, 3 (1) 69–85, 2002.

16. Arblaster, A., *Democracy*, Open University Press, Buckingham, 1994.

17. Birch, A.H., *The Concepts and Theories of Modern Democracy*, Routledge, London, 1993 [after Cook and Morgan, 1971].

18. Pacione, M., Public participation in neighbourhood change, *Appl. Geogr.*, 8 (3), 229, 1988 [after Jordan et al. 1976: 19].

19. Alty, R. and Darke, R., A city centre for people: involving the community in planning for Sheffield's central area. *Plann. Pract. Res.*, 3 (1), 7–12, 1987.

20. Bijker, W.E., Hughes, T.P., and Pinch, T. J., Eds., *The Social Construction of Technological Systems*, MIT Press, Cambridge Mass., 1987.

21. Burns, D., Hambleton, R. and Hoggett, P., *The Politics of Decentralisation: Revitalising Local Democracy*, MacMillan Press, London, 1994.

22. O'Doherty, R., Using contingent valuation to enhance public participation in local planning, *Reg. Stud.*, 30 (7), 667–678, 1996 [after Healey and Gilroy, 1990].

23. O'Doherty, R., Using contingent valuation to enhance public participation in local planning, *Reg. Stud.*, 30 (7), 667–678, 1996 [after Webster and Laver, 1991].

24. Kling, R., Can the "next generation Internet" effectively support "ordinary citizens"? *Inf. Soc.*, 15 (1), 57–63, 1999.

25. Smith, R.S. and Craglia, M., Digital participation and access to geographic information: a case study of UK local government. URISA Special Public Participation GIS Volume II, 49, 2003.

26. Wehn de Montalvo, U., Access to Geographic Information: towards a Trans-Atlantic Research Agenda, presented at 5th AGILE Conference on GI Science, Palma de Mallorca (Spain), April 25–27, 2002.

4 Metadata and Data Distribution

Bénédicte Bucher

CONTENTS

4.1 INTRODUCTION

Most geographic information applications are fed with geographical data produced in a different context than the application itself. These geographical data long were paper maps and are now often digital data. It can be, for instance, a topographical database built by National Mapping Agencies (NMA), earth imagery files, or databases produced by specific agencies working on located information. In other words, producing geographical data and using them often take place in separated contexts. Consequently, a data producer has to distribute its data for potential users to access them.

This access implies difficult tasks like discovering what data exist, understanding the information content of these available data, assessing the fitness for use of these data in the application context, selecting data sets for the application, and acquiring and using the data. These tasks are all the more difficult because storing geographical information in digital databases relies on different complex representation paradigms and on arbitrary choices. A user should be familiar with the various representations chosen by data producers, as well as by software used in his application, in order to chose the most relevant geographical data for his application. Standardization in the field of geographical information has greatly lightened user access. The remaining difficulties must be handled through specific access facilities as explained in this chapter.

Distributing information resources has received much attention due to the Internet growth. A crucial element in this context is the notion of metadata detailed in the following section. Distributing geographical data on the Internet should benefit from the general metadata effort on the Web. It is yet left to geographical information actors to define models for geographical metadata, possibly extending general metadata models, and to build and maintain geographical metadata bases. Applications dedicated to user access must also be designed, based on these metadata bases. Each of these elements is detailed in this chapter, as well as the associated remaining issues and ongoing research work.

4.2   GEOGRAPHICAL METADATA

This section introduces the notion of metadata and the importance of standard models of metadata in the context of data distribution. Section 4.2.2 describes existing geographical metadata models. Section 4.2.3 presents the production of metadata bases and the second part of this chapter will present what elements should be added to metadata bases to support user access.

4.2.1   METADATA

4.2.1.1   Definition

The prefix "meta" is used with strictness in knowledge engineering. In many other contexts, like the Semantic Web or geographic information science, "metadata" simply means data that contains information about a resource. On the Semantic Web, a resource is everything that can be uniquely identified. It may be a book, an idea, a service, or a person. A description of a resource is usually composed of a set of metadata. For instance, a data set may be described by the following set of metadata: name, size, format, abstract, location. In the field of geographical information, a resource may be not only a data set but also a feature, a data series, a model, or a service.

Practically, metadata are data that should support operations about a resource or a set of resources when it is difficult or impossible to perform these operations with the resources only. These difficulties or impossibilities are various. These operations may require input information that is not stored in the resource. For instance, the operation may be a user selecting a data set, and the required input information would be how often the data set is updated. To support user access to a data set, the data provider should thus document a specific metadata for each data set describing its maintenance frequency.

These operations may require input information that is ill-structured in the resource. For instance, finding a street on a city map can rely on a street index associated with the map.

These operations may be about a set of resources and require information that is ill-structured in the set of resources. For instance, buying aerial pictures, the extent of which intersects a specific river, and which are of good quality, is easily done through an index showing location of pictures and a database recording the quality of the pictures.

These operations may be about resources that cannot be handled. For instance, choosing a map on the Internet cannot rely on the maps themselves but on information about them that can be published on the Internet, like the spatial extent, the scale, and an overview.

To analyze the nature of metadata about information resources, [1] Kashyap et al. use several interrelated distinctions:

- There are content-dependent metadata, like the language of a book, and content-independent metadata, like the date of creation of a book.
- There are metadata that can be extracted from the resources, like the table of contents of a book, and metadata that cannot, like the glossary associated with a document.
- There are domain-dependent metadata, like the feature classification used in a data set, and domain-independent metadata, like the structure of a multimedia document.

4.2.1.2 Using Metadata in Catalogs

Metadata is not a new concept in cataloging; classical libraries or video stores already use metadata. For instance, a form describing a book by title, author, edition, the name of the person who last borrowed it, and a storage code is a set of metadata about the resource book. The Global Spatial Data Infrastructure group has studied the role of metadata in the distribution of geographical data [2]. They see three levels of metadata:

- Discovery metadata should provide support for answering the following questions: "What data sets exist? Which data set contains the data I am looking for?" They include, for instance, title, spatial extent, and feature catalog of a data set.
- Exploration metadata should provide support for assessing a data set's fitness for use. They include, for instance, quality elements.
- Exploitation metadata should provide support for the retrieval and use of a data set. They include, for instance, the projection used in the spatial reference system as well as the application schema.

Former metadata were specific to the system that owned the described resources and distributed them (e.g., a library, a grocery, or a video store). The distribution of geographical digital data is a different context — the cataloging service is not supposed to own the resources it describes. A resource may actually be cataloged by several services, and a service may catalog resources from various owners. It is up to the resource owner to provide metadata about his resource for catalog services to distribute it. This calls for the definition of metadata models so that resources of the same type can be described the same way, which eases the cataloging process. A metadata model should specify the structure of the description as well as how to document it, so that metadata stemming from various metadata producers will be homogeneous enough to be managed together in one catalog service.

Moreover, since catalogs may be dedicated to more or less specific resources, it has become important to have several metadata for the same resource, very generic metadata to index the resource in a generic catalog and more specific metadata to index it in a catalog dedicated to this type of resources. This has led to the definition of scaleable metadata models. The most generic model is the Resource Discovery Framework (RDF) that aims at describing on the Web everything that has an identifier [3]. A rather generic model is the Dublin Core, which is dedicated to textual documents and supports the description of html files [4]. A specific model is the Universal Discovery Description and Integration model (UDDI) for Web services [5].

4.2.2 GEOGRAPHICAL METADATA MODELS

In geographical information, metadata were first defined to support data exchange. Later on, metadata models were proposed to describe geographical data sets for cataloging purposes. In all these models, metadata information is organized into quite similar packages.

An early model, which has been partly reused by the next generations, is the Content Standard for Digital Geographic Metadata from the Federal Geographic Data Committee (FGDC CSDGM). The FGDC packages are: identification, quality, spatial properties, spatial reference system, entities and attributes, distribution, and metadata reference. The identification package contains the following elements: textual description, temporal and spatial extent, keywords, contact, access constraints, and technical information.

Organizations dedicated to normalization have also proposed metadata models for geographical data: the CEN TC287 in Europe and the ISO TC211 at an international level.

The CEN TC287 has led to a pilot implementation: the Geographical Data Digital Directory. It is a simple model that has been systematically documented for most geographical databases in Europe. Its elements are defined on the EuroGeographics GDDD Website (http://www.eurogeographics.org/gddd/INDEX.HTM) as follows:

- Overview: textual description, contact information (organization and individual)
- Commercial Information: geographical extension, conditions of sale, restrictions of use, format, etc.

Resource title	: Limiti Amministrativi
Owner	: Istituto Geografico Militare Italiano
Contact postal address	: Istituto Geografico Militare Italiano, Via Baracchini 59, 50127 Firenze, Italy
Contact person	: Sales department, Telephone: +39 55 41 04 10, Facsimile: +39 55 41 01 41, Email : igmiced@fi.nettuno.it
Unit of distribution	: the product is available as partial datasets : administrative regions
Supply media	: magnetic tape, floppy disc, CD-ROM
Supply formats	: MAPPA-83
Standard product type	: Digital Cartographic Model
Data capture technique	: Manual digitising
Geometric form	: Vector/no topology
Scale band	: 1:75 001 - 1:150 000
Geographic area	: ITALY
Extent status	: Dataset is strictly inside of the national boundary. Dataset is completed.
Features contained in the data set	: Administrative area, Administrative unit

FIGURE 4.1 Example of metadata describing a data set in the EuroGeographics GDDD model. This example has been summarized after the site http://www.eurogeographics.org/gddd/INDEX.HTM. Elements for which no information was available have been removed from this description.

- Technical Information: specifications, data source, content, update information, accuracy and other quality parameters
- Organizations: short description of the data provider. An example of a GDDD description is listed in Figure 4.1.

The ISO19115 model produced by ISO TC211 [6] is now an international standard. It is organized in the following packages:

- Metadata entity set information. This is the main package. A data set is described by one object MD_Metadata.
- Identification information. It contains information like an abstract and points of contact. Most elements from the GDDD should typically be mapped to elements in this package.
- Constraint information that describes restrictions associated to the data.
- Data quality information.
- Maintenance information.
- Spatial representation information. The main object of this package, MD_SpatialRepresentation, can be specified as MD_GridSpatial Representation and MD_VectorSpatialRepresentation.
- Reference system information.
- Content information. This package contains objects like MD_Feature Catalog Description and MD_CoverageDescription.
- Portrayal catalog information.
- Distribution information.
- Metadata extension information. This package contains elements that can be used to add new metadata information in a description.
- Application schema information. This package describes the schema of the data set.

An implementation of this object model in an XMLSchema is also proposed.

Finally, the interoperability consortium OpenGIS has included metadata in its specification, eventually adopting the ISO 19115 standard.

4.2.3 METADATA BASES

This section explores the process of building metadata bases compliant with the models presented above and lists the problems faced by a metadata producer. Examples refer to the ISO19115 standard.

It is not an easy task to understand the meaning of the metadata standards and how they should be used. For instance, the element MD_Metadata.hierarchyLevel has several possible values, among which are "series" and "model," but the difference between both is not always obvious; are the specifications of a data product a model or a series? For instance also, the element MD_Metadata.parentIdentifier is to refer to metadata of which this metadata is a subset. It is not easy to understand why its maximum occurrence is 1.

Besides, the metadata producer often has to specialize or extend the standard. For instance, in the ISO19115 standard, the definition of the browse graphic element is very fuzzy and needs specializing for the implementation. For instance also, the standards need extending elements for grid data and for services.

Discovery metadata will be queried in an iterative process to select resources of possible relevance to the user need. Defining a storage model that is adapted to the complex metadata model and that supports iterative queries is a difficult task. Most applications nowadays mix relational database techniques and XML storage.

Last but not least, acquiring metadata and maintaining a metadata base are difficult. Metadata information often exists in a nonstandard and possibly nonformal format. It can be, for instance, a text file storing the spatial extents of raster maps in a series. The person working on this file may change its structure over the year. A similar file may be stored on the next PC if his colleague is working on the same product. There may be no reference on either of these files (author, date, ...). Moreover, people holding such files may be unwilling to share "their" files and will probably be reluctant to commit themselves to maintaining them.

In the United States, it is mandatory for geographical data producers to document their data with the FGDC CSDGM. There is no such obligation in Europe. Yet, other initiatives favor the implementation of metadata such as the Spatial Data Infrastructures (SDI). SDI are presented in the last chapters of the present book. Regarding the production of metadata bases, SDI are useful to share the implementation efforts enumerated above, as well as to obtain a homogeneous implementation across various data producers.

4.3 GEOGRAPHICAL DATA DISTRIBUTION BASED ON METADATA

Once metadata become available for geographical data, applications enhancing user access to geographical data can be built based on these metadata. As explained in

Section 4.2.1, these applications should support different processes: resource discovery, exploration metadata, and exploitation.

These applications first focused on organizing metadata bases and querying these bases. These are geolibraries depicted in Section 4.3.1. They mainly support discovery.

The next step in data distribution is to enhance the interface between the user and the metadata, during the discovery, exploration, and exploitation processes. This is detailed in Section 4.3.2.

4.3.1 CATALOG SERVICES

In the context of the Web, specific digital libraries have been created to distribute located data, mostly environmental data. These are called geolibraries.

4.3.1.1 Existing Geolibraries

In a geolibrary, the main components supporting resource discovery are:

- Metadata bases structured after proprietary or standard format (and holding discovery metadata)
- A query model that can be very close to the metadata base structure
- Contextual databases like a thesaurus or a map

The expression of the user need relies on the query model. This is always a set of elements comprising the geographical area of interest, keywords or themes, and the data producer. Each element of the query model is associated to specification modalities, like the use of contextual databases to support the specification of the element in the user query. Typically, gazetteers and maps support the user specifying his geographical area of interest by writing down a place name or delineating a zone on a map. A thesaurus supports the specification of themes or keywords.

The obtained query is at last mapped into the formal query models of the metadata bases.

The most famous geolibrary is the Alexandria Digital Library (ADL, www.alexandria.ucsb.edu). The ADL query model is called the Bucket Framework [7]. The Bucket Framework specifies a set of buckets and the mapping between each bucket and corresponding elements in various standard metadata models. The buckets are: geographic locations, dates, types (e.g., map, aerial photograph), formats, assigned terms, subject-related text, originators, and identifiers.

Many geolibraries are dedicated to environmental data, such as the Environmental Services Data Directory of the U.S. National Oceanic and Atmospheric Administration grounded on a generic metadata model and the FGDC (http://www.esdim.noaa.gov/NOAA-Catalog/) or the UmWeltDataKatalog based on a specific metadata model and on a large thesaurus of environmental terms (http://www.umweltdaten-katalog.de/). To search such catalogs, the user should browse the thesaurus. For each selected keyword, the site proposes corresponding information resources, and the user may access online descriptions for these information resources.

4.3.1.2 Interoperable Catalog Services

An important step in cataloging is the definition of the OpenGIS® Catalog Services Implementation Specifications [8]. Indeed, this should support the interoperability of catalogs. A catalog will be able to query other catalogs and not only its own metadata store thanks to the OGC_Common Catalog Query Language. There already exist implementations of this OpenGIS specification [9].

4.3.1.3 The PARTAGE Experience at the Institut Geographique National (IGN)

In the French NMA, a network of metadata experts has been initiated to capitalize existing practices in the domain of metadata, as described in [10]. Such practices can be files or databases that are metadata, even if they are not called so. Such practices can also be access scenarios, possibly relying only on a person-to-person communication. The aim of this network is to explore enterprise portal solutions for IGN in order to capitalize and improve these practices. This experimental portal has been called PARTAGE, "sharing" in French. A major difficulty identified in PARTAGE is to have relevant actors involved in the metadata effort. There are two types of actors:

- The people who will provide and maintain the metadata
- The people who will use the metadata bases to access specific information

The PARTAGE approach formalizes existing practices, whether by building a metadata application that exploits an existing metadata base or by automating an existing user-access scenario through a metadata application. These applications are called dedicated testbeds. Dedicated testbeds are structured in a metadata database, an application programming interface (API) to this database, and a metadata service that often happen to be nonstandard. They should ultimately be integrated into the portal by standardizing the data aspect as illustrated on Figure 4.2. The data aspect of the application (i.e., the database and the interface to query it) is standardized without modifying the user interface to the service.

The final interface to the metadata store is a specialization and an extension of ISO19115. It describes not only data but also every kind of shareable geographic information resource. This includes derivation processes. In the PARTAGE API, all derivation processes are not described with the ISO19119 services description model. This model fails to describe generic processes that depend on various possible specifications of the context, such as "matching of two geographical data sets" or "geo-referencing user data." Indeed, these processes cannot be described as a sequence of operations. They should be described as a family of different sequences of operations yielding the same global result in different contexts as illustrated on Figure 4.3. TAGE, a model of tasks proposed in [11], is used to describe these generic derivation processes. A task describes a derivation process by two facets. The declarative facet describes the context: inputs and expected output variables of the derivation process. These variables are called roles. The operational facet or

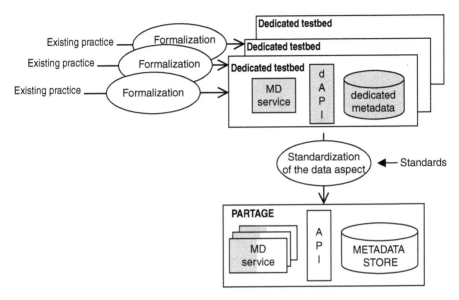

FIGURE 4.2 The approach to build an enterprise portal based on a standard metadata store, in PARTAGE.

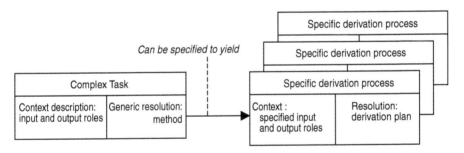

FIGURE 4.3 Tasks represent generic derivation processes in which resolution is not a fixed sequence of operations.

"method" describes how to realize the task. The method is not a sequence of operations, it is a structure of subtasks, and it can be specified to yield various derivation plans.

4.3.2 INTERFACING USERS WITH METADATA

Research on data distribution has so far focused on organizing descriptions of resources, the metadata, and on supporting the querying these descriptions in catalog services. The next step on the metadata services agenda is to improve the user interface to these metadata (i.e., to fully support resource discovery, exploration, and exploitation).

Supporting discovery implies translating a user need into a formal query involving standard metadata elements. Supporting exploration implies presenting to the

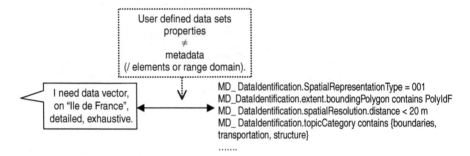

FIGURE 4.4 Distance between the expression of a user need for data and a formal metadata query.

user exploration metadata that are adapted to his need. All this is supported easily for a given need, with the help of expert people. Supporting it, in a formal way, for every kind of need, is much more difficult. Complementary solutions are needed. They are listed in the next sections. All examples refer to the ISO19115 standard. The following analysis of this issue distinguishes between the cases where a user expresses a need for data (i.e., an expression specifying properties of the needed data) and the cases where a user expresses a need for an information content. This needed information content may be explicitly represented in the searched data, or it can be user-specific information that should be derived from the raw data.

4.3.2.1 Answering a Need for Data

When a user expresses a need for data (e.g., vector data on Ile-de-France, detailed and exhaustive), the properties relevant to him may be different from these documented in the metadata. The catalog service has to translate user-defined properties into standard metadata. This is illustrated in Figure 4.4.

The difference may be in terms of range domain of the data properties. In the above example the property vector corresponds to the value "001" of the metadata element MD_DataIdentification.spatialRepresentationType. On Ile de France corresponds to the following values of the MD_DataIdentification.Extent.boudingPolygon: any polygon that contains the polygon "pIdF," where pIdF would be the boundaries of Ile de France expressed in the same spatial reference system as that used in the metadata base. The spatial extent could also be expressed in the metadata by a geographic entity or a bounding box.

Translating user-defined values into constraints on the corresponding metadata elements may rely on contextual databases, as already used by catalog services (see Section 4.3.1). This has been experienced in the project LaClef [12], where the Seamless Administrative Boundaries in Europe (SABE) database is used to map the user area of interest with a metadata constraint.

The difference can also be in terms of elements. For instance, the properties detailed and exhaustive should be mapped with several ISO19115 elements during discovery. The element MD_DataIdentification.SpatialResolution can be expressed as an equivalentScale or as a distance. The user is here looking for the largest scale or equivalently the smallest distance. The element MD_DataIdentification.topicCat-egory lists the themes of the resource content. The user is possibly looking here for the resource for which the list is the longest. He can also use the MD_FeatureCatalog Description.featureTypes element, which lists features contained in the representa-tion. The date, MD_DataIdentification.exTemp, is also a relevant metadata here because of the exhaustiveness requirement.

Translating user-defined data properties into metadata elements calls for a model of data properties relevant to users and a model of links between these properties and standard metadata used to describe data.

When it comes to the exploration process, the system should select metadata relevant to the exploration and present them to the user. In our example, it should present the MD_Identification_GraphicOverview entity and the MD_Application SchemaInformation and quality elements for accuracy and for completeness.

To conclude this section, in the case where the user is able to express a need for data, several types of elements are needed to truly support user access:

- Contextual databases to translate user range domain into standard range domain for specific metadata elements
- Links between data properties relevant to users and standard metadata elements
- User-friendly presentation of exploration metadata

The next two sections detail the case where the user expresses a need for an information content.

4.3.2.2 Answering a Need for Topographic Features

The user may express a need for topographic features (e.g., houses and small paths on Ile-de-France). These features are represented in the data, set but the user repre-sentation of geographic space may be different from those used in the geographical data sets. Corona and Winter [13] demonstrated these differences in the specific application domain of pedestrian navigation.

During the discovery process, the catalog service should interpret the terms "small paths" and "houses" into the metadata base, as illustrated in Figure 4.5. Moreover, the catalog should do this for various users and various data sets.

In this process, the catalog service should rely on a model in which a user may express which topographic features he is interested in. Having such a model for various possible users is a very difficult task, because there is no obvious model of geographic space that everybody would agree on. Instead, users communities build models adapted to their viewpoint. This issue is widely referred to as the definition of a geographical ontology. An ontology is a formal model of a domain that is shared by experts in this domain. Unfortunately, no ontology of commonsense geography is available, and the extensive research in this area failed to propose one.

The catalogs service also needs a list of features represented in the cataloged data sets to determine which data sets contain a representation of the needed features. This description is available in the ISO19115 model as MD_FeatureCatalog Description. An

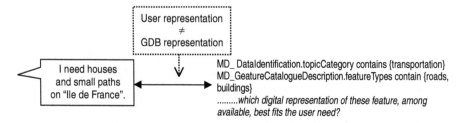

FIGURE 4.5 Distance between a user need for topographical features and a formal metadata query.

important issue is to propose a unified vocabulary (i.e., Feature Catalog) to list features in data sets, or to write down databases schema. An element of such a vocabulary would be, for example, a road object with a linear geometry described by a list of two coordinate points. So to speak, such a model would be an ontology of digital representation constructs for geographic space.

Lastly, the catalog service should rely on a model of links between the first model (that of topographic space in the users' mind) and the second one (that of topographic space in databases) to map the expression of a user need for geographical features into a set of digital features.

Once the discovery is performed, the exploration remains a difficult process. The user has to assess the fitness for use of the various resources proposed by the catalog, hence to understand the schema of the data sets and associated specifications. In the context of PARTAGE, for instance, a compared graphical representation of different IGN data products schemas has been proposed in the work of Gyorgyi Göder, a master student in cartography. The extracts shown in Figure 4.6 are meant to convey to the user the following representation notions.

The images on the upper cells tell that the representation of a river can be a surface or a linear object, depending on a width threshold. In the case where a river is a linear object, its represented width is limited to discrete values. The images in the cells below tell that there are two different thematic classifications of rivers, which are not equivalent. This helps the user to understand that there is a loss of information in the representation and that two rivers may have similar widths in one geographical database (GDB) and different widths in the reality or in another GDB. Moreover, in the same GDB, a river can be represented with a fixed width and another with a more precise geometry (a surface actually).

4.3.2.3   Answering a Need for User-Specific Features

The user may express a need for a feature that should be derived through geographic information system (GIS) functions (e.g., isolated industrial buildings near my place, my way home).

To map such a need with the relevant data set, the catalog service needs a model of links between user features and queries for data and functions.

The task-based TAGE model introduced in Section 4.3.1 has been initially designed to meet this issue. The expression of a user need can be supported by

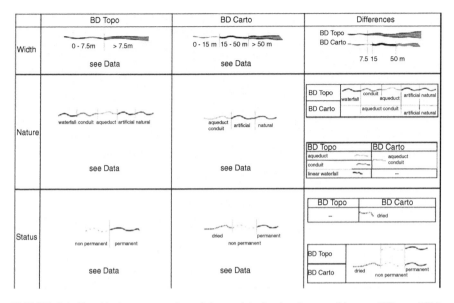

FIGURE 4.6 Graphical representation of the models for the river used in two different IGN data products (DBT$_{opo}$® and DBC$_{arto}$®).

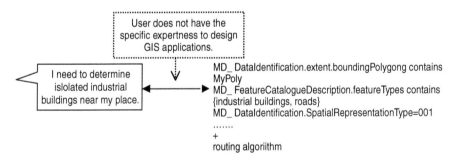

FIGURE 4.7 Distance between a user need for a geographical feature (not explicit in the raw data sets) and a formal metadata query.

specifying the corresponding generic task (e.g., "to determine the nearest beaches to my place" is expressed by specifying the task "to locate something"). This specification is the specification of a set of roles attached to the task as illustrated on Figure 4.7.

For instance, the user specifies the role "entity to locate" as follows: it is of feature-type "beach" and has a spatial relationship "nearest" with another user feature "my place." The spatial relationship "nearest" refers to a driving distance. The feature "my place" may itself be specified as follows: it is referenced by a postal address, it is inside the "Morbihan" (French administrative entity), or it can be referenced by the user through clicking on a map.

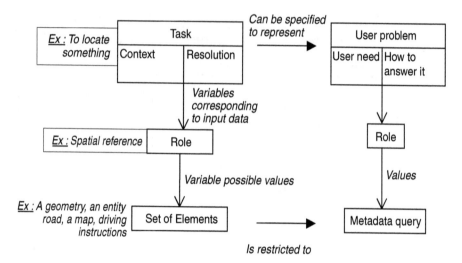

FIGURE 4.8 The use of tasks to translate a user need into metadata queries.

The specification of the task roles triggers the specification of its resolution. The specified resolution is actually an application plan with needed input data as illustrated in Figure 4.8. In our example, a possible metadata query would be of the form:

Spatial extent includes Morbihan
Feature catalogs includes {beach, all types of roads, building}
Spatial representation is vector
The roads have a topological representation
Resolution is higher than 50 m

4.4 CONCLUSION

Geographical data distribution should adapt to the current context. There are large amounts of data provided from various sources. These data are distributed by catalogs that do not own them. They are of potential interest for many different applications. User access should be as automated as possible. To meet this context, geographical data distribution should rely on a good management of metadata about geographical data.

Three key points can be listed. The first key point is the definition of an international standard metadata model for describing geographical data. This standard now exists. It is the ISO19115 standard. The next key point is the creation of databases based on this standard. In this area, there is much work to be done. The main difficulties are not technical but organizational and economical. People who can document metadata are not always willing to participate, and deciders that should launch these metadata projects are not always convinced of the stake of metadata and are afraid by the cost of building and maintaining metadata bases. The last key point is the development of catalog services based on available metadata. A first category of catalogs will soon give access to available metadata. They will be exploited by people who are already familiar with the geographical data they look for, or who are building a very simple application.

The remaining issues will be to improve the interface between users and such geolibraries to enlarge the number of users for these catalogs. It is a complex technical issue, as exposed in Section 4.3.2 of this chapter. It takes different techniques to solve it. Knowledge representation techniques are needed to build efficient models to automate the mapping between the expression of a user need and the determination of data sets that possibly answer this need and of the description of these data that should be given to the user. It will also take interface design techniques to support the presentation of the metadata to the user.

REFERENCES

1. Kashyap, V., Shah, K. and Sheth, A., Metadata for building the multimedia patch quilt, in *Multimedia Database Systems: Issues and Research Directions,* Jajodia, S., and Subrahmanian, V., Eds., Springer Verlag, New York, 1995, pp 277–319.
2. GSDI Technical Working Group, Developing Spatial Data Infrastructures: the SDI CookBook, v1.0, Douglas Nebert, Ed., 2000.
3. W3C, RDF Vocabulary Description Language 1.0: RDF Schema, W3C Working Draft, Brickley, D., and Guha R.V., Eds., 2002. http://www.w3.org/TR/
4. Weibel, S., Godby, J., Miller, E. and Daniel, R., OCLC/NCSA metadata workshop report, 1995. http://dublincore.org/workshops/dc1/report.shtml.
5. Januszewski, K. and Rooney, E. (Eds.), OASIS UDDI Specifications TC, UDDI Version 3.0, 2002.
6. ISO TC211, ISO 19115 Geographic Information—metadata, international standard, 2003.
7. Smith, T., Frew, J., Janee, G., and Hill, L., The Alexandria digital library project, paper presented at the 12th International Conference on New Information Technology— Global Digital Library Development in the New Millenium, Beijing, China, 2001.
8. OpenGIS, OpenGIS® Catalog services implementation specification, v1.1.1, OpenGIS® Project Document 02-087r3, 2002.
9. Muro-Medrano, P.R., Nogueras-Isos, J., Torres, M.P. and Zarazaga-Soria, F.J., Web catalog services of geographic information, an OpenGIS based approach in benefit of interoperability, in *Proceedings of the 6th AGILE Conference*, Lyon, France, 2003, p. 169.
10. Bucher, B., Richard, D. and Flament, G., A metadata profile for a National Mapping Agency Enterprise Portal—PARTAGE, in *Procceedings 11th GISRUK Conference*, London, 2003, p. 162.
11. Bucher, B., A Model to Store and Reuse Geographic Application Patterns, in Proc. 4th Agile Conference, Brno, Czech Republic, 2001, 289
12. Megrin, LaClef An Operational Model for Unlocking Public Sector GI through E-Commerce, Final report, Multipurpose European Ground Related Information Network (MEGRIN), 2001.
13. Corona, B. and Winter, S., Datasets for Pedestrian Navigation Services, in Proceedings AGIT Symposium, Salzburg, Austria, 2001, 84.

5 GI-Based Applications on Public Authorities' Web Sites and Their Nonprofessional Users

Mette Arleth

CONTENTS

5.1 INTRODUCTION

Public participation in the planning process requires well-developed communication between the authorities and the public. With the growth of the Internet, this communication has been enriched with the properties of the new digital media. This seems a great advantage for the public as well as the authorities, who see e-government as a means of offering more service to more people, while at the same time decreasing the number of employees in the administration [1]. In Denmark, much attention has been focused on e-government and digital management at all administrative levels of society during the last few years. This includes online access to a variety of GI-based services, with the aim of enabling citizens to service

71

themselves online, based on the geographic information offered. However, can we expect that a citizen who has no relevant professional basis for understanding the concept of geographic information should be able to use GI-based web services? Experiences gained from teaching surveyors and geography students about GI and GIS suggest that the concepts of geographic information (as opposed to maps), and the idea of layered information, are not intuitively understood. On the other hand, these Internet-based services are rather popular among those who have become regular users, such as farmers (e.g., applying for subsidies or for permission to increase their livestock), agricultural consultants, and property handlers. Unlike the average citizen, these regular users have some professional knowledge that enables them to comprehend the substance and context of the information in the interactive map. But the GI-based applications are meant as a service for all the citizens of the region, and they are — more or less — meant to replace the personal-based service. Earlier studies suggest that map-based services are popular among the majority of users, as long as they are not too complex or too technically demanding. Improving the usability of the GI-based web services obviously requires knowledge about the nonprofessional user's understanding and use of GI. This chapter discusses the need for investigations in this field and the methods relevant for such investigations.

5.2 E-GOVERNMENT AND DIGITAL MANAGEMENT IN DENMARK

Denmark has set the goal of being among the countries that are best at utilizing the global digital transformation to create growth and welfare. The ambition is to utilize the potentials of digital society across state, regional, and local levels of government to organize the public sector in a more flexible and efficient way and with higher quality of service for the citizens [1]. The core of e-government is to create better and more efficient solutions to administrative tasks through the use of information technology. The full digitalization of the public sector should ensure that work processes oriented toward paper handling and manual control are reduced, and that double work and unnecessary work processes are removed [2]. The goal is to reduce the costs in the public sector while improving citizen and company access to the public service. Digitalization of all major service areas and contacts with the public sector is seen as an important means to achieve these goals. Different stages in this digitalization process can be defined, moving from basic information published on the web, through increased interaction and integration, to high levels of direct public participation [1]. Similarly the United Nations (UN) and American Society for Public Administration define five stages of e-government [3]:

- Emerging: An official government online presence is established. Information is limited, basic, and static.
- Enhanced: Government sites increase. Content is updated with greater regularity.
- Interactive: Users can download forms, contact officials, and make appointments and requests.

- Transactional: Users can pay for services or conduct financial transactions online.
- Seamless: Full integration of e-functions and services across administrative boundaries.

According to a benchmarking report made by the UN [3], Denmark regarded as a nation is in the Interactive stage, but substantial differences occur between the different governmental levels and between the different institutions at each level [1].

The digitalization obviously includes all the processes involved in the management of rural and urban areas, processes that notoriously rely heavily on the use of maps and geographic information technology. In Denmark the vast majority of the rural and urban area management is carried out by regional and municipal authorities, respectively. During the last few years, all the regional and many of the municipal authorities have developed applications that enable the citizens to access various kinds of maps and geographic information via web sites. The impression one might get from browsing many of these sites is that the purpose and usefulness of the sites vary quite a lot, and one could be tempted to conclude that many of the sites are simply technology initiated: "technology is available, let's use it for something." While this motivation is valid for producing knowledge and experience with the potentials of the technology, it does not ensure a consistent, focused, and qualitative shift from paper-based management to digital management. To achieve such a goal as stated by the Digital Taskforce [2], a more comprehensive approach is needed.

5.2.1 THE LIGHTHOUSE IN NORTH JUTLAND

In the region of North Jutland, implementation of such a comprehensive approach has been sought. During the years 2001 to 2003, a regional project known as "the Digital North Jutland" or "the Lighthouse project" was performed. With a grant of 170 million DKK (approximately 23 million Euro) from the Ministry of Information Technology and Research, the project aimed at testing the potentials of the network society for all citizens in North Jutland. This was done through a large number of different IT-based projects, organized in the general themes "Digital Management," "Competence and Education," "Information Technology and Vocational Development," and "Information Technology and Infrastructure." Many of the projects under the theme "Digital Management" concern public access to digital geographic information on the local and regional authorities' web sites. Compared to what is offered at most public web sites, North Jutland wanted to broaden the spectrum of subjects and increase the level of interactivity in its GI-based applications on the web. For this purpose a service based on a map–metaphor user interface [4] was developed. At this moment eight different usages of the interactive map are suggested but not all implemented:

1. Self-service application for permission for digging up of wires and cables
2. Direct access to measurements of drinking water quality
3. Self-service application for permission to perform water catchments
4. Communication of the basis of the zonal administration

5. Involving the citizens of the region into an open and interactive planning process via the Internet
6. Self-service application for state subsidies for environmental agriculture
7. Self-service application for road neighborship administration
8. 3D-map covering the entire region (e.g., for the use of landscape visual-izations)

The intentions and the visions are high, and much focus is directed toward the development and change of the organization and the implications of the internal use of the new applications. So far, much less attention has been paid to the external users, the citizens. Will they have the skills and knowledge to use and comprehend the applications and the information without the facilitating guidance of a professional? If the digitalization process is to result in a reduction of the costs, the citizens must be able to service themselves. Self-service of an acceptable quality requires at least some understanding of the application in use.

5.2.2 DIFFERENT GROUPS OF USERS

Two major groups of users can be identified: professional users and nonprofessional users. The professional users are agricultural consultants, property handlers, wind-mill owners, municipal authorities, etc. The professional users use the GI-based web services as a tool in their work on a regular basis, and they have different kinds of professional knowledge that enable them to comprehend the substance and context of the information in the application. As opposed to this, the nonprofessional user might look up the application only a very few times and can, as a general rule, not be expected to have relevant professional skills or educational background. Between these two user groups a special group of nonprofessional but regular users can be identified: the farmers. Because farmers hold most of the areas that the regional authorities manage, they need to be kept informed about any kind of regulation, restriction, or zoning that involves their properties. Farmers that hold large properties often use GIS-like applications for managing their crops, dosage of fertilizer, and similar activities. The farmers must therefore be regarded as a more skilled user group than the average citizen, and furthermore, they have the opportunity to contact the agricultural consultants if they need assistance. The remainder of the article deals with the nonprofessional users.

5.3 WHAT IS KNOWN ABOUT THE SKILLS AND KNOWLEDGE OF THE USERS?

After identifying the target group of the investigation, it must be clarified which elements of these users' knowledge and understanding should be in focus. Many categorizations of knowledge as philosophical and psychological terms exist [5]. One way to categorize knowledge in context of a learning progression is given by Anderson et al. [5]:

- Factual knowledge ("knowing that")
- Conceptual knowledge ("knowing why")
- Procedural knowledge ("knowing how")
- Metacognitive knowledge (knowing of knowing)

These four kinds of knowledge should be regarded as a progression: factual knowledge being the most basic and metacognitive (including strategic knowledge and self-knowledge) the most abstract. Concerning the level of abstraction, there is some overlap between conceptual knowledge and procedural knowledge [5] in the sense that it is sometimes possible to "know how" to perform procedures and operations without "knowing why."

In the case of the nonprofessional users' understanding of GI, what is of interest is the users' factual and conceptual knowledge of GI and related issues. But these kinds of knowledge may be affected by the users' procedural knowledge concerning the use of ICT in general. The latter must therefore be clarified, at least on a basic level, as a precondition for the later analysis of the GI-related knowledge.

Regarding the general competence related to accessing and using the net, the developers can quite safely assume the users to be prepared for the digital management. The Danes are front-runners in the world when it comes to taking advantage of new technological opportunities, measured by citizen access to the Internet, mobile telephony, and computer use [1,2,6,7]. Obviously there are differences in skills across age and educational levels, but all citizens are encouraged to improve their qualifications in computer use, and public courses of computer and Internet use are offered. Some of these courses are directed toward specific groups such as retired people or women with little or no professional education that are known to have a slow adoption of information technology. As the access to computers and to the Internet is easy and as the concept of digital management penetrates more aspects and sectors of social life, the motivation and conditions for building up general skills in computer use may be considered good.

When it comes to forming an overview of the level of knowledge and understanding of maps and GI, the picture becomes less clear; sound data are scarce and divided. To this point it seems that very little effort has been directed toward a systematic collection of experiences with user reactions and use patterns on public GI-based web sites in Denmark. Some regional and municipal organizations use focus groups to evaluate and develop their web sites, but those efforts seem to focus primarily on the usability of the web sites. In the development of the "Active map of Aalborg" [8], the considerations of different user skills have been built into the concept, leading to an application with predefined functionality ("one-mouse-click") as well as extensive analytical functionality for advanced users.

Obviously, good usability and adequate cartographic design promotes understanding of the contents of the applications, but it does not guarantee a correct interpretation. An investigation of general map reading skills among different groups of Danes [9] found that, although the map user had access to a map legend, no more than 70% of the symbols of a topographic map were understood correctly. The test persons in the investigation represented a broad range of age and educational level, and four general levels of map reading abilities were defined:

1. Symbol-reading: ability to interpret simple iconic symbols such as blue = water, green = vegetation
2. Picture puzzle reading: knowledge of main groups of conventional symbols like roads, vegetation, urban areas, wetlands, contours, etc.; ability to read the map as a series of symbols and recognize simple patterns like hill, village and farm
3. Intermediate reading: more advanced ability to perform map reading moving toward ability to perform map interpretation; knowledge of most symbols and ability to look up the rest in the legend (but may not understand their meaning); recognition of known patterns in the landscape and ability to find new ones if the landscape is known and familiar
4. Content reading: map interpretation; ability to form an impression of new landscapes through a map; recognition of complex patterns

All the tested persons were found to be able to perform symbol reading. An average of 75% of the test persons reached the level of picture puzzle reading, but large differences were found between test persons with relevant education and those with only a short (elementary school) education. These differences become even more significant on the last two levels. Intermediate reading was performed by 50% of the test persons. On this level specific and general experience with different kinds of landscapes was found to be of considerable importance; knowing what a landscape could look like improved the interpretation process significantly. Less than 20% of the test persons reached the level of content reading. All the test persons in this group had either a long relevant education (academic degree) or extensive general and specific knowledge about landscapes.

The findings of the investigation might not be very surprising for researchers working with such matters. A general conclusion could be that most Danes have at least basic map reading skills. But on the basis of the above-mentioned results, one could easily get the impression that mapping professionals at the different administrative levels of Danish government generally overestimate the map reading skills of their users. There is no doubt that map-based services on the Internet, such as route planners and city information systems, are popular in Denmark. Moreover, most regional authorities report a high level of user satisfaction with the GI-based services on the web sites. Nevertheless, until now little has been done to systematically monitor the web site users and their understanding of the web site contents. Still, many of the GI-based web sites specifically stress that what is offered is geographic information — not necessarily maps. This implies that the users possess not only map reading skills, but also a basic (intuitive?) understanding of the concept of GI and layered information.

Experience gained from teaching surveyors and geography students suggests that the concepts of geographic information (as opposed to maps), and the idea of layered information, are not intuitively understood. Building an understanding of these concepts among the students is not a trivial task. It takes quite a lot of practical work and exercises with GI to enable one to recognize the difference between a map and geographic data that may or may not be part of a resulting map. Performing relevant analysis with GI presupposes understanding of the quality and structure of the data, as this determines the limitations of the usage. Whereas a citizen looking up the regional GI-based web site can be expected to have some map reading skills and supposedly have some general and specific knowledge about the landscape shown in the application, he can hardly be expected to have an intuitive understanding of the concept of GI. This forms two interesting research questions:

1. How does a user with no relevant professional education comprehend and understand GI?
2. Does an imperfect understanding of GI actually hinder qualified use of the GI-based web service?

5.4 HOW CAN THE USER'S UNDERSTANDING OF GI BE INVESTIGATED?

"Understanding" of concepts of GI are unlikely to be generally assumed from one person to the other, unless they receive formal teaching on the subject. A number of hardly generalizable aspects are involved in the basis of the understanding: level of education, spatial capabilities, and map reading skills, complexity of the information presented, situation and purpose of use, and more. This suggests the use of a highly qualitative method of investigation. On the other hand, to have legality and more general usefulness, the results need to be based on some kind of quantitative basis, to assure that the findings of the qualitative measure are sufficiently representative. Hence the subject calls for qualitative as well as quantitative measurements. The following methods can be relevant for the study [10,11,12,13,14]:

- Log-file monitoring and tracking of user behavior on GI-based web sites
- Questionnaires, online and/or paper based
- Focus group interview
- Personal interview
- Think-aloud study
- Video monitoring of user behavior and task solution in a test setting

5.4.1 LOG-FILE MONITORING AND TRACKING OF USER BEHAVIOR ON GI-BASED WEB SITES

Log-file monitoring as a means of evaluating usability in GI-related applications has been reported [13,14,15]. A necessary basis for evaluating the understanding of the user is to have a picture of how the users use the application: how do they navigate, how do they search the datasets, do they focus tight on a subject, or are they browsing the application? All user actions on a web-based application can be monitored using a log-file [13,14]. Depending on the kind of web site technology, much of the logging information will be stored in a server from which it can be extracted and analyzed. Furthermore, different commercial services exist that offer detailed tracking and logging of user behavior on a web site, such as the SiteCatalyst® (Omniture, Orem,

UT [16]). Systematic monitoring and tracking of user behavior will provide a more complete and detailed overview of the use of the web site than what can be achieved by asking users of their preferences or drawing conclusions based upon the feedback received from users. Such monitoring should be part of the systematic collection of experience for the purpose of evaluation and further development of the public GI-based web services. It also can provide a relevant basis for qualitative investigations through questionnaires and interviews, and for setting up realistic tasks for a video-monitored test setting.

5.4.2 QUESTIONNAIRES, ONLINE AND PAPER BASED

Analyzing log-files will provide information about how users behave and navigate on a web site, but explanations of the purpose of using the web service must rely on an interpretation of the use patterns. Mapping the different purposes of use as well as the user's expectations to the information at the web site is more easily done through questionnaires. Decisions must be made about how to reach the target group, in this case the nonprofessional users. Two different approaches can be followed: online questionnaires that users looking up information at the web site are encouraged to fill in and submit [17], or paper-based questionnaires distributed either by mail or manually at open house or relevant meetings. Distributing questionnaires by mail is a method that can immediately be eliminated because the costs are high and there is no way to ensure that the questionnaire reaches citizens who have actually used the web service. The regional authorities regularly participate in information arrangements of various kinds. At these arrangements users can try the web services and immediately thereafter be encouraged to fill in a questionnaire. The immediate and direct confrontation at such arrangements can be assumed to give a high response rate, but only a limited number of users can be reached in this way. As a supplement to this approach, therefore, the online questionnaire seems a good choice.

A combination of log-file analysis and questionnaires can form a sound quantitative basis for understanding what the users expect to find at the web site and how they use the web site. Such knowledge forms the basis for the creation of a test setting and decisions about the tasks the test persons should be appointed. It can also help to answer the second question stated above: whether or not the users are able to use the web site's applications correctly. But to get an impression of the user's actual understanding of the concepts of GI, more qualitative methods like different kinds of interviews or think-aloud studies must be used.

5.4.3 QUALITATIVE METHODS

Earlier in this chapter, various qualitative methods are mentioned: focus group interviews, personal interviews, think-aloud tests, and video monitoring. See Nielsen [11] for a general description and Van Elzakker [18] and Just [19] for specific reports on these methods. They can be used separately or in combination to throw light on the nonprofessional users' understanding of GI. Of particular interest here is getting an impression of how the users perceive what they see when they use the GI-based applications and eliciting the words and concepts they use to explain and describe it. It is assumed that the way the users verbalize their experience of the applications will tell quite a lot about how — and how well — they understand the concepts [12]. By analyzing the verbalizations, much can be found out about how the users understand the concept of GI, and obviously it will reveal any major misunderstanding. Implications of this method primarily involve the dependency of the test persons [11]. These must agree to participate in one or more interviews or tests, and they must be able to express their thoughts and doubts during the test. This puts strong demands on the test settings: the physical surroundings as well as the tasks and questions. Data capture methods must be chosen with great care to assure sufficient validity of data. Video filming, as an example, is a very effective qualitative way of capturing data in a test setting, but being filmed during tasks or discussions might make the test persons feel uncomfortable [11]. A way of overcoming this could be to use software that records the users on-screen actions and comments, as reported by Tobon [20]. An ultimate way of recording user focus and attention is eye-movement recording [21], a method currently used by the Danish National Mapping Agency to evaluate the cartographic quality of topographic maps.

5.4.4 POSSIBLE OUTCOMES OF THE INVESTIGATION

Apart from forming the analytical basis of an answer to the two research questions, an investigation consisting of a combination of log-file analysis, questionnaires, video monitored tests, and interviews will provide a vast amount of data useful in many ways to evaluate and further develop the GI-based services on the regional authority's web site. Three main conclusions of the analysis can be envisaged:

1. The users have sufficient understanding of the concepts to use the application correctly.
2. The users do not understand the concepts of GI, but are able to use the applications correctly anyway.
3. The users do not understand the concepts of GI, and are therefore unable to use the applications in a qualified manner.

Of these, the first two conclusions would imply that nothing other than general maintenance and development is applied to the web site, as the use of it apparently does not give rise to any problems for the user. Gradually adding more functionality and analytical capability to the applications would be a natural choice. But what should be done if the results of the investigation point toward the third conclusion? Such a conclusion could imply a different approach to the offering of online access to GI. Different approaches may be imagined, one being the adding of a number of facilitators outside the administrative authorities to help and service the citizens' use of the web services. Another approach might be to build up the web service in a different way, requiring less understanding of GI as a concept, and by using interactive maps only as a supporting tool.

5.5 GI-BASED APPLICATIONS AT THE REGIONAL WEB SITE

In this section some specific GI-based applications maintained by the regional authorities of North Jutland are described. The description is based on an interview of two developers working in the regional administration. Confronted with questions about the considerations of the users' skills and knowledge and the (above-mentioned) proposed investigational methods, they both pointed to the pragmatic and economic realities of their job. Obviously, between the ideal world of research, with its principle-based recommendations of "what should be done and how," and the organizational and political reality, there is often a gap. The budget for developing the GI-based applications at the region of North Jutland did not allow a large user-group investigation prior to development, nor does it leave room for extensive analytical evaluation of an application when it is up and running. The staff maintaining and developing the applications have to rely on their general experience and knowledge of the tasks and service that the regional authority usually attends. Supplemented with occasional log-file statistics and the feedback from users calling to get help, this is the basis for evaluating the existing applications and developing new ones. This situation must be assumed to be quite characteristic of most public GI-based web service administration. And after all, because the regional authorities are experts in their own field of administration and management, the in-house experience must be regarded as a sound basis for developing web-based applications for self-service. The developers do see a need to investigate further the skills and knowledge of the users, but mostly for the purpose of refining their interface design. While this is a valid and reasonable focus for the practitioners, the research should have a wider scope than this. In the case of this particular research project, being able to answer the questions in Section 5.3 and arriving at one of the conclusions mentioned above implies much more than making changes in an interface design, however needed that must be. It adds necessary input to the basis on which we build the most current development of our society, the growing field of e-government. If the transition toward e-government is to result in a more, and not less, democratic society, the general ability to participate in democratic processes must be taken into account. A recent survey reported in Remmen [1] showed that almost half of the Danish municipalities see "citizen's resistance to use of e-government" as a barrier to implementation of e-government. As mentioned in Section 5.3, the level of procedural knowledge concerning the use of ICT must be regarded as high. This could suggest that what the citizens consider a barrier in using digital services is comprehending the content of the service; that is, possessing the factual and conceptual knowledge needed to understand the applications. But one cannot decide if the level of abstraction in the application is too high or the information is too implicit for the intended user without examining the knowledge level of the users.

In what follows, some of the GI-based applications at the region's web site are described, and the regional authorities' experiences of the usage and users are mentioned. On the basis of this, three applications are chosen for the investigation of the nonprofessional users' understanding of GI.

5.5.1 THE PLOT OWNER INFORMATION SYSTEM

Of the eight applications mentioned in the second section, so far three have been launched. The first one is a general GI-based tool offering an overview of the regional area, with different search options (postal address and property/cadastral number). A large number of different themes concerning the management of the rural areas can be shown on a topographic basemap, or at large scales with an orthophoto as background. This application, known as "The Plot Owner Information system" was primarily meant as a service for farmers, to reduce the need for sending maps and paper-based information by mail to all farmers whenever changes in legislation or ownership made new maps necessary. Hence, the farmers are seen as the primary user group, a fact that is reflected in the terms used and in the grouping of the different themes. This grouping corresponds largely to the different kinds of applications the farmer could need to fill (e.g., applying for subsidies or for permission to increase their livestock). The terms used and the grouping of the themes might not be very easy to comprehend for nonprofessional users, but more general themes from the regional plan can be found as well. The application is widely used by different professional user groups, and presumably on a more occasional basis by nonprofessional users. Among the more surprising usages that have been reported by users are searching for hunting areas and studies of genealogy. The latter benefits particularly from the many place-names in the base maps.

5.5.2 THE RURAL AREA SELF-SERVICE APPLICATION

A further development of the plot owner information application resulted in the "rural area self-service" application. In this application, the users can indicate their property, either by property/cadastral number, postal address, or by sketching in a map. The application returns a list of all regulations, zonings, etc. that affects the property. This application was originally meant as a service to citizens applying for permissions to build in rural areas. The task of managing such applications used to be part of the regional administration, but lately the competence to allow or refuse such applications has by political decision been transferred to the municipal authorities. Hence the user groups of the "rural area self-service" today almost predominantly are the municipalities of the region.

5.5.3 THE DIGGING-UP SELF-SERVICE APPLICATION

The third and newest application on the web site is "self-service regarding application for permission to digging-up of wires and cables." This kind of activity often implies digging up roads, and to be allowed this, permission must be sought from the regional authorities. The application is targeted toward wire owners (usually telephone companies) and the entrepreneurs carrying out the actual task. The application works much as the above-mentioned rural area service; the area in question is indicated or sketched on a map. A GI-based analysis returns information about which roads are affected, and this information is used in the online application form. The "digging-up" service has been running for less than half a year, and according to the employees at the regional administration, so far some 25 to 30% of applications for permission to dig up wires are made online, and the number is growing.

Characteristic of the use of all three applications is that almost no inquiries are made about how to use the applications, but as soon as the applications break down (for technical reasons, etc.) many users contact the regional administration to find out what is wrong and when the problems might be solved. This clearly indicates a quite extensive use of the applications, a fact that might be taken as a proof that the users understand the philosophy and design of the applications. On the other hand, because no systematic log-file analyses are applied, there is no indication of the number of users that look up the applications but fail to use them.

5.5.4 THE 3D APPLICATION

As it will appear from the description of the different applications, they are generally targeted toward a specific usage and, thereby, a limited user group. This does not necessarily prevent other user groups from using the applications, but to some extent it does lower the demands on the applications in order to be understandable to all kinds of users. In October 2002 a 3D application was launched using the TerraExplorer® software (Skyline Software Systems, Inc., Woburn, MA). In this application, the users can fly above and investigate the entire region of North Jutland, visualized in 3D by an orthophoto mosaic draped on a digital elevation model (DEM). The application allows the user to "fly" from one address to another, circle around specified targets, and navigate freely in three dimensions. Buildings are extruded from the orthophoto as blocks, based on polygonal information from technical maps. This gives a rough yet realistic impression of the surroundings. The 3D application is targeted toward all citizens, and at this point it serves no specific use other than general information and entertainment. For this reason employees at the regional administration regarded it with some skepticism, but as the success of this application rises, the skepticism has been belied. The application is a flagship for the region. The attention it receives reflects on the other GI-based services, and for many users it has opened the door to thinking of the geography as an active part of their daily life. In this way, the application might help to promote interest in and understanding of GI as a general concept.

In the future the intention is to develop the 3D application further, building in more functionality and using it as a basis for visualizations of new installations (buildings, windmills, roads) that the public is encouraged to debate.

The investigation of the nonprofessional users' understanding of GI will focus on three GI-based applications: the 3D model, the plot owner information system, and the Active map of Aalborg. The latter is interesting to study, because it has already implemented many of the new ideas and approaches to usability for nonprofessional users, and it may as such work as a kind of reference or test of the usefulness of new ideas. The plot owner information system is preferred as an object of the investigation in front of the other self-service applications on the regional administrative web site because the latter are primarily meant for and used by professional or semiprofessional users. Finally the 3D application has been chosen due to its popularity and the development perspectives it holds. The three chosen applications present three very different user interfaces and very different cartographic designs as well. Obviously, when building the test, how these parameters influence the perceived usability and how this perception reflects on usage and understanding must be taken into consideration.

5.6 CONCLUSIONS

The digitalization project in the region of North Jutland ran until the end of 2003. Immediately afterwards followed an evaluation phase, building an overview of the progress of all the different projects and their success. For that use, a more detailed impression of the users and the usage of the GI-based web applications would be a valuable tool for evaluation and the strategy of further development. At this point, most of the GI-based applications are targeted toward relatively specific user groups whose general abilities to comprehend and use GI in specific tasks might not be known but can be estimated quite well. Still, if the ambition of reaching all citizens with the GI-based applications is kept, there is a need to get a clearer impression of the understanding of the concepts of GI among nonprofessional users. This chapter has suggested a number of investigation methods that, used in combination, might bring light to this area. Results of the study may prove to be valuable at a much broader range than just evaluating the GI-based applications at the regional and municipal web sites, as knowledge of users' understanding is crucial for the success of the transition toward digital management.

REFERENCES

1. Remmen, A., Images of e-Government — Experiences from the Digital North Denmark, paper presented at ICT and Learning in Regions, 2004, http://www.kommunikation.aau.dk/ddn/conference2004/index.htm.
2. The Digital Taskforce, Towards e-government — vision and strategy for the public sector in Denmark, January 2002, www.e.gov.dk.
3. United Nations, Division for Public Economics and Public Administration and American Society for Public Administration, Benchmarking E-government: A Global Perspective. Assessing the Progress of the UN Member States, 2002.
4. Cartwright, W. and Hunter, G., Enhancing geographical information resources with multimedia, in Cartwright, Peterson, and Gartner, Eds., *Multimedia Cartography*, Springer-Verlag, Berlin, Heidelberg, 1999, pp. 257–270.
5. Anderson, L.W. et al., *A Taxonomy for Learning, Teaching and Assessing. A Revision of Bloom's Taxonomy of Educational Objectives*, Longman, New York, 2001.
6. OECD; ICT statistics, available on the OECD web site: www.oecd.org.
7. Danish Ministry for Science, Technology and Innovation and Denmark's Statistics, Information Society Denmark, ICT Status, 2003.
8. The Active map of Aalborg (Det aktive Aalborgkort), www.detaktiveaalborgkort.dk.
9. Jørgensen, I., Korttolkning (Map Interpretation), KVL (The Royal Veterinarian and Agricultural University), Copenhagen, 1998.

10. Keller, C.P. and O'Connel, I.J., Methodologies for evaluating user attitudes towards and interactions with innovative digital atlas products, in *Proceedings, 18th ICA/ACI International Cartographic Conference*, Stockholm, Sweden, 3, 1243–1249, 1997.

11. Nielsen, J., *Usability Engineering*, Academic Press, Boston, 1993.

12. Heidmann, F. and Johann, M., Modelling graphic presentation forms to support cognitive operations in screen maps, in *Proceedings of 18th International Cartographic Conference*, Stockholm, Sweden, 1997, pp. 1452–1461.

13. Arleth, M., Using log-file analysis for testing cartographic web based applications, in *Proceedings of 20th International Cartographic Conference*, Beijing, China, 2001, pp. 2313–2319.

14. Todd, P., Process tracing methods in the decision science, in Nyerges et al., Eds., *Cognitive Aspects of Human-Computer Interaction for Geographic Information Systems*, Kluwer Academics, Norwell, Mass., 1994.

15. Møller-Jensen, L., Monitoring user responses to web-based GI interfaces, paper presented at 2nd AGILE conference on Geographic Information Science, Rome, 1999.

16. Sitecatalyst, www.omniture.com.

17. Wherrett, J.R., Issues in using the Internet as a medium for landscape preference research, in *Landscape Urban Plann.* 45, 209–217, 1999.

18. Van Elzakker, C.P.J.M., Map use tasks in regional exploratory studies, in *Proceedings of 20th International Cartographic Conference*, Beijing, China, 2001, pp. 2496–2505.

19. Just, L., Designing new maps adapted to user's needs, in *Proceedings of 19th ICA/ACI International Cartographic Conference*, Beijing, China, 2001, pp. 1056–1063.

20. Tobon, C., Usability testing for improving interactive geovisualization techniques, CASA working paper series www.casa.ucl.ac.uk.

21. Brodersen, L., Quality of maps — measuring communication, in *Proceedings of 19th ICA/ACI International Cartographic Conference*, Beijing, China, 2001, pp. 3044–3051.

6 Geographic Information as an Economic Good

Alenka Krek

CONTENTS

Geoinformation is an intangible economic good determined by the high fixed cost of production and low marginal cost of reproduction. Its characteristics and the properties of the geoinformation market cause difficulties in applying neoclassical economic principles of pricing to geoinformation. We present cost-based pricing that is still used by many data providers and value-based pricing. Value-based pricing is pricing according to the value the buyer attaches to the characteristics of the product. It represents an alternative approach in pricing of geoinformation and is generally independent of the geoinformation production cost.

6.1 INTRODUCTION

Imagine that the producer wants to sell geoinformation. How much are the potential buyers willing to pay for it? What would be the right pricing policy, how much should a producer charge for geoinformation? These are the questions that many producers of geoinformation address while preparing their marketing strategies. Price is an important element of trade and can, if set in the right way, persuade the potential buyer to buy. It is defined as "an amount, usually in money, for which a thing is offered or exchanged." A "thing" may be a product, service, or money. Possible pricing techniques range from full cost recovery to no-cost dissemination. Selection of an efficient pricing strategy depends mostly on the characteristics of the market and the properties of the good offered on the market.

Pricing of geoinformation is complicated because of the nature and economic characteristics of geoinformation, its cost structure, and the structure of the market that makes applying standard neoclassical economic models difficult. We review the basic characteristics of a geoinformation market in Section 6.2 of this chapter.

Section 6.3 of this chapter is devoted to cost-based pricing, which is still often used by the producers of geographic data and providers of geoinformation. The price according to this strategy is set in such a way as to recover the cost of producing the data. Such pricing can lead to relative high prices and therefore prevent the development of new geoinformation products along the geoinformation value chain and, with this, an efficient exploitation of geographic data sets. We show in this chapter why this strategy might not guarantee profitability to the geoinformation producers.

An alternative to cost-based pricing is value pricing that is introduced in Section 6.4. It is based on a strong belief that price needs to be related to the value the potential buyer attaches to the product and not to the production cost. The economic value the buyer attaches to the product and its characteristics reflects his or her preferences and needs. Certain conditions have to be fulfilled for a successful implementation of the value pricing. The producers have to be able to identify the characteristics of the product that have an economic value for the buyer, segment the potential buyers into groups with similar information needs and willingness to pay, and design their products in such a way as to satisfy the need for varieties. We review the requirements that have to be fulfilled for a successful implementation of the value pricing of geoinformation.

We propose metric conjoint analysis methods for analyzing the potential buyer's decision-making process and preferences, sorting the users according to these preferences, and the design of differentiated geoinformation products. A special form of value pricing can be applied when the users cannot be segmented into groups with certain similarities. This approach is based on the self-selection principle that we present in Section 6.4.

The disadvantage of value pricing is price dispersion, which is a variation in prices charged for the same good and can be a consequence of value pricing. It may create perceptions of unfairness among potential buyers of geoinformation if they are able to share information about price. We conclude the chapter with a discussion and suggestions for further work.

6.2 CHARACTERISTICS OF A GEOINFORMATION MARKET

Geographic information is intangible, difficult to quantify, and lacks transparency. It is not a standard public good such as, for example, national defense, because it is not possible to define whether it is excludable and nonrival or not. It has a similar cost structure to other information goods determined by the high fixed cost of production that is sunk cost and low marginal cost of reproduction. In this section, we review the specific characteristics of geoinformation, stress the importance of considering the geoinformation transaction cost, and analyze the market structure of a geoinformation market.

6.2.1 PROPERTIES OF GEOINFORMATION

In this section we look at some properties of a public good as defined in economic literature and analyze whether they are significant for geographic information. Examples of public goods are national defense, traffic lights, clean air, or street signs. One of the characteristics of a public good is that it is nonrival, which means that the consumption of the good does not diminish the amount available to other users. Whether geographic information is nonrival or not depends on the situation of use and technical characteristics of the media through which this information has been offered to the users. The Internet allows for the same information to be used many times, by different users at the same time without being used up. This might also cause problems; when too many users try to use the database via Internet at the same time, too low connection capacity might prevent some users from using the information from the data set.

A public good is nonexcludable, which means that one person cannot exclude another from consumption (for example, from using a geographic data set) and extract information from it. It is practically impossible to control the use of geoinformation acquired from the data. Geoinformation offered over the electronic network can be used by different users at the same time, and it is difficult to prevent copying, reproducing, and disseminating collected data sets by unauthorized users. Whether geographic data is excludable or not depends on the legal regime (terms of agreement between the producer and the user), protection of the intellectual property (copyrights, patents, and trade secret protections), and the technologies used to disseminate or control the access to the data [1].

Geoinformation is an intangible good, which means that one cannot hold it or touch it, and it is not easily countable as, for example, other standard economic goods. It lacks transparency, which means that it is not easy to see what the user will get (will it work on the computer, will it be possible to visualize the data with the existing software, will the user be able to extract the information he needs from the delivered set of the geographic data). This effect is in economic theory known as "imperfect information." The potential buyer is not perfectly informed about the characteristics of the good that is a subject of trade. Geographic data is an experience good as defined by Nelson [2], which means that one has to try, test, and experience it in order to assess its usability and the economic value.

6.2.2 Cost of Geoinformation

The cost of geoinformation consists of the cost of producing and delivering geo-information to the user, called transformation cost and the transaction cost. The transformation cost is the cost of transforming resource inputs, in this case datasets, into the physical attributes of a good. The transaction cost is the cost of inquiring about the characteristics of the product or service and the cost of negotiating the conditions of exchange such as, for example, price, amount, time of delivery, or the quality of the product.

6.2.2.1 Transformation Cost

Transformation cost of a data set, sometimes also called production cost, is transparent and relatively easy to determine. It includes the investment in the creation of the first copy of the data set. The cost of making another copy of the geoinformation product, or marginal cost, is, for the geographic data set, very low or even zero and can be neglected. Experience indicates that data collection accounts for 60 to 80% of the total cost of a full operational geographic information system [3]. This cost is fixed and is high mostly because of the high labor cost of capturing the data from data sources or acquiring them with measurement techniques, the cost of data integration, transformation, analysis, and modification. The share of this costs has been reduced in the last few years due to well-established technology for data capturing and new techniques that enable faster, efficient, accurate data capturing with less resources needed, compared to the case ten years ago. Nevertheless, the fixed costs of producing and delivering geoinformation stay very high, which is a general characteristic of information goods [4] such as, for example, CDs, books, or newspapers. The high fixed cost of producing a geographic data set is a sunk cost, which implies that it is not easily recoverable if the production is altered.

6.2.2.2 Transaction Cost

Transaction cost is the cost associated with choosing, organizing, negotiating, and entering into contracts and the cost of acquiring information about the product or service that is a subject of trade [5]. It arises because information about the good is costly and asymmetrically held by the parties to exchange. North [6] distinguishes between the measurement and enforcement cost. Measurement cost is the cost of measuring the valuable attributes and characteristics of what has been exchanged. Measuring the quality of geoinformation, searching for the right data set, and acquiring the information on the level of quality and usability for the specific application is very costly; it requires time, energy, and knowledge on usage of the data set. Enforcement cost is the cost of protecting rights, policing, and enforcing agreements. It includes resources involved in defining, protecting, and enforcing the property rights of the products and services, which are the right to use, the right to derive income from the use of, the right to exclude, and the right to exchange.

Transaction cost of geoinformation is not transparent and is rather difficult to measure. It was up to now neglected, because neoclassical economic theory completely ignores transaction cost. The buyer is, according to the assumptions made in the theory, perfectly informed about the good that is a subject of exchange. In the case of perfect information, we have an example of zero transaction cost. We estimate that transaction cost represents a substantial share of the geoinformation buyer's budget. Future research should consider several classes of transactions [7,8], which are affected in different ways along the geoinformation value chain. The role of electronic networks in relation to transaction cost should be analyzed.

6.2.3 GEOINFORMATION MARKET STRUCTURE

What price should be charged for the products depends also on the whole structure of the geoinformation market. Traditionally, a "market" was a physical place where buyers and sellers exchanged the goods. Discussions on the definition of a market (see [9], p.13) will be ignored for the purpose of this chapter. In general, economists describe a market as a collection of buyers and sellers who transact over a particular product, product class, or service. Marketers view the sellers as constituting the industry and the buyers as constituting the market [10].

The market structure itself directly influences the price strategy for the goods offered on the selected market. Standard economic classification distinguishes among perfect competition, monopoly, monopolistic competition, and oligopoly. These market structures differ in the number of producers on the market, type of products, the extent to which an individual firm controls its price, how easily can the firms enter the market, and the extent to which the companies compete on the basis of advertising and differences in product characteristics, rather than price. Table 6.1 gives an overview of the main characteristics of the market structures.

The market of the data sets produced by the National Mapping Agencies has characteristics of a monopoly, with high barriers to entry into the market due to the copyrights they preserve on certain data sources and rather high prices set for the digital data. On the market of the geoinformation software, we can observe characteristics of an oligopoly. A few producers offer standardized or somehow differentiated geoinformation software. The market for geoinformation applications has a very similar structure, with few producers, rather high barriers to entry due to the high initial investment required and with some power over the pricing strategy preserved. The question that we face is what is then the market structure for geoinformation? Can we describe it with the standard characteristics known for the market structures (see Table 6.1)? Additional research has to be devoted to the issues related to the implications of electronic networks to geoinformation market structure. There is a need to better understand the evolutionary impact of trade over the electronic network on the existing geoinformation market structures and the role of the companies involved in a geoinformation value chain.

6.3 COST-BASED PRICING

The data set producers are aware of the cost of geoinformation, and they often try to cover them with cost-recovery pricing. This is still the most common pricing

TABLE 6.1
Characteristics of the Market Structures

Market Structure	Examples	Number of Producers	Type of Product	Power of Firm over Price	Barriers to Entry	Nonprice Competition
Perfect Competition	Parts of agriculture	Many	Standardized	None	Low	None
Monopolistic Competition	Retail trade	Many	Differentiated	Some	Low	Advertising and product differentiation
Oligopoly	Computers, oil, steel, GIS software	Few	Standardized or differentiated	Some	High	Advertising and product differentiation
Monopoly	Public utilities, national mapping agencies	One	Unique	Considerable	Very high	Advertising

Source: Mansfield, E., *Managerial Economics*, W.W. Norton & Company, New York, 1993.

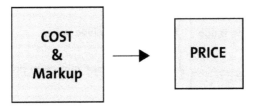

FIGURE 6.1 Cost-plus pricing.

practice on the geoinformation market. In this section we discuss the basic theoretical concepts of the cost-based pricing technique and its applicability in the geoinformation market.

6.3.1 COST-PLUS PRICING

Cost-plus pricing is a pricing technique used by a large number of firms and can appear in several different forms. It is used primarily because it is easy to calculate and requires rather limited information. The typical calculation consists of two steps. First, the firm estimates the cost of production and calculates the cost per unit of output. The estimation of the level of output can be based on past data or predicted according to the current trends of the sell. Second, the firm adds a markup to the estimated average cost. The markup includes the return on the investment and certain overhead costs, is usually expressed in percentage, and represents a constant number. The price is then calculated as follows:

$$\text{Price} = \text{Cost} \ (1 + \text{Markup})$$

The cost is the sum of average variable cost and percentage allocation of fixed cost. In some cases the companies incorporate the fixed cost allocation directly into the markup percentage (Figure 6.1).

The markup is, in general, influenced by several factors including market structure, special offers such as discounts, and pricing strategy that also depends on the stage of the product in a lifecycle. The pressure to lower the markup is stronger in highly competitive markets. A seller might decide to lower the markup for special offers such as a discount where he expects to raise his revenues by selling higher quantities of the good. The skimming and penetration prices where the producers or sellers enter the market also require a lower markup. A product in the early stages of the life cycle needs a lower markup percentage to help establish demand for the product.

Another form of cost-plus pricing is activity-based pricing. According to such pricing, every activity is linked to the resources it needs and uses and requires more precise and specific determination of costs. According to the product life-cycle pricing, the seller adjusts the markup to the phase of development of the product, depending on the position of the product in the life cycle.

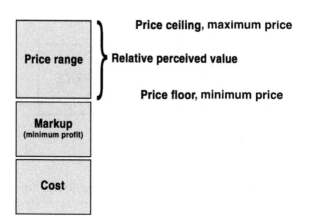

FIGURE 6.2 Adaptive cost-plus pricing.

6.3.2 Adaptive Cost-Plus Pricing

Some companies use what we call here adaptive cost-plus pricing. Such pricing adapts the level of the markup according to the market conditions, structure, and relative perceived value. If the producer is maximizing the profit, the markup is determined by the price elasticity of demand for a product. Figure 6.2 shows the price floor, which is the level of the price that includes the cost and a minimum profit, and the price ceiling where the producer sets the price higher according to the relative perceived value of the product to the user. This happens if he expects high valuation of the product in a certain segment of the market.

6.3.3 Target Rate of Return Pricing

Some firms set up a target rate of return that they aim to earn. A target rate of return is a figure that determines the markup. Under target rate of return pricing, a price is set equal to [11]:

$$P = L + M + K + F/Q + (Pr \times A) / Q$$

where P is price, L is unit labor cost, M is unit material cost, K is unit marketing cost, F is total fixed or indirect cost, Q is the number of units the firm plans to produce during the relevant planning period, A is total gross operating assets, and Pr is the desired profit rate on those assets expressed in percentage. General Motors, for example, used cost-based pricing and stated the objective of earning a profit of about 15% after taxes on the total invested capital [11].

6.3.4 Applicability to Geoinformation

Cost-plus pricing has an advantage that it is relatively easy to calculate and administer, the price is based on the cost and estimated output, and it requires minimal information. The increase in prices can be justified when costs increase, but this is possible only if the price also includes other changes on the market such as changes in the income level of the potential and current buyers, changes in the prices of the competitive products, etc.

Cost-based pricing cannot easily be applied to geoinformation. One of the problems represents the structure of the cost with high fixed cost of data collection and rather low level of output at the beginning of the geoinformation value chain. The question of estimation is crucial for the calculation. If budgeted costs are overestimated or the level of estimated sell of the products is underestimated, this might result in too high a selling price, which may lead to lower demand, higher costs, and lower profit. This can prevent further development of the geoinformation market, and it does not bring the highest possible profitability to the geoinformation producer. In this case he can sell his products only to the users with high willingness to pay, neglecting the other potential segments of the market. The estimation of cost ignores the size of marginal cost, and the sunk cost that appear if the investment is altered. It also completely neglects the transaction cost of geoinformation.

An additional disadvantage of this technique is that it tends to ignore the role of the user of the geoinformation product and the elasticity of demand. Taking the elasticity of demand into consideration, it might be possible to charge a higher or lower price to maximize profits, depending on the responsiveness of the user to changes in price. A further disadvantage is that it ignores the role of competitors and their pricing strategies. This is often an issue that appears further along the geoinformation value chain. The competition has little impact at the beginning of the geoinformation chain, especially for the so-called "raw data" that are produced by the National Mapping, and other agencies at the state level, where we can observe the characteristics of natural monopoly due to the high investment cost and restrictions on copyrights.

6.4 VALUE-BASED PRICING

Value pricing is pricing according to the value of the product perceived by the potential buyer. The value the potential buyer attaches to the product reflects his preferences. Understanding the concepts of value and the demand for the products helps producers in setting a successful pricing policy. In this section, we review the concepts of value, as known in economic theory, and the conditions that have to be fulfilled in order to be able to successfully apply value pricing to geoinformation products. We list the following three conditions: knowing the user's preferences, sorting the users according to these preferences, and the design of differentiated geoinformation products.

6.4.1 VALUE IN ECONOMIC THEORY

The issues of value, its concepts, and value formation in the human brain are complicated, and economic theory does not have a unique value theory. In spite of that, several groups of economists contributed to a better understanding of the concepts of value. Adam Smith [12] held scarcity to be the source of value in his early work. Later, he considered labor required in the production of a good to be the source of value. His approach is important because he has made a distinction between the "value of use" which is perceived by the buyer, and the "value of exchange" that is the price the potential buyer is willing to pay for the product.

Neoclassical economists neglect the issues of value formation and its importance for defining the price of the product. They consider it irrelevant for economics to study how patterns of value are formed in the human mind. According to their assumptions, it is identical to price in the state of the market equilibrium, and the buyer's preferences are given, known, and well defined.

Veblen [13,14] criticized neoclassical preoccupation with static equilibrium, abstracted from sociocultural changes and their impacts on economic activities. He belonged to the group of institutionalists who argue that economics has to focus on the analysis of the different processes by which modern societies develop their valuation systems. North [6] shows that social institutions and values have profound impacts on the behavior of consumers.

The cognitive approach in the value theory studies the development of consumers' value patterns, consumers' judgments of utility, and consumers' decisions in consumption activities [15]. Woo [15] proposes a new value theory that consists of a theory of value formation, a theory of consumer choice, a theory of cost formation, and a theory of consumer–supplier interaction.

Porter [16] introduces the concept of a value chain as a tool for analyzing the firm's competitive advantage. "The value chain desegregates a firm into its strategically relevant activities in order to understand the behavior of costs and the existing and potential sources of differentiation. A firm gains competitive advantage by performing these strategically important activities more cheaply or better than competitors [16]." Value is "the amount buyers are willing to pay for what a firm provides to them [16, p. 38]." It is measured by the total revenue that is a reflection of the price of a firm's product and the number of products sold on the market.

The concept of value moved the focus from the product or a firm to the customer in the late 1980s [17]. Hanan and Karp [18] define the value as "the added competitive advantage" the seller brings to the customers. For example, if the product helps to reduce the cost of the customer this adds to the competitive advantage of the customer as a low-cost supplier. Their major contribution is the idea to add value to the customer and not to the product. Brandenburger and Nalebuff [19] introduce the concept of "added value" in an economic situation considered as a game. Added value measures what each player brings to the game. Intuitively, what a player can get from the game is limited by what he brings into the game, that is, his added value. Tapscott [20] suggested the provision of value to be something that is generated through an ever-changing open network. The digital infrastructure sets the foundation for the creation of fundamentally new and different kinds of value. Kotler [10] sees the value primarily as a combination of quality, service, and price (QSP), called the customer value triad, where value increases with quality and service and decreases with price.

6.4.2 Concept of Value Pricing

Value pricing, sometimes referred to as "smart" pricing [21], is a market-based pricing technique where the producer sets the price of the product according to the value the product has for the potential buyer. The economic value the potential buyer attaches to the product reflects his preferences and is directly related to his needs and willingness to pay for certain characteristics of the product. Value pricing is often suggested as the most economically efficient pricing of information products [4,22]. One of the reasons for that is the significant cost structure of information products determined by the high fixed cost of producing an information product and low marginal cost of reproduction.

Value pricing can be economically efficient for geoinformation products, because it can better match, in combination with product differentiation, the varieties of information needs and the varieties of the willingness to pay for the information. It enables the producers and sellers to gain higher revenue serving new markets that would otherwise not be served.

6.4.3 IMPLEMENTATION OF VALUE PRICING

Value pricing can be successfully implemented if the producer knows the preferences of the potential buyer and is able to segment them into different groups. Conjoint analysis can help to understand why consumers choose certain products. It is concerned with quantitative descriptions and methods and can be used to identify the attributes of the product that have an economic value for the buyer, determine the contributions of certain attributes to consumer preferences, and predict consumers' behavior. Other conditions for an efficient implementation of value pricing include market segmentation and product differentiation. We describe them within this section.

6.4.3.1 Conjoint Analysis

Conjoint analysis is a generic term coined by Green [23,24] and refers to a number of paradigms in different research areas that are concerned with the quantitative description of consumers' preferences or value trade-offs [25]. It involves the use of modeled choice situations to examine consumer behavior, measure his preferences, and predict his choices among several alternatives [26]. It is based on the assumption that the potential buyer can evaluate multi-attribute alternatives on a category rating scale [27].

Theoretic background for a metric conjoint analysis is partially given by Lancaster's characteristics approach [28–30], where he argues that a product consists of several attributes called characteristics. These characteristics directly influence the buyer's decision whether to buy a product or not. He argues that it is the color, the model, or the material of the product that attracts the potential buyer and directly influences his decision-making process. His approach can be applied to geoinformation products [1].

A conjoint analysis deals only with characteristics of the product that can be identified and measured. The researcher or the producer defines and selects the characteristics of the products that are assumed to have an impact on the potential buyer's valuation and his decision-making process. These characteristics are then classified into numerical categories and combined into product profiles. Each profile

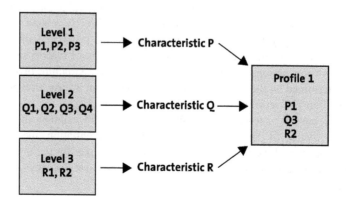

FIGURE 6.3 Characteristics, levels, and a profile. (Adopted from Dijkstra, J. and Timmermans, H.J.P., *Proceedings of the Second Conference on Computer Aided Architectural Design Research in Asia*, Liu, Y.-T., Ed., Hu's Publishers, Hsinchu, Taiwan, 1997, p. 61. With permission.)

is a combination of characteristic levels for the selected characteristic. Different product profiles are then presented to the potential buyer, who expresses the degree of preference for these profiles or chooses between them. Figure 6.3 shows the relationship among levels, characteristics, and a profile.

In order to establish a valid model of the buyer's judgments, the researcher estimates so-called "utilities" or strength of preferences for the various characteristic levels. The analysis concentrates on a general approach of estimation of partial and joint evaluations. Valuation of the partial or characteristic strength of preferences is often called part worth [26,27] and represents a numerical expression of the value that consumers place on each characteristic level. Joint evaluation is a profile evaluation that is an overall strength of preferences of a product profile, which is calculated by summing up all partial values of characteristic levels defined in that profile. This approach is known as functional measurement, because the partial measures of interest are those that "function" in models of human information processing [27], and is based on information integration theory, and a theory of human information processing.

Conjoint analysis is one of the most celebrated tools in marketing, and academics distinguish among several variations of this analysis, for example traditional conjoint analysis, adaptive conjoint analysis, and choice-based analysis. It is usually based on experiments, statistical calculations, and evaluations. Its advantage is that it is a metric computational method that enables a systematic approach in analysis of consumer behavior in the marketplace. It can help researchers or product producers to examine the direct trade-offs among competing geoinformation products or product varieties, and perceptual mapping, which assesses the benefits of different products that may not be direct substitutes for one another seeking to identify the benefits that no other product offers. There are many issues to explore (see [31,32]), especially those related to the applicability of these methods to geoinformation products. The most challenging issue is the identification of the characteristics of geoinformation that can possibly have an economic value for the potential buyer. Additional research is to be devoted to these issues.

6.4.3.2 Geoinformation Market Segmentation

In order to be able to price according to the value the potential buyers attach to the product, the researcher or the producer segments the market into groups of buyers with similar geoinformation needs and similar willingness to pay for the information. Such segmentation helps him to identify the groups of individuals with corresponding geoinformation needs and to design the geoinformation products that can satisfy these needs. For example, the producer of a tourist application can differentiate between a city tourist interested in cultural heritage of the city and a tourist interested in hiking in the mountains. Their needs for geoinformation, the technology that delivers this information, and their willingness to pay for the geoinformation differ substantially.

In general, economists distinguish between exogenous and endogenous market segmentation [11,33]. Segmenting is exogenous if the potential users can be sorted according to some observable characteristics such as age, location, time, or income. The producer can, for example, define driving tourists, city tourists, and hiking tourists as separate groups. Another market segmentation of tourist user's groups could make a distinction among families, couples, and individual tourists. The segmentation can also be based on time, for example, weekend tourists, summer tourists, and winter tourists. Every group has certain characteristics of use and needs for geoinformation.

Segmenting is endogenous where the potential users cannot be sorted by observable characteristics. In this case, the producer can develop varieties of a product, differentiate them according to some characteristics, and price them according to the value these characteristics have for the buyer. The potential buyer can then "self-select" the product that satisfies his information needs and his willingness to pay. In this case, the potential buyer segments himself with his choice of the product.

Conjoint analysis can be also useful for segmentation purposes [32,34]. North and de Vos [34] suggest segmenting the potential buyers on the basis of strength of preferences. Such simulations can be viewed as segmentation analyses that group buyers according to their most preferred product profile among other substitute product profiles or competitive products. The applicability and usability of conjoint analysis for the purpose of geoinformation market segmentation should be analyzed, tested, and studied.

6.4.3.3 Differentiated Geoinformation Products

Product differentiation is concerned with how a firm competes on the market. The company differentiates itself and its products from its competitors when "it provides something unique that is valuable to buyers beyond simply offering a low price [16]." What differentiates the products are the characteristics they possess, and differentiated products are both; similar and different. These differences are grounded in the preferences of the buyer. Geoinformation products can be differentiated according to the quality of the sources of geoinformation, completeness of the application, time, form, and format of delivery, copyright, etc. Frank and Jahn [35] show an example of quality differentiation on a street network data set. The list of possibilities for geoinformation product differentiation is endless and depends on the imagination of the producer and his or her business perspectives.

The term product differentiation was first introduced by Chamberlin [36] in Chapter 4 of his book *Theory of Monopolistic Competition*. We can find some aspects of the differentiation in Hotelling's paper [37], in which he uses a spatial component in the economic analysis and represents the foundation of so called "spatial," sometimes referred to as "locational," economic research investigated by numerous authors [38–40]. Lancester [28] further developed the idea and modeled an economic good as a bundle of different characteristics, where the buyers have preferences over characteristics.

Standard economic literature on product differentiation distinguishes between vertical and horizontal differentiation [41–43], sometimes referred to as address and nonaddress approach. Products are vertically differentiated if the consumers can rank them according to the quality index, and they would all rank them in the same way. This means that the buyers can objectively agree that the product A is better than the product B for a particular use. A typical example is quality. Most agree that higher quality is preferable. For example, a smaller and more powerful computer is preferable to a larger, less powerful one. Products that cannot be objectively ranked by the quality index are modeled as horizontally differentiated. For some characteristics, the optimal choice depends on the particular buyer. An obvious example is the case of colors, another example is location [9].

The advantage of product differentiation is that the differentiated products aim at better matching the buyer's preferences and needs, and the producers hope to be able to serve new markets. The existence of product differentiation may imply that firms retain some market power, especially because it might represent a barrier for other producers to enter a certain market.

6.4.4 Self-Selecting Geoinformation Products

The self-selecting principle can be used for geoinformation products. This principle is known in economic literature [44,45] and implied in situations where the buyer self-selects the product he is willing to pay and reveals the value the product has for him through the selection of the product. The seller makes several products available, differentiates them according to one or several characteristics, and prices them according to the value they might have for the potential buyer.

An illustrative example of a self-selecting pricing is pricing in movie theatres. The visitor of a movie theatre can decide in which row she wants to sit which also reflects the price he is willing to pay for the characteristics of his choice. She has a choice of paying less and sitting in the first few rows of the movie theater or paying more for a better product. The self-selecting principle naturally leads to value pricing where the buyer pays the price for the product according to the value he

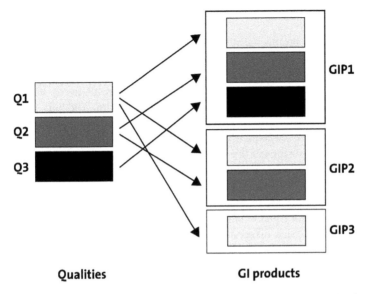

Qualities   **GI products**

FIGURE 6.4 Geoinformation products as a composition of qualities. (Krek, A., An Agent-Based Model for Quantifying the Economic Value of Geographic Information Institute for Geoinformation, Technical University Vienna, Vienna, Ph.D. thesis, 2002. With permission.)

attaches to it. In equilibrium and under the assumption that all characteristics have equal value weights, the product with several valuable characteristics will be charged at a higher price than the compositions with fewer valuable characteristics. The self-selection principle is a special case of second-degree price discrimination [44,45].

The producers of geoinformation products can construct price-quality packages, introduce them to the potential users, and give the possibility to self-select the option that satisfies their information need.

The options can be as follows [1]:

$$GIP\ 1 = \{(Q1,\ Q2,\ Q3)\}\quad p1$$

$$GIP\ 2 = \{(Q1,\ Q2)\}\qquad p2$$

$$GIP\ 3 = \{(Q1)\}\qquad\qquad p3$$

where GIP is a geoinformation product, Qi quality, and pi price of composition of the qualities where p1 > p2 > p3. Figure 6.4 shows a composition of three qualities that can be combined in three different products.

The user usually needs a particular composition of qualities that he considers as optimal; higher is more expensive, and lower does not satisfy the information need [46]. In such a way the producer encourages the self-selection principle by adjusting the quality of the good. The buyer self-selects the option that satisfies his information need and his willingness to pay.

6.4.5 PRICE DISPERSION

Value pricing is a promising strategy for pricing of information products. When applied properly, it can result in a higher profit for the producers through serving the markets that would otherwise not be served. The buyer pays the price that he is willing to pay and can choose the product that satisfies his needs without paying for all additional characteristics that have no value to him. The only problem with value pricing represents price dispersion, which is a variation in prices for the same good and can be a consequence of value pricing. It may create perceptions of unfairness among consumers if they are able to share information about prices. Such comparisons can be easily done over the Internet. Price search engines, often called "shopbots," are sites that enable comparison of selected product offerings from multiple producers or sellers [47]. They allow the buyer or potential buyer to directly compare the prices of equal or similar products. Clemons, et al. [48] found empirical evidence for wide price dispersion among online travel agents, with ticket prices varying by up to 28% for the same customer request, and up to 18% after accounting for ticket quality and route differences. Sinha [21] claims that value pricing achieved through versioning or mechanisms like auctions can be extremely risky in the long term. The producers of geoinformation have to study the ways of applying the value pricing technique in practice in order to avoid price dispersion and perceptions of unfairness among the buyers.

6.5 CONCLUSIONS

The geoinformation producers have problems with the selection of an economically efficient pricing policy. The reasons for that are the historical development of the data sources supply, the nature of a geoinformation market, and specific characteristics of geoinformation as an economic good. The market of geographic data produced by National Mapping Agencies (NMA) has characteristics of monopoly. An inappropriate pricing strategy for the basic geographic data can prevent further development of the geoinformation market.

Many data providers use cost recovery pricing in order to recover the transformation cost by setting a price on the cost basis. This strategy might induce several difficulties. The major problem is that the producers can serve only particular groups of geoinformation users that need such products, are able to use them, and are willing to pay the price set for these products. All other groups, for example, less knowledgeable users, or the user who needs only geoinformation and not the whole set of data, are often not served by these providers. Cost-based pricing at the beginning of the geoinformation value chain can also prevent further development of the geoinformation market if there are not enough intermediate companies that use the "raw data" and transform them into new geoinformation products. Pricing of information goods should always be considered together with the analysis of what has been offered to the user and not strictly related to the cost of producing the information good.

Value pricing represents a promising alternative to cost-based pricing. This pricing technique is market-based and considers the potential buyer's needs and preferences. In order to price according to the value, the producers of geoinformation products should be able to identify which characteristics of the products have an economic value for the buyer.

In this chapter we show how conjoint analysis can help the producers to examine consumers' behavior, measure their preferences, and predict their choices among several alternatives. These methods are metric and enable a systematic approach in analysis of consumer behavior. In performing conjoint analysis, the researchers in geoinformation can face several difficulties; selection of attributes of geoinformation is not a trivial task, and the potential buyers of geoinformation might find it difficult to indicate which attributes they considered and also how they combined them to form their overall opinion.

Value pricing also requires product differentiation and segmentation of the market according to certain characteristics of the potential buyers. Product differentiation is involved in production of the products that can satisfy different geoinformation needs. The producers should try to avoid price dispersion, which can create a perception of unfairness among the potential buyers and result in a lower trade of the geoinformation products.

This chapter provides a short overview of our research on pricing of geoinformation. In our future work we will continue working on the issues of value pricing, which is still rather a theoretical approach and has not been implemented by many geoinformation producers. The most promising is a metric conjoint analysis that has to be further investigated. It would be necessary to improve our knowledge of these methods and execute some experiments with different geoinformation products. This would contribute to a better understanding of the potential buyers and their preferences for geoinformation, as well as to designing differentiated products that can satisfy the variety of information needs. More attention should be paid to adaptive pricing, which is pricing based on algorithms that respond to business fluctuations by adjusting changes in real time. It is a natural expansion of a demand-driven pricing and offers a degree of automation that lowers the cost of administration and calculation. Adaptive pricing is in the early stages of development. The researchers in geoinformation science should investigate whether it can be efficiently applied to geoinformation products. Research on economic issues of geoinformation is still limited to a very small group of researchers. What is missing is a strong research agenda that would encourage more grounded basic research on geoinformation economy.

ACKNOWLEDGMENTS

Most of the research presented within this chapter was done during my stay at the Technical University Vienna, Institute for Geoinformation, and has been partially published elsewhere. I thank my Ph.D. supervisor Professor Dr. Andrew U. Frank for his guidance and valuable suggestions. This research would not be possible without him.

REFERENCES

1. Krek, A., An Agent-Based Model for Quantifying the Economic Value of Geographic Information, Institute for Geoinformation, Technical University Vienna, Vienna, 2002.
2. Nelson, P., Information and consumer behavior. *J. Political Econ.*, 78, 311, 1970.
3. Bernhardsen, T., Geographic Information Systems, Viak IT and Norwegian Mapping Authority, Arendal, Norway, 1992.
4. Shapiro, C. and Varian, H.R., *Information Rules, A Strategic Guide to the Network Economy*, Harvard Business School Press, Boston, 1999.
5. Williamson, O.E. and Masten, S.E., *Transaction Cost Economics, Theory and Concepts*, Edward Elgar, 1995.
6. North, D.C., Institutions, *Institutional Change and Economic Performance*, Cambridge University Press, Cambridge, UK, 1990.
7. Sarkar, M.B., Butler, B., and Steinfield, C., Intermediaries and cybermediaries: a computing role for mediating players in the electronic marketplace, *J. Comput. Mediated Commun.*, 1(3), 1995.
8. Adelaar, T., Electronic commerce and the implications for market structure: the example of the art and antiques trade, *J. Comput. Mediated Commun.*, 5(3), 2000.
9. Tirole, J., *The Theory of Industrial Organization*, MIT Press, Cambridge, Mass., 1995.
10. Kotler, P., *Marketing Management*, 11th Edition, Prentice Hall, May 2002.
11. Mansfield, E., *Managerial Economics*, W.W. Norton & Company, New York, 1993.
12. Smith, A., *The Wealth of Nations*, reprint, Penguin Books, New York, 1986.
13. Veblen, T., Why is economics not an evolutionary science? *Q.J. Econ.*, 12, 373, 1898.
14. Veblen, T., *The Theory of the Leisure Class: An Economic Study of Institutions*, Rev. ed. (orig. 1898), Viking Press, New York, 1953.
15. Woo, H.K.H., *Cognition, Value and Price*, The University of Michigan Press, Ann Arbor, Mich., 1992.
16. Porter, M.E., *Competitive Advantage, Creating, and Sustaining Superior Performance*, The Free Press, New York, 1985.
17. Fletcher, T. and Russell-Jones, N., *Value Pricing, How to Maximize Profits Through Effective Pricing Policies*, Kogan Page, London, 1997.
18. Hanan, M. and Karp, P., *Competing on Value*, AMACOM, American Management Association, New York, 1991.
19. Brandenburger, A.M. and Nalebuff, B.J., *Co-opetition*, Bantam Doubleday Dell, New York, 1996.
20. Tapscott, D., Ed., *Creating Value in the Network Economy*, Harvard Business School Press, Boston, 1999.
21. Sinha, I., Cost transparency: the net's real threat to prices and brands, *Harvard Bus. Rev.*, 43, March/April, 2000, pp. 43–50.
22. Varian, H.R., Differential pricing and efficiency, *First Monday,* peer-reviewed journal on the Internet, Issue 2, 1999, available at: http://www.firstmonday.dk/issues/issue2/different/.
23. Green, P.E. and Srinivasan, V., Conjoint analysis in consumer research: issues and outlook, *J. Mark. Res.*, 8, 355, 1971.
24. Green, P.E. and Tull, D.S., *Research for Marketing Decisions*, Prentice-Hall: Engelwood Cliffs, New Jersey, 1978.
25. Timmermans, H.G., Consumer choice of shopping centre: an information integration approach, *Reg. Stud.*, 16, 171, 1982.

26. Dijkstra, J. and Timmermans, H.J.P., Exploring the possibilities of conjoint measurement as a decision-making tool for virtual wayfinding environments, in *Proceedings of the Second Conference on Computer Aided Architectural Design Research in Asia*, Liu, Y.-T., Ed., Hu's Publishers, Hsinchu, Taiwan, 1997, p. 61.

27. Louviere, J.J., *Analysing Desision Making, Metric Conjoint Analysis, Quantitative Applications in the Social Science*, SAGE Publications, Newbury Park, CA, 1988.

28. Lancaster, K.J., A new approach to consumer theory, *J. Political Econ.*, 74, 132, 1966.

29. Lancaser, K., *Consumer Demand: A New Approach*, Columbia University Press, New York, 1971.

30. Lancaster, K., *Modern Consumer Theory*, Edward Elgar Publishing, Hants, UK2, 1991.

31. Bradlow, E.T., A unified approach to conjoint analysis models. *J. Am. Stat. Assoc.*, 97(459), 674, 2002.

32. Bradlow, E.T., Current Issues and a "Wish List" for Conjoint Analysis. 2003, http://www.statisticalinnovations.com/articles/bradlow.pdf.

33. Frank, R.H., *Microeconomics and Behavior*, McGraw-Hill, New York, 2000.

34. North, E. and de Vos, R., The use of conjoint analysis to determine consumer buying preferences: A literature review, *J. Fam. Ecol. Consumer Sci.*, 30, 2002, pp. 32–39.

35. Frank, A.U. and Jahn, M., How to sell the same data to different users at different prices, in *Proceedings of the 6th AGILE Conference*, Lyon, France, 2003.

36. Chamberlin, E.H., *The Theory of Monopolistic Competition*, Harvard University Press, Cambridge, Mass., 1933.

37. Hotelling, H., Stability in competition, *Econ. J.*, 39, 41, 1929.

38. Andersson, A.E., Beckmann, M.J., Loefgren, K.G. and Stenberg, A., *Economics of Space and Time, Scientific Papers of Toenu Puu*, Springer-Verlag, Heidelberg, 1997.

39. Greenhut, M.L., *Spatial Microeconomics*, Edward Elgar, 1995.

40. Loesch, A., *The Economics of Location*, Yale University Press, New Haven, Conn., 1954.

41. Beath, J. and Katsoulacos, Y., *The Economic Theory of Product Differentiation*, Cambridge University Press, Cambridge, UK, 1991.

42. Constantatos, C. and Perrakis, S., Vertical differentiation: entry and market coverage with multiproduct firms. *Intern. J. Ind. Org.*, 16 (November), 81, 1997.

43. Norman, G., Spatial pricing with differentiated products, *Q.J. Econ.*, 98 (2), 291, 1983.

44. Varian, H.R., *Intermediate Microeconomics: A Modern Approach*, 4th ed., W.W. Norton, New York, 1996.

45. Norman, G., Ed., *The Economics of Price Discrimination*, Edward Elgar Publishing, London, 1999.

46. Krek, A. and Frank, A.U., Optimization of quality of geoinformation products, in *Proceedings of the 11th Annual Colloquium of the Spatial Information Research Centre (SIRC'99)*, Dunedin, New Zealand, 1999.

47. Elbers, A., Barwise, P., and Hammond, K., The impact of the Internet on horizontal and vertical competition: market efficiency and value chain reconfiguration, in *The Economics of the Internet and E-commerce*, Vol. 11, pp. 1–27, Baye, M.R. Ed., *Advances in Applied Microeconomics*. Amsterdam. JAI Press, 2002.

48. Clemons, E., Hann, I. and Hitt, L., Price dispersion and differentiation in online travel: an empirical investigation, *Manage. Sci.*, 48(4), 534, 2002.

Part II

GIS Research Perspectives for Sustainable Development Planning

7 Advanced Remote Sensing Techniques for Ecosystem Data Collection

Alexandr A. Napryushkin and Eugenia V. Vertinskaya

CONTENTS

7.1 INTRODUCTION

The problems of monitoring and ecological control of ecosystems of different natures are becoming more and more urgent. Monitoring of the Earth's surface has a multidisciplinary character and allows a wide spectrum of issues to be solved. The ecosystem components involved in monitoring are manifold and include, among others, surface waters, soils, vegetation canopy, and anthropogenic landscape components. The latter represent the man-made and man-changed ecosystems and are of primary interest in the context of monitoring and management problems due to degradation of recent ecological conditions [1].

One of the most important issues solved in the monitoring process is representation of its results as a series of thematic maps indicating the spatial structure of complex ecosystem components [2]. The basic concern of thematic mapping is graphical modeling of ecosystems and providing the information on their conditions for efficient natural resources management. The geoinformation provided by the thematic maps is used for analysis and assessment of natural resource conditions, recording and accounting destructive natural phenomena, studying natural and man-made ecosystems interaction, revealing anthropogenic impact to environment, and assessing its consequences [1,3].

Initial information used for ecosystems thematic mapping is acquired by means of terrestrial and remote monitoring techniques. The former characterize only 1 to 5% of surface and are not efficient to provide sufficient information on large ecosystems. Moreover, when detailed research is conducted, personnel, equipment, and time costs increase dramatically. Remote monitoring techniques provide a number of advantages over the terrestrial techniques, allowing the limitations of the latter to be overcome. In the literature, the concept of remote monitoring or surveying is referred to as remote sensing (RS) [4]. The RS techniques involve detecting and measuring electromagnetic radiation or force fields associated with terrestrial objects located beyond the immediate vicinity of recording instruments, such as radiometers or radar systems mounted on an aircraft or satellite. Remote monitoring, unlike the terrestrial one, allows a large-scale ecosystem to be surveyed with a short repeat cycle. The latter in most cases is a crucial criterion for ecosystem-change research. Generally, RS data represent images much like photos of the sensed surfaces of the objects under surveillance, and in the literature, RS images are often referred to as aerospace imagery [5].

Recently, thematic mapping of ecosystems has been widely implemented through employing geographic information systems (GIS) characterized by advanced capabilities for spatial information storing, manipulating, and processing [6]. Modern GIS provide wide capabilities for both computer-aided thematic mapping and spatial analysis of mapped features and phenomena, allowing derivation of complex quantitative characteristics indispensable for ecosystem conditions modeling and forecasting. Commonly, GIS facilities are oriented mainly for vector data handling, while RS-based thematic mapping methodology requires supporting functions of raster image processing. This fact makes urgent the problem of developing efficient and highly integrated software means enabling GIS to implement aerospace imagery processing and facilitate the thematic mapping technologies with use of RS data.

In this chapter, the methodology of RS-based thematic mapping is introduced. The implementation of the methodology is based on application of a vector GIS and original image processing and interpretation system "LandMapper" [7], developed at Tomsk Polytechnic University (TPU). The main distinction of the system from its counterparts is adaptive classification procedure (ACP), making the process of image interpretation more flexible and efficient in comparison with existing recognition techniques. The chapter considers the basic methodology of image processing and interpretation adopted in the "LandMapper" system and gives the results of its application for solving problems of mapping two anthropogenic ecosystems with the use of multispectral imagery acquired from the Russian satellite RESURS-O1.

7.2 RS-BASED THEMATIC MAPPING METHODOLOGY

7.2.1 GENERAL CONCEPT

Today, thematic mapping technologies making use of RS monitoring data and modern GIS-based tools are of great value, especially when significant interest is taken in research of various aspects of anthropogenic ecosystems. The wide range of anthropogenic issues that can be solved by means of RS-based thematic mapping involve urban areas monitoring [2], land use mapping, anthropogenic load of petroleum-production territories assessment, snow cover surveying, and flood forecasting. Recently joint use of GIS and thematic maps designed with aerospace imagery proved to be an efficient approach to creating and employing comprehensive models of anthropogenic ecosystems that were indispensable for decision-making.

Designing thematic maps with the use of RS imagery consists of a number of steps, including complicated processing of initial imagery, and is, as a rule, a nontrivial task to accomplish. Figure 7.1 illustrates the general scheme of thematic mapping of landscape ecosystems with use of remotely sensed images. According to Figure 7.1, in the methodology of RS-based thematic mapping, the stages of preliminary and thematic processing of imagery may be distinguished.

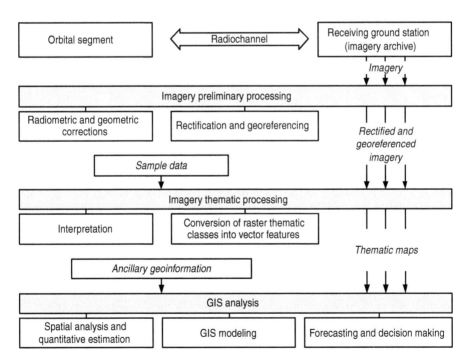

FIGURE 7.1 Thematic mapping with use of remotely sensed imagery.

Initially, imagery acquired from a satellite or aircraft is exposed to multilevel preliminary processing in order to make it usable for comprehensive analysis and facilitate transition from a simple raster image to a complex thematic map model. The preliminary processing involves solving the tasks of geometric and radiometric error correction. The tasks include compensation of radiometric distortion caused by atmospheric effect and instrumentation errors, correction of geometric distortion due to the earth curvature, rotation, and panoramic effect, noise reduction, image registration in a geographical coordinate system (georeferencing) through its rectification, and visual properties enhancement by histogram transformation [8].

The thematic and geometric information defining the application domain of the final thematic map is extracted at the stage of imagery thematic processing [5]. In thematic processing, very significant attention is paid to the image interpretation issue. Image interpretation provides revealing thematic knowledge about a studied ecosystem component and its spatial relationships by identifying image features and assigning them appropriate semantic information such as, for instance, landscape cover type.

Commonly, two main approaches can be adopted for image interpretation. One is referred to as photointerpretation and involves a human analyst/interpreter extracting information by visual inspection of an RS image [5]. In practice, photointerpretation is a very laborious and time-consuming process, and its success depends mainly upon the analyst effectively exploiting the spatial and spectral elements present in the image product. Another approach involves the use of a computer to assign each pixel in the image semantic information (land cover type, vegetation, or soil class) based upon pixel attributes. This approach deals with the concept of automated image interpretation–classification. Commonly, the approach appears to be most efficient when applied to multispectral imagery [4] having several bands of data acquired in different not overlapped spectral ranges.

In practice, classification is often carried out in so-called supervising mode, requiring the classification procedure to be trained beforehand. Training of the classification procedure relies upon selecting a set of representative elements (pixels) in the image for each informational class (land cover type) and forming training sets to be used further by the procedure as prototypes of extracted classes. Forming training data for supervised classification is one of the important issues in imagery thematic processing. This is carried out by gathering ancillary sample data that helps obtain a prior knowledge of the properties of ecosystem components present in RS imagery. Practically, sample data is acquired from different sources of information about the studied ecosystem — site visit data, topographic maps, air photographs, or even results of initial imagery photointerpretation.

The final product of the thematic processing stage is a raster map, each pixel of which is labeled with an appropriate code (label) corresponding to a landscape thematic class. Thus, different groups of equally labeled pixels in a thematic map represent thematically uniform objects recognized in imagery by the classification procedure.

Imagery thematic processing is followed by transferring the resultant thematic map into GIS, where it can be integrated with other data acquired from various informational sources, and comprehensive spatial analysis of the data can be conducted. Since many GIS software packages basically manipulate vector information, the stage of transferring a thematic map into GIS is performed through conversion of the raster map into a set of vector features thematically grouped in layers, each representing a specific class of ecosystem components — water surfaces, vegetation canopy, urban areas. The automated raster–vector conversion is not a straightforward procedure and is implemented by means of applying complex algorithms using "running window" and "tracing contour" principles as well as line generalization techniques [7].

In GIS the extracted vector features are assigned the additional attributive information. At that stage, the resultant vector thematic map is becoming a valuable informational model of the ecosystem. Such a model can be used efficiently for visualizing, measuring, and analyzing various characteristics of ecosystem components imaged in initial imagery. In cases when time-series RS imagery has been used for ecosystem thematic mapping, the resultant informational model allows acquiring knowledge for revealing trends of ecosystem change and forecasting its behavior.

The RS-based thematic mapping methodology described above is quite common and may be readily adopted in anthropogenic ecosystem research. However, the methodology of RS imagery processing and further thematic analysis can be very specific and can differ considerably in various case studies. In the remainder of this discussion, the imagery thematic processing approach elaborated in the GIS laboratory of TPU is considered.

7.2.2 IMAGERY INTERPRETATION APPROACH

The problem of automated imagery interpretation is still one of the most complicated among those of RS data processing. Among the general problems of automated RS data interpretation, that of efficient image classification techniques synthesis should be addressed. Classification efficiency is commonly defined by the accuracy and computational complexity of the recognition procedures that allow image objects to be categorized and depends on two main factors — conformity of classification decision rule and optimality of feature space.

The statistical classification decision rule (CDR) may be represented as function $m(X)$ allowing unambiguous assigning image pixels defined in P-dimensional feature space by respective feature vectors $X = \left\{ x_j, j = \overline{1,P} \right\}$ to one of M nonoverlapped classes $\omega_i, \left(i = \overline{1,M} \right)$. Commonly, $m(X)$ returns the index of the class for which X membership was proved through finding the largest discriminate function $\phi_i(X)$ defined for each class $\omega_i, \left(i = \overline{1,M} \right)$ [9]. The overall efficiency of a statistical decision rule is determined by *a priori* knowledge of the imagery classes, classification optimality criterion $R(m(X))$, and type of discriminate functions adopted.

For decision rule synthesis, it is common to employ a Bayesian approach to determining the discriminate functions calculated as a product of the class conditional probability density function (PDF) $p(X|\omega_i)$ and its *a priori* probability $p(\omega_i)$, with which class ω_i membership of X can be guessed before classification [5]. The crucial parameter $p(X|\omega_i)$ used in the Bayesian rule may be estimated in different ways, allowing a few CDRs to be derived. The applicability of the derived CDRs may differ, depending on feature vectors X distribution low, as well as the amount and quality of training data used for PDF estimations. The relatively fast parametric Bayesian CDR, making use of the Gaussian (normal) distribution hypothesis, produces good results with only unimodal distributions, whereas nonparametric CDRs, being free of normality constraints, can be efficient with distributions of any form, but at the expense of great computational complexity. In other words, finding a universal CDR effective by accuracy and performance for an arbitrary RS imagery is a big concern.

Endeavoring to solve the problem, an idea of adaptive classification approach has been proposed [7]. The approach is based upon employing a few CDRs in the classification procedure and an adaptive decision rule allowing an optimal CDR, in terms of accuracy and performance, to be chosen for classification. In the ACP, synthesis of $m(X)$ rests upon adopting a Bayesian rule that makes use of an empirical risk minimization criterion, $R(m(X))$, showing the probability of wrong pixel classification.

In practice, a common approach for probabilistic description of RS image classes is making an assumption of normal form of PDF $p(X|\omega_i)$ for each of M classes and using Gaussian parametrical PDF estimate in the Bayesian decision rule given by:

$$\hat{p}\left(X|\omega_i\right) = \left(2\pi\right)^{-P/2}\left|\hat{\Sigma}_i\right|^{-1/2}\exp\left\{-\frac{1}{2}\left(X-\hat{\mu}_i\right)^t\hat{\Sigma}_i^{-1}\left(X-\hat{\mu}_i\right)\right\}, \quad i=\overline{1,M} \quad (7.1)$$

in which $\hat{\mu}_i$ is sample vector of means, and $\hat{\Sigma}_i$ is sample covariance matrix of class ω_i.

The approach making use of the parametric estimate (1) is effective when probability distributions are unimodal and/or close to those of normal form that is usually achieved with large training sets. Practically, these constraints may not always be overcome due to lack of prior information and non-normal form of a class features distribution. In such cases, more accurate classification may be obtained with use of a nonparametric approach to multivariate conditional PDF $p(X|\omega_i)$ approximation. As a nonparametric estimate, the ACP employs the multivariate analog of Parzen function [10] given by:

$$\hat{p}\left(X|\omega_i\right) = \left(n_i\prod_{v=1}^{P}c_v^i\right)^{-1}\sum_{s=1}^{n_i}\prod_{v=1}^{P}\Phi\left(\frac{x_v-x_v^s}{c_v^i}\right), \quad i=\overline{1,M} \quad (7.2)$$

in which n is the number of training samples, P is the number of features, c_v is a smoothing parameter; and $\Phi(u)$ is a kernel function.

It should be noted that the efficiency of the Bayesian approach depends on PDF estimation techniques requiring large training sets to be available. Practically, when the training set size is too small for PDF function to be estimated properly, a simpler decision rule of minimum distance is used by the ACP that does not utilize probabilistic description of the RS image classes.

The adaptive decision rule includes a set of discriminate functions $\phi = \{\phi^1(X),$ $\phi^2(X), \phi^3(X)\}$ corresponding to Bayesian CDR with Gaussian PDF estimate (1), CDR with Parzen PDF estimate (2), and CDR adopting minimum distance principle, respectively. Assuming that $\phi^*(X)$ is the most effective CDR, the adaptive decision rule $m(\phi^*(X))$ can be expressed as follows:

$$m\left(\phi^*(X)\right): \phi^*(X) = \arg\min_{i=1,3}\left\{\hat{R}\left(\phi^i(X)\right)\right\} \tag{7.3}$$

The adaptive decision rule (3) allows the ACP to choose the most accurate CDR $\phi^*(X)$ of three functions $\phi^1(X)$, $\phi^2(X)$, $\phi^3(X)$, using minimum empirical risk criterion. Ambiguity between those CDRs having relatively equal values of the $\hat{R}\left(\phi(X)\right)$ parameter (different by any accepted measure of inaccuracy) is resolved through choosing the fastest one. Thus in the classification stage, the ACP reveals the most effective CDR by accuracy and performance for an imagery with arbitrary characteristics independently of training set size, and so doing the ACP adapts to the data to be classified, in order to obtain the most accurate results in the shortest time.

Unfortunately, the adaptability principle employed in the ACP cannot predefine the overall efficiency of the procedure, since classification success also depends to a large extent upon optimality of the feature space used. Commonly, feature space of an RS imagery is formed by considering the intensity (brightness) values of its pixels in different bands of electromagnetic spectrum (in the case of multispectral imagery) as the components of a multidimensional feature vector. It has been shown that feature space formed by only spectral features allows obtaining accurate classification results for the image areas with relatively uniform intensity distribution [11]; otherwise, the produced classification contains high-frequency noise caused by misclassified pixels. In some works [12] it has been proved that in a RS image the neighbor pixels are spatially correlated, which makes reasonable the idea of using information about pixel context for its classification. So self-descriptiveness of the spectral feature vectors can be improved through extending them with complementary components representing the image texture descriptors calculated within the context of the classified pixels.

In order to account for image textural information, the ACP utilizes an extended feature space (EFS) when performing classification. The EFS is formed through calculating a textural component of initial image by means of Haralick's textural analysis approach [12]. The initial image is sequentially scanned by running windows of odd size $b \times b, (b = 3, 5 \ldots, Z)$ and textural feature sets $X^{TX} = \left\{X_{3\times3}^{TX}, X_{5\times5}^{TX}, X_{Z\times Z}^{TX}\right\}$ are generated. The elements of each textural feature set $X_{b\times b}^{TX} = \left\{T_{b\times b}^1, T_{b\times b}^2, \ldots, T_{b\times b}^S\right\}$, $(b = 3, 5 \ldots, Z)$ are computed as the first and second statistical moments of intensity function of initial image pixels falling into current running window of odd size $b \times b$. Since the textural feature sets computed with windows of different size do not contribute equally to discriminating the RS image classes, the ACP performs the feature selection procedure, improving computational efficiency of the EFS classification. The procedure selects the features that are more significant (informative) for classification and excludes the rest, using the image classes pairwise separability criterion of Jeffries-Matusita [11].

An original particularity of the ACP is that, once the EFS is built, the further classification of its textural and spectral components is performed separately in an iterative manner. Classification starts from processing textural component of the EFS, in the course of which the different scale textural feature sets $X^{TX} = \left\{ X_{3\times3}^{TX}, X_{5\times5}^{TX}, X_{Z\times Z}^{TX} \right\}$ are classified sequentially in iterative manner, going from coarser feature sets (calculated in bigger running window) to finer ones. At every iteration, the classification results represent posterior probability maps [5] computed for current textural feature set $X_{b\times b}^{TX}$. The probability maps acquired for feature set $X_{b\times b}^{TX}$ are transferred to the next iteration, to be used as prior probabilities for classifying finer scale feature set $X_{(b-2)\times(b-2)}^{TX}$. The iterations are repeated until the finest feature set is classified. The completion phase of the classification is processing of the spectral feature component of the EFS with use of posterior probability maps calculated at the stage of textural component processing. At each iteration while classifying the image, the ACP employs an adaptive decision rule, finding the best CDR for the data currently processed in order to obtain the most accurate classification in the fastest way.

The principle of the EFS iterative processing adopted in the ACP allows the procedure to overcome the shortcomings of the traditional stacked vector approach for employing textural features for image classification, in which the extended feature vectors are formed by stacking textural and spectral features together [5]. Adopting this approach faces the problem of losing fine spatial details in the resultant thematic map, which makes the approach not very practical, whereas the EFS iterative processing preserves the finest details in the resultant thematic map.

Thus, by employing extended feature space processed in an iterative manner and an adaptive decision rule, the ACP produces better classification results compared to traditional image interpretation techniques, as is shown in the following application examples.

7.3   THEMATIC MAPPING METHODOLOGY IMPLEMENTATION

7.3.1   THE RS IMAGERY PROCESSING AND INTERPRETATION SYSTEM "LANDMAPPER"

The thematic mapping methodology based on improved imagery interpretation approach has been implemented in the framework of the "LandMapper" system of imagery processing and interpretation developed in the GIS laboratory of TPU. The "LandMapper" system is a software package, which is launched as an additional unit for a vector GIS (MapInfo Professional®, MapInfo Corporation, Troy, New York) providing it with image processing functionality. The general structure of the "LandMapper" system is given in Figure 7.2.

As can be seen from Figure 7.2, "LandMapper" is based upon vector-raster architecture comprised of two components, Raster (RC) and Vector (VC), respectively. The RC provides means for raster data visualization in a GIS environment and implements functions of RS imagery preliminary and thematic processing. The

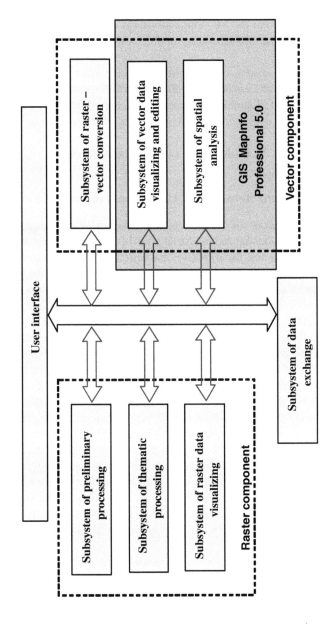

FIGURE 7.2 General structure of the "LandMapper" system.

supported functions solve the problems of image spectral and geometric correction, visual enhancement, georeferencing, and projection transformation, as well as comprehensive imagery interpretation. The spatial analysis subsystem, which allows complex quantitative estimations, and the vector data visualization and editing subsystem are implemented by means of a vector GIS, which together with a raster-vector conversion unit form the VC. The subsystems of "LandMapper" developed as original software in the GIS laboratory of TPU are shadowed with light gray in Figure 7.2.

The "LandMapper" system can be applied to solving different problems of RS-based thematic mapping and can be an essential tool in GIS research. In the following section, two examples of "LandMapper" applications are given, which consider the issues of anthropogenic ecosystems mapping with use of remote sensing imagery.

7.3.2 APPLICATION OF "LANDMAPPER" FOR ANTHROPOGENIC ECOSYSTEMS RESEARCH

7.3.2.1 Mapping Hydro Network and Urban Areas of Tomsk City

Tomsk City is the capital of the Tomsk region situated in the southeastern part of Western Siberia. The residential and industrial areas of the city, together with natural landscape components such as the Tom River and surrounding forestry, form a typical anthropogenic ecosystem. Thematic mapping was implemented with the purpose of updating topographical information on urban areas and the hydro network as well as assessing the ecological condition of water bodies. In the research, the imagery acquired in July 2000 by a domestic RESURS-O1 satellite (sensor MSU-E, resolution 30×45 m, three spectral bands) was used, allowing the thematic map of 1:50,000 scale to be produced.

Georeferencing of the initial imagery was carried out by means of the "Land-Mapper" imagery preliminary processing subsystem, making use of an obsolete vector map (1994) of the Tomsk area hydro network. On the base of the 30 ground control points clearly distinguished both in the map and in initial imagery, a triangulation network was designed linking map and imagery coordinate systems to each other. The triangulation network was used for performing imagery linear rectification, allowing imagery local geometric errors to be compensated (Figure 7.3), and then the resultant georeferenced imagery was assigned the Gauss-Kruger projection. Imagery rectification was followed by the thematic processing stage. First, the set of training samples was formed, relying upon the reference data acquired from the site visit information as well as from topographic and landscape maps of the Tomsk area.

Imagery interpretation was performed by means of the ACP, which computed a few different scale textural sets (with various running window sizes) for every spectral band of the initial imagery and selected the most informative textural sets in each scale by the Jeffries–Matusita separability criterion.

The generated textural component of the EFS comprising informative textural feature sets was classified by the ACP in iterative manner, and the resultant posterior probability maps were then used for classifying spectral components of imagery

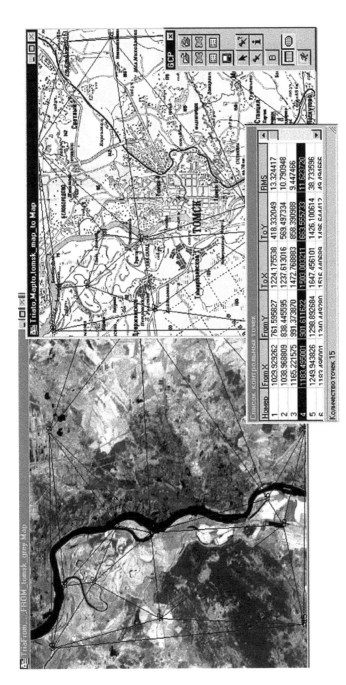

FIGURE 7.3 Forming a triangulation network linking Tomsk imagery and map with use of ground control points.

TABLE 7.1
Tomsk City Imagery Classification Details

Iteration	Window Size	Features Selected	PDF Estimate Adopted in Bayesian Rule	Accuracy, %
1	7 × 7	5	Parzen	85.3
2	5 × 5	5	Parzen	92.4
3	—	3	Gaussian	94.5

involving three initial spectral features. The iterative classification details, such as number of textural or spectral features used, type of PDF estimate chosen by the ACP in Bayesian rule, as well as thematic map overall classification accuracy acquired at each iteration, are given in Table 7.1.

Figure 7.4 illustrates some thematic mapping results for the Tomsk City area produced by the ACP implemented in the "LandMapper" thematic processing subsystem. In Figure 7.4a and 7.4b the initial multispectral imagery of the Tomsk City area as well as the obsolete topographic map of the area (1994) superimposed by updated thematic data are shown. The highlighted thematic layers (Figure 7.4b) correspond to water bodies and urban constructions (scale 1:50,000). Comparative GIS analysis of the obsolete map and the updated thematic layers revealed that boundaries of the urban areas and water objects have changed considerably in the course of time. A good example of hydro network change detection is given in Figure 7.4c, 7.4d, and 7.4e, depicting the Um river bed area. In addition to change detection outcomes, the resultant thematic map showed clearly the contaminated conditions of the Tom River in the northern part of Tomsk City caused by power station. Together with on-ground measurements data, this map is a valuable information source for making a decision on ecological conditions improvement.

7.3.2.2 Landscape-Ecological Research of Pervomayskoe Oil Field

Pervomayskoe oil field is situated 180 km southwest from Strezhevoy City and belongs to the Vasyugan oil-producing area of the Tomsk region. The "LandMapper" system was applied for landscape-ecological mapping of the oil field with use of the multispectral imagery acquired in July 1998 by a RESURS-O1 satellite (sensor MSU-E, resolution 30 × 45 m, three spectral bands). The purpose of landscape-ecological mapping was to reveal various environmental changes caused by petroleum production in the field area.

The preliminary processing stage involved imagery georeferencing and its visual properties enhancement to enable effective visual inspection. To perform thematic processing, a number of training sets was selected making use of reference data acquired from photointerpretation results, aerial photographs (1997 and 2001), and topographic maps of Pervomayskoe oil field describing its geomorphological structure and degree of anthropogenic influence. The training sets cover all general landscape types and consist of anthropogenic objects (roads system, well clusters, and settlements), old deforested areas and quarry, and natural objects (lakes, swamp,

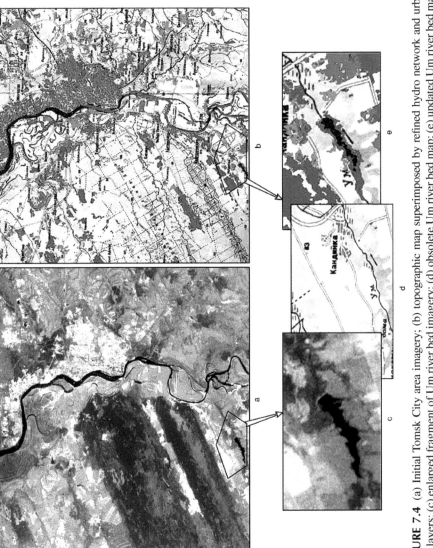

FIGURE 7.4 (a) Initial Tomsk City area imagery; (b) topographic map superimposed by refined hydro network and urban area layers; (c) enlarged fragment of Um river bed imagery; (d) obsolete Um river bed map; (e) updated Um river bed map.

TABLE 7.2
Pervomayskoe Oil Field Imagery Classification Details

Iteration	Window Size	Features Selected	PDF Estimate Adopted in Bayesian Rule	Accuracy, %
1	11 × 11	7	Gaussian	77.2
2	7 × 7	4	Parzen	79.1
3	5 × 5	5	Parzen	84.3
4	—	3	Gaussian	89.5

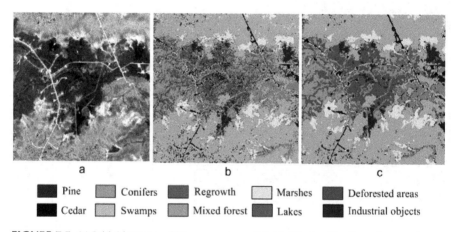

a   b   c

■ Pine   □ Conifers   ■ Regrowth   □ Marshes   ■ Deforested areas
■ Cedar   □ Swamps   ■ Mixed forest ■ Lakes   ■ Industrial objects

FIGURE 7.5 (a) Initial imagery of Pervomayskoe oil field; (b) classification using maximum likelihood technique; (c) classification using the ACP.

and forest). The class representing forest area was split up into four subclasses: pine sphagnous forest, mixed cedar sphagnous forest, coniferous sphagnous forest, and mixed mossy forest. The class representing swamp was set to subclasses representing upper sphagnous swamps and marsh areas.

The imagery interpretation stage was implemented by means of the ACP. First, the informative EFS was generated in the manner described in Section 7.2.2. Feature selection was performed among textural feature sets generated with running windows of 11 × 11, 7 × 7, and 5 × 5 pixels size. Then classification was conducted involving four iterations. Table 7.2 gives the information on classification details in a similar way to the example described in Section 7.3.2.1. Overall classification accuracy reached with use of the ACP is about 90% which is almost 15% higher than accuracy given by the traditional maximum likelihood classification (MLC) technique.

Figure 7.5a shows the fragment of initial imagery of Pervomayskoe oil field, and Figure 7.5b and 7.5c illustrate the resultant landscape-ecological thematic map fragments produced with use of the MLC technique and the ACP, respectively. The comparison of the two classification fragments demonstrates the advantage of employing the ACP for RS-based thematic mapping. As it can be seen from Figure

7.5b, the map produced with MLC is heavily noised, and the mapped classes have quite fuzzy boundaries. This effect is due to strong similarity and, as a result, mixing of some landscape cover types in the spectral feature space of the initial multispectral imagery (marsh areas and upper swamps; coniferous sphagnous forest, mixed mossy forest, and deforested areas). By contrast, the map in Figure 7.5c does not contain any noise, and all classes have clear boundaries thanks to incorporation of textural information within the EFS and using an adaptive decision rule in the ACP.

The acquired raster thematic map produced with the ACP was converted into vector features, and the feature layers corresponding to different landscape types were designed. After assigning appropriate attribute information to the mapped features, the resultant vector landscape-ecological map was applied, together with vector GIS analysis tools, for computing areas of marshes that appeared close to industrial objects and for defining the areas of forest devastation. The quantitative estimations obtained in the GIS analysis allowed the overall anthropogenic load within the oil field to be assessed.

The designed landscape-ecological map is an essential means for both qualitative and quantitative statistical analysis of anthropogenic ecosystem structures of the Pervomayskoe oil field, which is of great importance for supporting management decision-making on the oil field environment enhancement and ecological situation forecasting.

7.4 CONCLUSION

The chapter has introduced a methodology of RS-based thematic mapping, based on an advanced approach to image classification. The approach makes use of image extended feature space and an adaptive decision rule selecting an optimal (in terms of accuracy and performance) classification algorithm during the interpretation process and allows the quality of the data extracted from a RS image to be improved. The proposed methodology has been implemented on the base of original image processing and the "LandMapper" interpretation system functioning in the framework of a vector GIS. Two sample applications have demonstrated the efficiency of using the "LandMapper" system while solving problems of collecting data on two anthropogenic ecosystems (Tomsk City and Pervomayskoe oil field) with images acquired from a domestic RESURS-O1 satellite. The average accuracy of the thematic maps produced in the applications with use of "LandMapper" adaptive classification procedure is about 90%, which is almost 15% higher than that reached by traditional MLC technique with the same training sets. Along with the accuracy improvement, the adaptive classification procedure is more time consuming than traditional MLC technique, due to involving new textural components in the image feature space and an iterative manner of classification. However in practice, the computational efficiency factor, as a rule, is considered as less important than classification accuracy and thus can often be sacrificed to obtain more accurate thematic maps.

In conclusion, it should be noted that later investigations of the work considered in the chapter will focus on research on applicability limits of the adaptive classification procedure for solving issues of anthropogenic ecosystems data collection and thematic mapping with use of space imagery from SPOT, LANDSAT, and QUICK-BIRD satellites.

ACKNOWLEDGMENTS

The authors would like to thank Novosibirsk Regional Center of Data Acquisition and Processing for providing medium-resolution multispectral RESURS-O1 imagery of the Tomsk region. The work described in the publication has been carried out with financial support of the Russian Foundation of Basic Research (grant number 03-07-90124).

REFERENCES

1. Vinogradov, B., *Foundations of Landscape Ecology,* Geos, Moscow, 1998, 418 (in Russian).
2. Markov, N. and Napryushkin, A., Use of remote sensing data at thematic mapping in GIS, in *Procceedings of the 3rd AGILE Conference on Geographic Information Science,* AGILE, Helsinki, 2000, 51.
3. Vinogradov, B., *Aerospace monitoring of ecosystems,* Science, Moscow, 1984, 320 (in Russian).
4. NASA's RS tutorial, The Concept of Remote Sensing, 2003, http://rst.gsfc.nasa.gov/Intro/Part2_1.html.
5. Richards, J. and Xiuping, J., *Remote Sensing Digital Image Analysis: An Introduction,* Springer, Berlin, 1999, 400.
6. Star, J. and Estes, J., *Geographic Information Systems: An Introduction,* Prentice-Hall, Englewood Cliffs, N.J., 1990.
7. Markov, N. and Napryushkin, A., Self-organizing GIS for solving problems of ecology and landscape studying, in *Proceedings of the 4th AGILE conference on Geographic Science,* AGILE, Brno, 2001, 462.
8. Moik, T., *Digital Processing of Remotely Sensed Images,* NASA, Washington, D.C., 1980.
9. Duda, R. and Hart, P., *Pattern Classification and Scene Analysis,* Wiley, New York, 1973.
10. Lapko, A. and Chenzov, S., *Nonparametric systems for information processing,* Science, Moscow, 2000, 350.
11. Markov, N. et al., Adaptive procedure for RS images classification with extended feature space, in *Proceedings of the 9th International SPIE Symposium on Remote Sensing,* Vol.4885, SPIE, Bellingham, 2002, 489.
12. Haralick, R. and Joo, H., A Context Classifier, in *IEEE Trans. Geoscience Remote Sensing,* N24, 1986, 997.

8 Spatiotemporal Data Modeling for "4D" Databases

Alexander Zipf

CONTENTS

8.1 INTRODUCTION

Conventional GIS are usually quite static, as they do not cover dynamic aspects of geo-objects in their data model. The information on the modeled domain is usually separated into models of geometric space (2D/3D) and thematic aspects (attributes). But if someone wants to develop a system that is capable of modeling objects of the environment including their history, presence, and future, most available systems lack expressive power. It has been demanded that a temporal GIS (TGIS) needs to provide functionality for spatiotemporal data storage, data handling, and analysis as well as visualization. These functions are usually more complex than in conventional GIS and are still an area of active research.

Within the Deep Map/GIS project a flexible and extensive temporal object-oriented model had been developed [1–3]. The aim was to allow the management of 3D geo-objects of urban areas over historic epochs and act as a basis for the data management components of temporal 3D-GIS ("3D-TGIS" or more colloquial "4D-GIS") to be developed in the future. Since the temporal part of this model is a self-consistent OO-model for temporal structures, it can also be used with 2D geodata.

The proposed framework is a contribution toward the development of a temporal 3D-GIS by offering guidelines on how to model the time in a sophisticated way. It also shows how to integrate these temporal aspects of geo-objects along with their 3D spatial (topological) and thematic aspects. Working prototypes have been realized that implement these models in an object-oriented and an object-relational database management system (DBMS), showing the applicability of the proposed concepts. The model has been demonstrated within the domain of 3D historic city models for an urban information system.

8.2 SPATIOTEMPORAL DATA MODELING

A geo-object or feature in general consists of the aspects theme, geometry, topology, and time [4,5]. Still, today's GIS don't handle all aspects equally well. The temporal dimension is an important aspect of most real-world phenomena. Nevertheless, databases or GIS delivered only a snapshot of the real world. Therefore there was a need for new data models that allow the handling of temporal data [6,7]. In recent years, a range of temporal models was also developed in the field of object-oriented databases [8–10], presenting possibilities for an object-oriented integration of temporal models into 2D-GIS [11–14].

To represent the basic elements of the temporal framework, some important concepts are defined briefly. The period of the physical process used to measure time is called "chronon," while the duration of the period is described as a "granularity." A temporal framework should provide means for representing arbitrary calendars. Further aspects of time are explained in more detail by Krüger [15].

8.3 TOPOLOGICAL MODELING OF THREE-DIMENSIONAL GEO-OBJECTS

The development of the data model for 3D geometry is largely influenced by the model of Molenaar. It combines the geometry and topology of 3D geodata and allows retrieval of multiple topological properties directly from the model.

The basic concepts include the primitives node (point), arc (line), and face (area). Thematic attribute data are attached using feature identifiers. Molenaar extended earlier models by the new primitives edge and body to model the third dimension (Figure 8.1, [16]).

The topology of the 3D primitives has been modeled through several 1:n relationships between the five primitives:

- For every arc there exists exactly one start- and endpoint (node).
- A node can belong to several arcs.

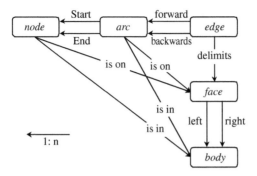

FIGURE 8.1 Topological relationships between the 3D primitives (after [16]).

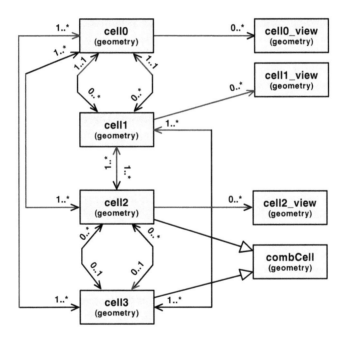

FIGURE 8.2 Class diagram for 3D topological geometry information.

- A face can only margin two bodies, while one body can have several faces.
- There are links between arcs and nodes to the face they belong to or the body they are part of.
- Face and body both consist of several nodes or arcs.

A unified modeling language (UML) class diagram that models the geometry model of the framework is depicted in Figure 8.2.

The data model introduced so far describes the topology of up to three-dimensional objects. The actual geometrical data is integrated by relating multiple versions

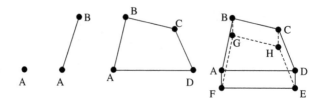

FIGURE 8.3 Graphical representation of the primitives 0-Cell, 1-Cell, 2-Cell and 3-Cell.

of geometry to the primitives. In the case of *nodes* these are the actual coordinates; for an *arc* these are the coordinates of the vertices (points in between nodes, representing geometry).

The body primitive does not need further geometric information, because it is described by the constituting faces. The classes for the geometry were realized similarly according to the 3D model using the primitives *Point, Face,* and *Body.* They shall be called 0-Cell (cell0), 1-Cell (cell1), 2-Cell (cell2), and 3-Cell (cell3) according to their dimensionality (see Figure 8.3).

Within the spatiotemporal model, only the primitives 2-Cell or 3-Cell have been used.

The realization of the relationships between the spatial and temporal parts of the model has been achieved using coupling classes. This class is called *combCell.* Both primitives *2-Cell* and *3-Cell* inherit properties from that. Modeling these relationships using coupling classes offers the following benefits: first, redundancy is minimized, and secondly, the geometrical components can be coupled in a more flexible way with temporal aspects, as the individual parts of the model can be exchanged or altered freely. If another class also inherits from *combCell,* it can replace the spatial model we used with a different one easily.

8.4 MODELING OF THEMATIC DATA: THE EXAMPLE OF THE HISTORY OF A CITY

The structures describing the thematic aspects of the features (geo-objects) are also realized using an object-oriented model. The thematic model cannot be generic but is oriented toward the application domain. In the case of the Deep Map project, this was a city information system, where individual buildings with their visible parts (from outside) and other man-made structures within a city are modeled. Other geographic domains can also be applied by extending or exchanging the thematic model.

The most important three-dimensional real-world objects are in our case *buildings, monuments, bridges, fountains, gates,* and *roads.* Parts of such 3D objects may belong to the classes *body* (of a building), *stair, tower, roof, wall,* or *yard.* But as it is likely that more complex 3D objects need to be represented, it seems sensible to be able to aggregate such objects to a more complicated semantic unit (Figure 8.4). This is realized through the relationships between the class *threeD_Obj* and *part_threeD_Obj.* This allows assembling several parts of a 3D object together within the thematic model. An example is the definition of an object "southern wing" (e.g.,

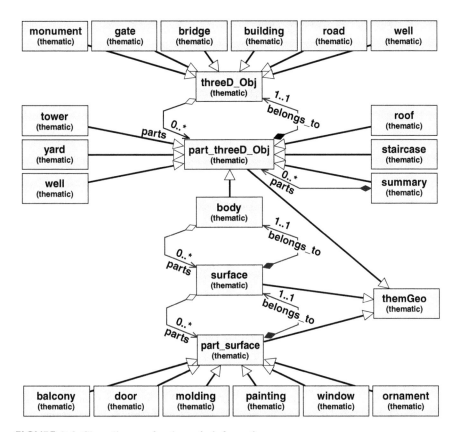

FIGURE 8.4 Class diagram for thematic information.

of the building "Villa Bosch") by combining the objects "*body*" (of south wing), *roof* (of south wing), and further parts of the south wing. These objects can also be used in queries to the database.

A further requirement on the data model was that it should allow queries to details of a facade of a building, like "Which parts belong to the northern facade of an object?" or "What are the properties of the window next to the entrance door?" In order to allow this, the main elements of a facade are modeled explicitly. This includes classes for *balcony, door, molding, painting, window,* or *ornament,* which all can be attached to a part of the facade. So just as there are 3D objects and their parts, there are surfaces that can be separated in several parts of a surface that can be addressed independently.

A part of the thematic model for buildings is depicted in Figure 8.4: The classes *threeD_Obj, part_threeD_Obj, surface,* and *part_surface* are used for realizing the corresponding main aspects of a *threeD_Object,* part_*threeD_Object, surface* (facade), and *part_surface.* Using these classes the properties of the corresponding subtypes are modeled.

As already explained, generalization allows not only minimizing redundancy when defining subtypes, but also results in a well extensible structure. The integration

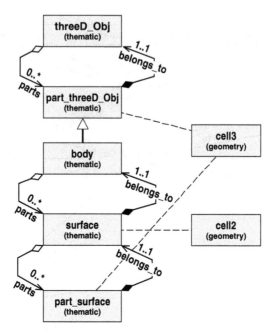

FIGURE 8.5 Class diagram for the relationship between thematic aspects and geometry.

of new subtypes can be achieved by defining inheritance relationships to the corresponding main class (Figure 8.5).

In order to model the relationships between 3D objects and their parts, base bodies, and their facades as well as between facades and the objects belonging to a facade, bidirectional 1:n relationships are being used.

In the realized prototype, only parts of 3D objects, facades, and parts of facades are linked to spatiotemporal data structures. This is very application specific. In order to change this relation easily, these kinds of relations are modeled using an extra class, which is called "*themGeo.*" This improves flexibility, for example, to exchange the geometry model with a different description (e.g., GML, [17]).

Figure 8.5 shows the realized relationships between thematic and geometric data within the "4D" model explained. The geometric description for the thematic classes *part_threeD_Obj* and *part_surface* is realized using the class *cell3* (*body*), while for the thematic class *surface* the class *cell2* (*face*) is used. When there is only 3D information available for a part of a 3D object or for parts of a facade, this can be expressed by the modeler through the usage of the hierarchical structure of the spatial model by using 3-*Cells* that only use a 2-*Cell* (*face*). The geometry of a 3D object is represented through the geometries (3-*cells*) of the parts of the 3D object.

8.5 AN OBJECT-ORIENTED MODEL FOR TEMPORAL DATA

The object-oriented paradigm has also been used for the modeling of a general time framework. The range of possible different applications puts quite complex requirements on temporal support. First it is necessary to identify the dimensions limiting the modeling space of a general temporal model. Further, the components and properties have to be determined in order to be able to define an adaptable structure that fulfils the various requirements. From these a framework for building temporal models was developed using the identified components. It supports design alternatives by the provision of a range of classes and accompanying properties. These temporal classes can be integrated with the models for the geometry and for the thematic aspects already introduced to a composite model for temporal 3D geo-objects.

Regarding time, one can distinguish the following general aspects:

- *Temporal Structure* defines a structure using temporal primitives, domains, and structures concerning temporal determination (certain or uncertain representations).
- *Temporal Order* describes the possible types of orders of temporal structures.
- *Temporal History* describes the semantic meaning of the different states of the object.
- *Temporal Representation* describes how to represent calendars and granularities.

8.5.1 TEMPORAL STRUCTURE

The temporal structure defines, through its parts, a base for the temporal model (Figure 8.6). This temporal "structure" can have the following properties [18]:

1. *Temporal primitives* are represented either as absolutes (anchored, "date" [e.g., 5-9-1999] or relative (unanchored, "period of time" [e.g., 30 days]).
2. *Temporal domain:* It is possible to distinguish discrete and continuous domains. In the field of temporal databases a discrete time domain is usually used.
3. *Temporal determination:* In the deterministic case complete and exact knowledge is available for temporal primitives. On the other hand, these are not determined exactly in indeterministic cases [18] (e.g., fuzzy temporal borders).

The topmost level of the temporal structure-model consists of absolute (anchored) and relative (unanchored) temporal primitives. The next hierarchical level supplements the structure with domains, being either discrete or continuous. The deterministic and nondeterministic primitives form the last component. A temporal structure consists of a combination of all of the represented temporal primitives. Through the combination of the different properties offered within the three levels of the hierarchy to model temporal aspects of the world, it is possible to distinguish eleven temporal types as "temporal primitives" (the twelfth one is only a theoretical combination, because "nondeterministic continuous time points [instants]" are not possible because of contradicting properties). The temporal primitives represent the fundament for representing temporal data. Further, it is necessary to distinguish between the logical and physical representation of a time value. If the time value is described by means of a calendar, it is a logical representation.

One can define a broad range of operations for the suggested data types. Langran [12] defines a range of categories for the operations according to their purpose and the types of arguments and results, as stated below. Krüger [15] explains the realized operators within our model in more detail:

- *Build-in-functions* allow the type conversion between temporal data types as well as combination or comparison functions.
- *Arithmetical Operators* offer the corresponding adaptation of the basic arithmetic functions.
- *Comparison operations* give back a Boolean value (they are used for checking the correctness of selection criteria).
- *Aggregation functions:* The well-known aggregation functions from SQL like COUNT, SUM, AVG, MAX, and MIN can also be adapted for temporal data types.

8.5.2 TEMPORAL REPRESENTATION

The proposed temporal primitive data types offer a basis for the representation of temporal data. For a temporal value that is represented by an instance of such a temporal data type, it is necessary to distinguish between logical and physical representations of this value. If the value is represented using a calendar, it is a logical representation. While supporting multiple logical calendars, the value of temporal data types is stored independent from a calendar within the implemented framework. This means that the point of time is stored as a chronon of the base watch. But within the framework, there are classes for different calendars available, which define the logical representation through the definition of usable granularities, referencing the chronons of the base watch. They also offer functions for converting between the physical and logical representations of the temporal objects.

8.5.3 TEMPORAL ORDER

The course of the time can be classified as linear, sub linear or branching. In both cases time is generally regarded as running linearly from past to future. They only differ regarding the handling of subordinate spatial basic types (primitives). In the linear case overlapping borders of temporal primitives are forbidden, while they are possible in the sub-linear case. A sub-linear order can also be used for managing indeterministic temporal phenomena. This can for example be used for the temporal description of the changes of an object that are only known roughly.

The concept of branching order time allows time to be to regarded as linear only up to a certain point of time. A typical example would be town planning, where different planning alternatives can be managed in different branches of the resulting temporal tree. In each of the branches of that tree a partial order of time is defined.

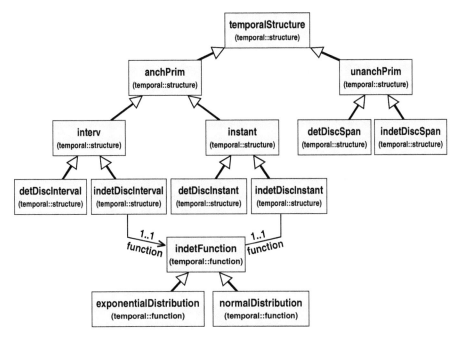

FIGURE 8.6 Class diagram of the prototypical implemented temporal structures.

8.5.4 TEMPORAL HISTORY TYPE

One of the fundamental requirements for a temporal model is to represent the development of real world objects over time regarding geometric, topological, or thematic attributes. This is a basic functionality needed within a TGIS. This development of the object — the set of observations of these attributes within time — forms the "*temporal history*" of the object and can be distinguished in *valid time* and *transaction time* [6].

The valid time describes the time for which an entity of the real world is "valid" or a statement about the entity is true (e.g., "Fountain A is in front of building B").

The transaction time deals with the points of time when a value is inserted into the database. A database management system that supports both aspects of time is generally called "bi-temporal" (Figure 8.7).

8.6 PUTTING THE COMPONENTS TOGETHER

The focus of the following discussion is on the relationships between the components defined through the design alternatives for a temporal model that have been explained so far.

A temporal model can support either one or many histories of the types *valid time*, *transaction time*, *event*, or *user-defined histories*. Each of these histories consists of a set of temporal orders (either linear, sublinear, or branching). For example, the

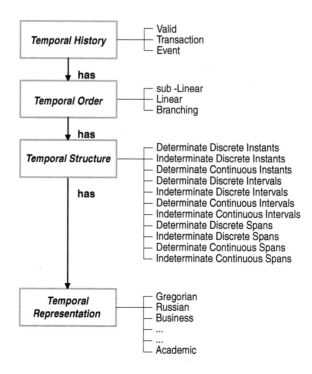

FIGURE 8.7 Basic data types for a temporal model.

borders of structures that belong to linear orders cannot overlap. These represent a total temporal order. On the other hand, it is possible for sublinear or branching orders to have (absolute) overlapping temporal primitives. In the case of branching orders, they may represent multiple partial temporal orders. Each of these temporal orders includes a temporal structure which consists either of all or a subset of the eleven temporal primitive types that have been introduced (Figure 8.7). For each of these temporal primitive types, it is necessary to define a function to convert between the physical representation (real, integer, …) and one of the logical representations (e.g., "March 11, 1971, 8:22:45") and vive versa. In order to allow a maximum degree of flexibility, it is possible to define different calendars that can be related to the temporal primitives. This relationship is represented in Figure 8.8 through a "has" relationship. Figure 8.8 shows a summary of the different alternatives for modeling a temporal structure.

8.7 INTEGRATING GEOMETRY, THEMATIC AND TEMPORAL MODEL

As explained earlier, object-oriented modeling allows modeling of the different aspects of geo-objects within their own class hierarchies and then combining these by defining

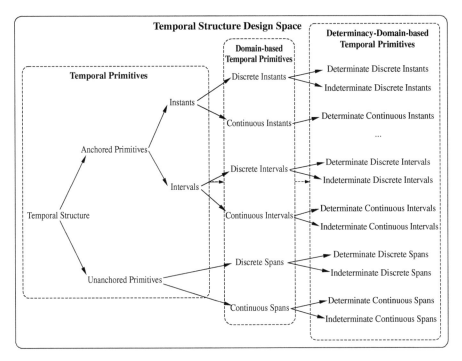

FIGURE 8.8 Design space for temporal structures.

relationships between the classes. Figure 8.9 shows the relationships between the main classes of the resulting temporal 3D model. So how can this be applied?

When defining a temporal object within the proposed framework, the first step involves the definition of the thematic structure. Each thematic structure can be linked with a temporal order to model the change of the (spatial) data of that object over time. Within each order the spatiotemporal relationship is expressed using objects of the class *combTempCell*.

These allow the aggregation of spatial and temporal information of the represented objects.

In order to make this work as explained, it seems sensible to introduce structures (classes) for the coupling of the three data models into a common model. This is done using the new classes *themGeo* and *combCell*. These do model the relationship between thematic objects and temporal orders, on the one hand (*themGeo*), and the relationship between temporal and spatial parts of the model within a spatiotemporal structure, on the other hand.

This way, it is possible to model the classes of the thematically or spatial (partial) models through the definition of inheritance relationships of the classes *themGeo* or *combCell*. Through the decoupling of the structure that links the thematic with the spatial data model, it is possible to replace one part of the model through a different one quite easily. This might be useful for adaptations to other application domains.

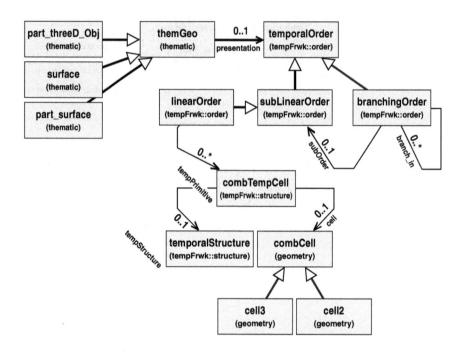

FIGURE 8.9 Class diagram for the relationship between thematic, spatial, and temporal data.

8.8 OBJECT- VERSUS ATTRIBUTE-TIME-STAMPING

Through assigning a temporal order to every time variant feature or attribute, it is possible to model temporally changing spatial data. Object-time-stamping can be used to describe the changes of a complex object (e.g., GML-feature) over time. In this case, the whole object (feature) will be time-stamped by adding a reference to an order.

Zipf and Krüger [2001] [20] illustrate how such a link could be realized using an XML representation of geographic entities like GML. This shows that the temporal framework can not only be coupled with 3D objects as explained before, but also with other representations of geo-objects and their spatial as well as nonspatial attributes, offering a high degree of flexibility.

One does not have full control in all cases of all available class packages that model domain issues, but often it is necessary to extend just existing software libraries. Aspect-oriented programming now gives the possibility to do that without editing the actual code of the original software. This results in even greater freedom to mix domain models that deal with different aspects of the real world into a new and richer domain representation. Zipf and Merdes [21] propose that these benefits can even be realized in a more automated way by using existing formal ontologies to derive aspect descriptions. This will be explained in the following section.

8.9 DYNAMICAL EXTENSIONS OF SPATIAL CLASS HIERARCHIES WITH "ASPECTS"

Let a class hierarchy for features be given (e.g., one based on OGC standards). Now we want to extend this through a "time" aspect, which can be modeled in more or less sophisticated ways. In order to combine both, it is necessary to modify the original model by either enhancing the super-class or defining coupling classes. This might not be desirable or possible in all situations. In such cases a new paradigm of software engineering called "aspect-orientation" (AO) offers help to enhance existing class-libraries to enrich them with new (cross-cutting) aspects dynamically. But what is the benefit over conventional approaches?

Object-oriented programming has become mainstream. There, in short, the classes assembling software have well-defined responsibilities. But often some parts cannot be traced down as being the responsibility of only one class. These cross-cut several classes and affect the whole system. One can add code to each class separately in order to handle such parts, but that violates the basic rule that each class has well-defined responsibilities. Here comes Aspect Oriented Software Development (AOSD) (http://www.aosd.net) into play; AOSD defines a new program or language construct, called an "aspect." This allows capturing cross-cutting aspects of software in separate program entities. This new concept has recently been added to several programming languages as an extension. In Java it is called AspectJ (http://www.aspectj.org). Aspects (in AspectJ) have much in common with classes. They can have methods and fields, extend normal Java classes, implement interfaces, and may be abstract. They also can extend other aspects and can contain new constructs called pointcut and advice. Pointcuts provide a mechanism for specifying join points (i.e., well-defined points in the execution of the program). Examples for join points include object initialization, method calls, and field access. When defining a join point related to a method call, it is possible to use powerful wildcards semantics for the method signature including name, arguments, and return type, as well as target object. The definition of an executable piece of functionality is called advice. An advice is defined with respect to a pointcut and can be run in a variety of ways (e.g., before, after, or even instead of a method call). Elements of the surrounding nonaspect code, such as method call parameters, can be made accessible within an advice. AspectJ also offers a mechanism for adding elements (fields, methods) to existing classes and changing the inheritance and interface structure. This mechanism is called introduction. Introduction effectively changes the static structure of a program at compile time as opposed to the dynamic nature of join points. An example for combining spatial and temporal models on the fly is presented using Java and AspectJ notation (see Listing 1 in Table 8.1).

First, a standard Java interface named TimeDependent is defined. This is greatly simplified for the sake of clarity. Then an AspectJ aspect named TimeDependency is declared. This aspect will contain all program elements relevant to the temporal modeling. These elements are static introductions such as parent, constructor, method, and field introductions, as well as pointcut and advice definitions. The aspect will affect all members of the two separate class hierarchies with the root classes

TABLE 8.1
Listing 1: TimeDependency.aj

```
1  package aspectExample;
3  import org.eml.modell_4d.temporal.structure.interv;
4  import org.eml.deepmap.gml.objects.*;
6  // simplified example interface for time dependency------
7  public interface TimeDependent {
9  public long getStart();
10 public long getEnd(); }
13 aspect TimeDependency {
15 //static introduction for interface implementation------
16 declare parents: (Feature || Geometry) implements TimeDependent;
18 // ------field introduction ------
19 public interv (Feature || Geometry).validTime = null;
21 //------ constructor introduction ------
22 public (Geometry+ || Feature+).new(interv interval) {
24 //call to existing constructor
25 this();
26 // ------additional initialization ------
27 this.validTime = interval; }
30 //------ method introductions ------

31 public long (Feature || Geometry).getBegin() {
33 return this.validTime.getBeginChronons(); }
36 public long (Feature || Geometry).getEnd() {
38 return this.validTime.getEndChronons(); }
41 //------ the pointcut definitions ------
42 pointcut AllGetterMethods(TimeDependent object):
43 call( public * get*(..)) &&
44 target( object);
46 pointcut DesiredGetterMethods(TimeDependent object):
47 AllGetterMethods( object) &&
48 !call( public long getBegin()) &&
49 !call( public long getEnd());
51 //------ the around advice ------
52 Object around(TimeDependent object): DesiredGetterMethods(object) {
54 long now = System.currentTimeMillis();
56 if (( object.getBegin() < now) && ( object.getEnd() > now))
57 return proceed( object);
58 else return null; } }
```

Source: Adapted from Zipf, A. and Merdes, M., AGILE Conference Proceedings, Lyon, France, 2003.

Geometry and Feature. Therefore, new parents are being introduced into both classes in line 16. These implement the interface TimeDependent as if it was declared in the original source code. In order to be valid, it is necessary to introduce both methods of the interface into both classes (lines 31–39). Both methods reference an instance variable named validTime of type interv (which was introduced into the Feature and Geometry classes in line 19).

The constructor implementation additionally initializes the introduced instance variable validTime. This constructor is introduced into all members of both class hierarchies individually. This allows the new constructors to be used as if they were declared within the respective class definitions: Box box = new Box(new interv()). Boxes can then be created with a time interval constructor argument just like any other subclass of Geometry or Feature. These static introductions change the class structure, hierarchy, and dependencies at compile time. They do this in a crosscutting manner, which means that they affect a lot of different and (potentially) unrelated files from a single aspect definition.

In the aspect some additional pointcut and advice definitions can be found. These modify the behavior of the classes at run time. In the example given, a composite pointcut named DesiredGetterMethods is being defined (lines 46–49). It selects certain getter-methods of objects of type TimeDependent, that is, instances of Geometry, Feature, or any of their subclasses.

Now the pointcut DesiredGetterMethods can be used to define a piece of advice (i.e., the functionality that is to be executed before, after, or instead of the methods selected by the pointcut). In the example, the former behavior of all getter-methods (that is, all access methods) is being replaced by a new, time-dependent behavior. This is done through an around advice (lines 52–58). For the sake of simplicity, it is only checked if the time of the method invocation is within the time span defined as valid. If not, null is returned. Otherwise, the keyword *proceed* signals to proceed as usual (line 57).

This illustrated some advantages of using aspects to enhance GIS data models and class libraries, such as:

- Extension of the functionality of an existing GIS library without modification of its sources
- Combination of an existing GIS library with an unrelated library from a different domain
- Modification of the behavior of classes scattered over many places in the source code in a single aspect

Similarly, it is possible to enhance other existing non-spatial-aware domain-models with spatial "aspects" in order to spatially enrich them. More classical examples include "weaving in" middleware-style features (e.g., logging, tracing, performance monitoring, persistence, error-handling, security, distribution) into existing GIS class libraries.

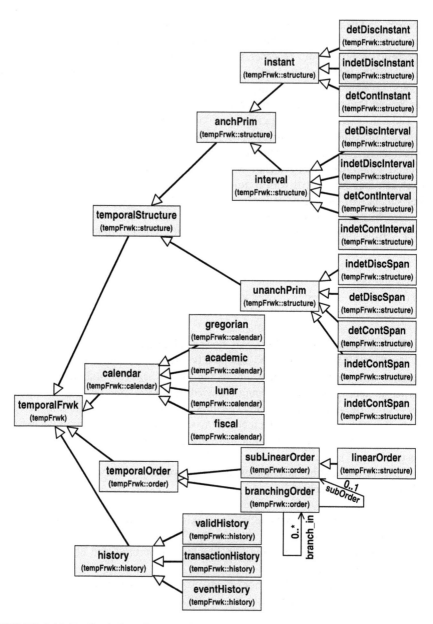

FIGURE 8.10 Realized class diagram of the temporal framework.

8.10 CONCLUSIONS

For a multitude of geographic applications, and in particular for an information system covering the aspects of town history, the management of changes in feature data (in our case, buildings) is of particular importance. But since the temporal model introduced is very generic, it can be used in different domains. The framework

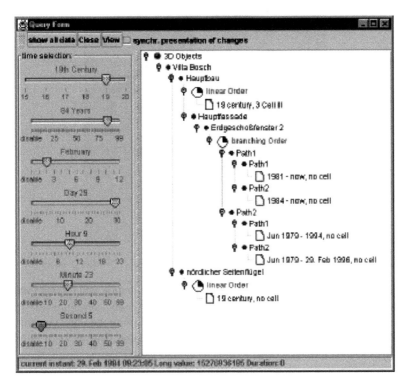

FIGURE 8.11 GUI of temporal query component of prototype.

is able to cover both valid-time as well as transaction-time in a flexible way. This is done by providing the necessary building blocks for temporal structures such as *intervals,* time *spans,* or *instants.* Each of these can be subdivided being either *continuous* or *discrete* on the one hand, and *determinate* or *indeterminate* on the other hand. By supporting the definition of additional application-specific calendars, with their respective granularities, the framework supports all the notions and notations of time that can be considered relevant for present practical applications (Figure 8.10).

Further, a prototype has been developed that allows editing and querying spatiotemporal features through a graphical user interface (Figure 8.11 and Figure 8.12).

The development of flexible and efficient temporal 3D-GIS is an attractive but demanding task for further GIS research. A "4D"–GIS that covers all aspects of GIS functions from data handling, analysis, and visualization equally well will not appear within a short time. Spatiotemporal analysis possibilities should include and perform the analysis of attributes, geometry, and topology equally well. For example, there is a lack of research on the possible changes of topological relationships in "4D" space-time in particular within vector-oriented GIS. Similarly the sparse availability of functions for inter- and extrapolation within vector-oriented "4D"-GIS is not satisfying. Because existing GIS use proprietary data models that could not be extended easily, it soon became clear that only development from scratch could

FIGURE 8.12 3D window for visualization of geometry view of the query result.

cover all the requirements for a "4D"-database. This situation will change with the spreading of more open data models. Further research and developments are of course necessary, ranging from the technical side, regarding the implementation of multidimensional indices for the efficient access to large data sets, to the query languages for 4D queries to represent moving objects.

ACKNOWLEDGMENTS

This work has been undertaken in the context of the project Deep Map at the EML, supported by Klaus Tschira Foundation (KTS), and the BMBF project SMART-KOM. Special thanks go to Sven Krüger, now at Quadox AG, Germany, for contributing considerably to the modeling and implementing the model.

REFERENCES

1. Zipf A., Deep Map, ein verteiltes historisches Touristeninformationssystem. Inaugural-Dissertation, Geographisches Institut, Universität Heidelberg, 2000.
2. Zipf, A., DEEP MAP — A prototype context sensitive tourism information system for the city of Heidelberg, in *Proceedings of GIS-Planet 98*, Lisboa, Portugal, 1998.
3. Malaka, R. and Zipf, A., Deep Map — Challenging IT research in the framework of a tourist information system, in *Information and Communication Technologies in Tourism 2000*. (Proceedings of ENTER 2000, 7th International Congress on Tourism and Communications Technologies in Tourism, Barcelona, Spain), Fesenmaier, D. Klein, S. and Buhalis, D., Eds., Springer Computer Science, Wien, Pp, 15-27, 2000.
4. Bill, R. and Fritsch, D., Grundlagen der Geoinformationssysteme, Bd. 1, *Hardware, Software und Daten*, Wichmann, Heidelberg, 1991.
5. Streit, U., Geoinformatik online, 1998, http://ifgivor.uni-muenster.de/vorlesungen/Geoinformatik/index.html.

6. Snodgrass, R. and Ahn, I., A taxonomy of time in databases, in *Proceedings of ACM SIGMOD International Conference on Management of Data*, 1985, pp. 236–246.

7. Tansel et al., *Temporal Databases*, Benjamin/Cummings Publishing, Menlo Park, CA, 1993.

8. Skjellaug, B., Temporal Data: Time and Relational Databases, Research Report 246. Department of Informatics, University of Oslo, 1997.

9. Skjellaug, B., Temporal Data: Time and Object Databases, Research Report 245. April 1997. Department of Informatics, University of Oslo, 1997.

10. Skjellaug, B. and Berre, A.-J., Multi-dimensional Time Support for Spatial Data Models, Research Report 253, May 1997. Department of Informatics, University of Oslo, 1997.

11. Breunig, M., On the way to component-based 3D/4D geoinformation systems. Lecture Notes in Earth Sciences, Vol 94, Springer. Heidelberg, 2000.

12. Langran, G., *Time in Geographic Information Systems*, Taylor & Francis, London, 1992.

13. Wachowicz, M., *Object-Oriented Design for Temporal GIS*, Taylor & Francis, London, 1999.

14. Worboys, M., Object-oriented approaches to geo-referenced information, *Int. J. Geogr. Inf. Syst.*, 8(4), 385–399,1994.

15. Krüger, S., Konzeption und Implementierung eines temporalen objekt-orientierten Modells für 3-dimensionale Geo-Objekte im Deep Map-Projekt. Diplomarbeit, institut für Praktische Informatik und Medieninformatik, Technische Universität Illmenau, 2000.

16. Molenaar, M., A formal data structure for three-dimensional vector maps, in *Proceedings of the 4th International Symposium on Spatial Data Handling*, 1990, pp. 830–843.

17. Zipf, A. and Krüger, S., TGML — Extending GML by Temporal Constructs — A Proposal for a Spatiotemporal Framework in XML. ACM-GIS 2001. *The Ninth ACM International Symposium on Advances in Geographic Information Systems*, Atlanta, GA, 2001.

18. Snodgrass, R., The TSQL2 Language Design Committee. *The TSQL2 Temporal Query Language*, Kluwer, Dordrecht, 1995.

19. Dyreson, C.E. und Snodgrass, R., Valid-time Indeterminacys, in *Proceedings of the 9th International Conference on Data Engineering*, 1993, pp. 335–343.

20. Zipf, A. and Krüger, S., Ein objektorientierter Framework für temporale 3D-Geodaten. AGIT 2001, Symposium für Angewandte Geographische Informationsverarbeitung, 04–06 Juli 2001, Salzburg. Austria, 2001.

21. Zipf, A. and Merdes, M., Is Aspect-Oriented Programming a new paradigm for GIS development? On the relationship of geoobjects, aspects and ontologies, AGILE Conference Proceedings, Lyon, France, 2003.

9 Spatial Multimedia for Environmental Planning and Management

Alexandra Fonseca and Cristina Gouveia

CONTENTS

9.1 INTRODUCTION

Effective environmental management aims to achieve goals for optimizing resource use and minimizing environmental impact, while at the same time maintaining economic growth and viability. Environmental management as it is used here includes not only formal management processes, but also a range of environment-related activities of individuals and groups and those interested in environmental programs, policies, and outcomes. The word environment is directed not only to purely physical environmental factors, but also to the understanding that their effective management must take into account the social and economic factors.

Environmental management activities are strongly associated to the nature of environmental problems, which are characterized by a high level of complexity resulting from their multiplicity of components, interrelationships, and spatiotemporal variability. Being multisensory, environmental problems have strong visual and audible features that should be taken into account within environmental management approaches.

The spatial nature of environmental problems has favored the use of geographic information systems (GIS) and associated technologies, such as GPS or remote sensing. GIS have been used mainly for data exploration and visualization, improving communication among stakeholders. Since their early developments, GIS have been used within environmental applications to support the analysis of alternative uses that compete for space.

One of the most important changes to have occurred within the geographic information (GI) domain has been the increase in users' diversity. GI and GIS are no longer used by a limited number of professionals, but they have now become present in citizens' daily life. The popularity of map channels on every WWW portal, the existence of GPS receivers included in mobile phones, or car navigation systems are some of the multiple examples of how GI and GIS are present in daily life activities. Furthermore, citizens not only have become GI users, but they also demand tools to manipulate such data and support their activities.

Nevertheless, GIS are still considered an elitist technology [1], and several obstacles remain. For a complete review of the major barriers to a more democratic use of GI and GIS, please refer to the public participation GIS literature, for example [2–4]. Within the scope of this chapter, two types of barriers are underlined: (1) the difficulty of using GIS and associated tools due to poor interfaces and (2) the abstraction level required to decipher the representations of the world produced by cartographers, urban planners and environmental engineers.

The development of multimedia systems, together with spatial data-handling capabilities, have been proposed by Fonseca et al. [5] and Raper [6], among others, to surpass some of the barriers to the use of GIS. This approach has been preferentially applied to environmental management and planning, due to the multisensory and spatial nature of environment [5–7]. Accordingly, spatial multimedia systems have been explored to disseminate information to different types of audiences and to provide access and manipulation tools for environmental management processes.

Spatial multimedia provides new rich forms of multidimensional geo-representation, taking advantage of the exploration of realistic representations and easy-to-use tools. Raper [8] argues that the key challenges have been to give multimedia environments geographic qualities and to show that the use of such representations has had significance and validity. Nevertheless, spatial multimedia systems have benefited from technological developments in computer science and telecommunications, making such systems more pervasive in citizens' daily life. The emergence of the Internet and, more recently, the mobile communication and computing developments have created new platforms for spatial multimedia systems, creating opportunities to explore such tools for environmental management. This context is shaping current and future research activities to explore the use of spatial multimedia systems tools for environmental management.

This chapter starts by presenting spatial multimedia key concepts and the technological developments that have been determining its evolution. It goes on to analyze the use of spatial multimedia for environmental management, considering the spatial and multisensory nature of environmental problems. Some case studies illustrate the use of spatial multimedia within environmental management activities.

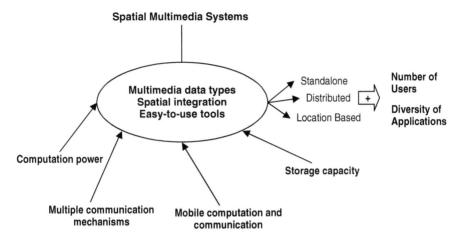

FIGURE 9.1 The contribution of spatial multimedia systems to increase GI users' diversity.

Finally, some research questions are identified that are shaping future developments in this domain.

9.2 SPATIAL MULTIMEDIA KEY CONCEPTS

In recent years, several technological developments associated with computer science and telecommunications have supported the development of multimedia systems. The increases in computation power, storage capacity, and miniaturization associated with the spread of multiple communication media have allowed integration of data of different types and have made them available through easy-to-use tools and interfaces. Multimedia software includes electronic games, hypermedia browsers, and authoring and desktop conferencing systems [9]. Multimedia systems have been defined by Steinmetz et al. [10] as systems that deal with processing, storage, presentation, and manipulation of independent information from multiple time-dependent and time-independent media.

Within the GI field, such developments have been used to create spatial multimedia systems where location is used as a key variable to integrate and explore the multimedia data. Spatial multimedia systems create new possibilities for multidimensional representation of a more direct nature [11] and may help to overcome the existing barriers to the use of GI and GIS. Figure 9.1 summarizes the contribution of the technological developments for the creation of spatial multimedia systems and their impact in the number of users and diversity of applications.

Spatial multimedia includes a multiplicity of definitions that tend to favor one of its specific characteristics, for example, the integration capability, the dynamic nature, or the structuring role (for a review please refer to Fonseca [12]). Raper [6] defines spatial multimedia as the use of hypertext systems to create webs of multimedia resources organized by theme or location. Spatial multimedia, according to this author, has the capability of exploring multimedia data types for spatial analysis and modeling.

TABLE 9.1
Major Spatial Multimedia Data Types

Data Types	Observations
Nontemporal	
Alphanumeric data	Includes text or numbers of variable size and structure. Within GIS, alphanumeric data are usually associated to graphical data. Alphanumeric data may be associated to a location or to a theme.
Still images	They can be bitmaps or raster images. Within spatial multimedia systems, still images are geo-referenced and include ground photos, aerial photos, and satellite images.
Temporal	
Still and animated computer-generated graphics	Vector data, such as centerlines, are included in this data type. Spatial multimedia systems, in general, use graphics and animated sequences intensively as interface elements and to visualize data. 3D models of landscapes and associated fly-overs are one of the most common examples.
Audio	Audio can be synthesized or captured and replayed. Spatial multimedia systems may use spoken and nonspoken sounds. The first are mostly used as interface to the system (data input is one example), while the last are used more as earcons or as alerts. The use of stereo sounds provides a notion of space.
Video or moving frames	The most demanding multimedia data type concerning storage. The proliferation of inexpensive video acquisition systems, such as web cams, has favored the use of videos as a source of data. Video can be used both in the form of airborne device or ground sensor device [9].

The creation and management of spatial multimedia information systems imply the consideration of the requirements that the different media types have in terms of storage and data transfer. Media types can be divided into two major groups, temporal and nontemporal [13]. The data types that can be used in spatial multimedia systems are presented in Table 9.1.

Most spatial multimedia systems are hypermedia structured, presenting a web of nodes and links. Such a structure may contribute to the development of applications where the user can intuitively explore a set of data [14]. The data structures within spatial multimedia systems are closely related to the authoring and hypermedia development environments used (for a review on spatial multimedia authoring environments and data structure please refer to Fonseca [12] and Raper [8]).

The concept of linking videos with maps was pioneered in the mid-1980s by the BBC Domesday project [15–17]. However, it was in the beginning of the 1990s that spatial multimedia applications started to flourish. Two major perspectives have been historically associated with spatial multimedia information systems: (1) the incorporation of multimedia data types into GIS software, and (2) the integration of spatial functionality into multimedia software and hardware environments [12]. Nevertheless, the technological developments have blurred this distinction, enabling the creation of more open architectures where data and services are exchanged.

Initiatives like the Open GIS Consortium or GRID computing are examples of developments that contribute to create web services and languages that facilitate data sharing and integration, avoiding the traditional black box systems. On the other hand, the pervasive use of data types such as aerial photos within GIS applications and the increasing users' demands to include more visualization tools brought GIS closer to spatial multimedia applications.

Spatial multimedia systems have a powerful level of flexibility, allowing the development of multimedia operations that can be useful in different contexts. Multimedia operations may be associated with different media types, and their use may have different purposes. Multimedia operations can target data access, exploration, spatial analysis and visualization, and data presentation and communication. Table 9.2 presents examples of spatial multimedia operations.

The characteristics of spatial multimedia applications have changed, taking advantage of the technological developments that have occurred since then. Spatial multimedia applications can be stand-alone, distributed, and more recently, mobile (Figure 9.2).

The decision to develop stand-alone, distributed, or mobile spatial multimedia systems has to consider the goals of the application and the current advantages and disadvantages that each platform presents. Stand-alone applications are more difficult to update, and the storage capacity is restricted to the capacity of the storage device (CD-ROM, DVD, among others). Nevertheless, stand-alone applications are not limited by network access, which facilitates the use of highly demanding media types, such as videos. Additionally, stand-alone applications may take advantage of stand-alone software and are not constrained by the Internet architecture and protocols. However, with the proliferation of broadband access and Internet developments, such as XML, these advantages have become less important. Nevertheless, stand-alone applications have the advantage of being more permanent than distributed versions where their lifetimes are determined by application providers. It has been estimated that the average web page lifetime is around 100 days [25].

Distributed applications are easier to update (both in content and functionality). The Delft Hypermap and the Geoexploratorium are examples of early distributed spatial multimedia systems [8]. Additionally, they may reach a wider group of people without having to create a distribution channel. The use of Internet protocols and architecture enables the user to access through the Web browser heterogeneous sources of data and services. This advantage has supported the change, within the GI community, from data-centric to service-centric approaches, making it possible to reach more users and satisfy diverse needs.

Mobile applications refer to access and use of spatial multimedia data and operations through mobile and wireless devices such as personal digital assistants (PDA) and cellular phones. Recent developments within mobile computing and communication have supported the creation of location-based services (LBS) and multimedia data processing and acquisition tools such as the ones that support moblogging [26]. Examples of applications of such developments can be found in several domains such as environmental management and monitoring, urban planning, journalism, and transportation planning and tourism [27–32].

TABLE 9.2
Examples of Spatial Multimedia Operations

Spatial Multimedia Operations	Examples of Application
3D modeling	3D graphics improve realism of many spatial data and rendering techniques allow the representation of volumes or surface objects. Architecture projects are one of the most common areas of application of 3D models.
Animated sequences	Animated sequences of temporal series of aerial photographs, which may provide a dynamic view of land cover changes.
Annotations	Annotations can be made using sounds, images, graphics or videos over maps, images or videos. The use of spoken sounds to annotate maps and images can be an interface to collect public's preferences and opinions ([18] and [5]).
Data search and access	The possibility to query the information system using images or sounds may allow for more intuitive interfaces. Multimedia systems, in general, use point-and-click tools to facilitate data search and access interfaces.
Filters	The use of filters over multimedia data allows changing the views of the objects viewed through that operator. In virtual environments, filters can be used to focus on a certain area while flying in a virtual world.
Flyover	The aerial photographs together with the topographic 3D model may be used to simulate a flyover through real-time animation. Flyover aerial photographs may be used as an interface to explore and manipulate spatial data.
Morphing	It is an animation technique used to dynamically blend two still images by creating a sequence of in-between images. When the images are played back rapidly, the first image gradually metamorphoses into the second. Fonseca et al. [5] proposed the use of such technique to visualize environmental impacts.
Navigating images	The use of 360-degree circular views from specific points and to access to video or sound sequences in specific directions.
Overlay	The overlay of graphics to aerial photographs is a traditional GIS operation. Within spatial multimedia systems, it is also possible to overlay synthetic videos, graphics, or images to videos and images.
Simulation models	Words, images, and sounds may be used in modeling. Linguistic and pictorial approaches to modeling environmental problems have been proposed by Câmara et al. [19,20]. Another approach is to combine modeling and spatial information representations. Fedra and Loucks [21] and Loucks et al. [22] have pioneered such concept within environmental modeling.
Sketching	Drawing is one the most intuitive methods for human communication. The use of sketching tools applies the principles already tested in successful drawing software. Fernandes et al. [23] proposed the use of sketching tools to interact with mosaics of aerial photos. Nobre and Câmara [24] proposed the use of sketching to define and simulate the evolution of graphical objects in a background.

Source: Adapted from Fonseca, 1998. With permission.

FIGURE 9.2 The evolution of information and communication technologies and its impact within spatial multimedia systems.

Within GIS, Peng and Tsou [33] define major mobile users as fieldworkers and location-based services customers. Mobile spatial multimedia applications have additional design constraints, not shared by the other alternatives, namely the size and quality of the screen and the multidimensionality of the mobile phone's keyboard — most keys stand for one number and three letters. On the other hand, they make location and spatial multimedia more pervasive within daily activities. LiveAnywhere traffic [32] is one example of a mobile spatial multimedia system, where data from traffic cameras, road sensors, and text databases are processed and sent to users, automatically formatting the data to be correctly displayed on the mobile device being used.

Examples of spatial multimedia systems exist across several areas of application: urban planning, tourism, fleet and transportation management, coastal zone management, environmental management and education, and GIS education. Such applications have different characteristics and range from spatial multimedia databases and visualization systems to decision support systems or collaborative systems.

9.3 ENVIRONMENTAL MANAGEMENT AND SPATIAL MULTIMEDIA

Environmental management requires an effective access to environmental information. Although it is a broad subject, it includes several environmental management processes such as impact and strategic assessment, auditing, monitoring, decision-making, and public participation. These processes tackle environmental problems in different ways, involving diverse data requirements, methods, and tools.

Environmental problems are characterized by a high level of complexity resulting from their multiplicity of components, interrelationships, and spatiotemporal variability (Table 9.3). On the other hand, most environmental phenomena have strong visual and audible features and involve large data sets. The dynamic nature is possibly the major feature of environmental problems and stresses the importance of being able of portraying it.

Within environmental management, information and communication technologies (ICT)-based tools have been used to improve activities such as data acquisition and management, data exploration and analysis, and data presentation. For example, distributed database systems have been used for environmental monitoring, and the

TABLE 9.3
Main Features Characterizing Environmental Problems

Environmental Problems	Description	Comments
Complexity	Multiplicity of components, variables, criteria, or points of view under which environmental problems can be considered. The involvement of different social groups is an additional factor to be taken into account.	Complexity of environmental problems requires massive sets of data.
Dynamic nature	Temporal dimension, associated to duration, reversibility, time of occurrence, and also in social and cultural terms.	Simulation models can have an important role in dealing with the dynamic nature. Models are used to enhance the understanding of the system dynamics.
Spatial dimension	Spatially varying phenomena with no clear borders.	GIS can contribute to better deal with this characteristic and has been used to perform spatial analysis of environmental problems.
Large data sets	Environmental data sets are large, often containing many time steps, representing changing conditions.	Data formats vary, contributing to additional problems in terms of data handling.
Visual and audible nature	Most environmental phenomena can be seen or heard. Visual and audio representations can be very helpful in this context.	Symbolic and realistic representations should be explored within the available tools.

Adapted from Fonseca, 1998. With permission.

issues involved within data integration from different sources have been documented by Rosen et al. [34], Hale et al. [35], and McLaughlin et al. [36].

Spatial information systems such as GIS have been used for several environmental applications. In fact, the environment drove some of the earliest applications of GIS and was a strong motivation for the development of the very first GIS in the mid-1960s, a system that aimed to support policies over the use of land [37].

The environmental applications of GIS support operations such as handling of coverage data (e.g., land use, land cover, geology, and soils), handling of network-oriented data (e.g., streams, water distribution systems, and sewerage systems), and terrain information, or handling of subsurface information (e.g., for groundwater modeling). These applications can go from simple inventory and management tasks to sophisticated analysis and modeling of spatial data.

These systems may include models linked to GIS visualization tools and also the exploration of advanced spatial visualization tools, such as virtual reality and multimedia animation. For a review of the use of collaborative visualization environments, refer to MacEachren [38]. For example, geo-visualization developments generated tools to interactively explore maps and associated data [39], allowing, in some cases, the possibility of building alternative future scenarios within environmental and planning activities. Examples may be found in Kingston et al. [40] and Peng [3].

Spatial multimedia may help enhance the analysis and comprehension of environmental phenomena and to improve the communicability of results to the public (e.g., within the environmental impact assessment [EIA] process). It can also be explored to improve cooperative work between the different professionals involved in environmental assessment activities, contributing to enrich environmental problems analysis and to increase the level of integration among the different fields of knowledge.

The concepts of nonsequential association and presentation, characteristic of multimedia technologies, are easily connected to planning and environmental management activities. Aspects of the growth and change of urban areas or the detection of land use changes are not sequentially connected, and most environmental problems are interrelated. The ability to cross-reference and compile information from different sources, provided by hypermedia systems, can be most adequate for planners or environmental managers. As argued by Wiggins and Shiffer [41], the ability to link together many disparate pieces of information on a single screen, of different types (text, graphics, images, videos, sound), communicating different things and views of the same subject, is one of the main advantages. Additionally, multiple representations of environmental problems can be obtained, and the possibility of viewing the information in several different contexts increases the potential for generating alternative approaches and solutions.

Accordingly, spatial multimedia tools may improve several environmental management tasks (Table 9.4) that go from collection, inventory, and analysis of environmental information and phenomena to the evaluation, negotiation, and voting of alternatives or scenarios.

Examples of environmental spatial multimedia applications exist in stand-alone, distributed, or mobile platforms. These applications include examples in domains such as environmental planning and management, natural phenomena representation, or public participation in decision-making and can preferably support data acquisition, data manipulation, data distribution, or data visualization/representation. For a review on the spatial multimedia systems applied to environmental management and planning, please refer to [8, 9, 12, 42].

One of the first spatial multimedia developments applied to environmental planning is the Collaborative Planning System (CPS) [7]. CPS was developed as a stand-alone spatial multimedia system that combined the activities of tool usage, information access, and collaboration. It was intended to increase access to relevant information, leading to greater communication amongst participants in a group planning situation, exploring the use of multiple representation aids. The CPS was used in public hearings and allowed a hypermedia navigation to access spatial and multimedia data on the study area. Multimedia tools such as on-screen video navigation, animation techniques associated with models, the superimposition of video sketching onto a video, or the access to graphics with sounds were included in the system. A multimedia brainstorming component, based on the sketchpad metaphor, and a multimedia, multicriteria evaluation and map-based audio annotation were created to be used during meetings.

TABLE 9.4
Examples of Projects That Use Spatial Multimedia Tools for Environmental Management Tasks

Environmental Management Tasks	Access to Multimedia Data and Navigation Tools	Spatial Analysis with Visualization and Other Multimedia Tools	Collaborative and Communication Tools
		Spatial Multimedia Tools	
Environmental data collection and production	The use of video to measure geo-phenomena such as air pollution [43]	GIS based multimedia database in Pielsen project [39]	Map-based annotation tool developed in Virtual Slaithwaite [40]
Organize and explore the information about environmental problems	Fly over realistic representations of the terrain in Portugal Digital project [44] and Virtual Tejo [45]	Interactive maps in Descartes [46]	Data input in Internet-based database in Naturdetektive, (http://www.naturdetektive.de/2002/dyn/1407.htm); schoolchildren can input their nature observations (including photos), and the data are represented over satellite images
Analyze environmental problems and phenomena	The Visualizer tool within the EXPO '98 Environment Exploratory System [12]	Access to animated maps, which are .gif images, showing atmospheric ozone levels for U.S. and Canada ozone in the Mapping Project by the USEPA (http://www.epa.gov/airnow/)	Emails are sent automatically to registered users anytime ozone reaches unhealthy levels in the Sacramento county (http://www.sparetheair.com/)
Share and discuss problems, solutions, and ideas and generate alternatives	Playing the sound of aircraft taking off as heard from different locations in CPS [42]	Animation techniques associated with models; superimposition of video sketching onto a video; access to graphics with sounds, in the CPS [42]	The use of argumentation maps as a way to summarize and discuss the different alternatives [47]; argumentation maps use text, maps and images
Evaluate, negotiate, and vote alternatives/scenarios	CPS [42] allows access to a wide range of multimedia data by selecting resources from a geo-referenced image map base	Multicriteria evaluation techniques in the Open Spatial Decision-Making on the Internet system [48]	Multimedia online survey for measuring visual preferences, in Piesen Project [39]

The Web-based Public Participation System (WPPS) [37] is a distributed application of the spatial multimedia concepts to environmental planning. It is an online spatial collaborative system designed to enhance public participation in the planning and decision-making process, allowing users to evaluate, comment, select, and above all formulate their own alternatives. It is based on a webmapping tool to access and explore multimedia data on the study area, including basic spatial analysis functions. It also includes tools to allow users to add local data to the system and online editing and drawing tools for scenario building.

With the developments associated with mobile computing, new applications to environmental management are emerging. One example is Software Tools for Environmental Field Study (STEFS, http://web.mit.edu/envit/www/), an integrated system for data collection on mobile computers. STEFS uses a GPS and a water-quality sensor to collect data, which are sent through a wireless network to a database server. Mobile mapping software records and maps the exact locations where the environmental readings are taken [31].

The following three examples of spatial multimedia applications to environmental management illustrate the stand-alone, distributed, and mobile approaches. These examples have been developed within research projects that were intended to explore the use of information and communication technologies to improve environmental-management-specific activities, for example to support the involvement of the public in the Environmental Impact Assessment process and to promote the use of environmental data collected by concerned citizens.

9.3.1 EXPO '98 ENVIRONMENTAL EXPLORATORY SYSTEM: A STAND-ALONE APPLICATION

The development of a spatial multimedia application to allow access to the environmental information of the EXPO '98 site in Lisbon is presented herein. The research involved in the development of this spatial multimedia application brought together most of the concepts previously discussed and functioned as a case study to investigate the use of spatial multimedia for environmental impact assessment [12].

Expo '98 was held in a place that was totally renovated between 1993 and 1998. The EXPO '98 intervention area presented several environmental problems that had to be solved and for which a considerable amount of environmental information had to be collected and produced. The area included, among others, an oil industry complex covering 50 hectares, an Army depot, several harbor facilities, Lisbon's slaughterhouse, a wastewater treatment plant, a landfill and a solid waste treatment plant, and a highly polluted river on the northern border.

The environmental data collected at the EXPO '98 site (water, air, and soil parameters, land use, meteorological, energy, fauna and flora) were stored in an environmental information system that presently continues to accommodate the data from the environmental monitoring programs. This information system represented one of the data sources used in the development of the spatial multimedia application, designated as the EXPO '98 Environmental Exploratory System, that aimed to support decisions and at the same time promote communication with the public and the media. The developed spatial multimedia application system includes technical

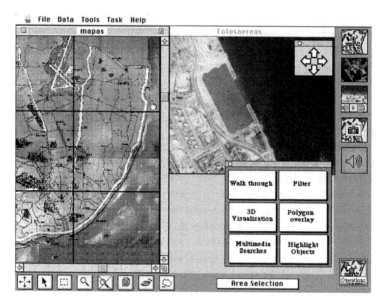

FIGURE 9.3 Prototype 1 — EXPO '98 Multimedia Spatial Information System for EIA.

information, but a more educational and entertainment approach can also be explored.

Three prototypes were developed for the EXPO '98 CD-ROM application, using different user interface design approaches [12]. In all three prototypes, the hyper-media structuring model was followed, where the nodes included multimedia information related to the EXPO '98 site (maps and aerial photos, photographs, digital video, synthetic three-dimensional images, animated sequences, audio, text), and the links enabled the user to explore the application. Three types of structural links are available in the system: spatial, temporal, and thematic. The user can navigate through maps, aerial photos, and three-dimensional walkthrough models, for each of the four temporal stages: baseline conditions, construction phase, exhibition phase, and postexhibition phase. The thematic option allows the access to physical and chemical characteristics of the environment (e.g., water and air quality), biological features (e.g., fauna and flora), and cultural characteristics (e.g., recreation).

The system includes associative links that give the user the possibility to access multiple representations for certain spatial, temporal, or thematic options, and referential links that link all the technical information included in the system to an environmental glossary.

Figure 9.3 presents the first prototype user interface developed in 1994, which aimed to bring traditional environmental impact assessment studies into a multimedia platform. This prototype provides multiple views of the same area or detail of the EXPO '98 intervention area and includes some demos of tools to explore, visualize, and simulate existing environmental problems.

The second prototype was developed in 1995 and followed a more flexible and exploratory approach, which resulted in a much more compelling interface (Figure 9.4). A traveling metaphor was used, were the user can explore the area controlling

FIGURE 9.4   Prototype 2 — EXPO '98 Multimedia Environmental Exploratory System.

a transparent air bubble/vehicle through the use of a control panel that gives access to different operations available in the upper part of the screen.

The third prototype (Figure 9.5) was finished in 1997 and resulted from the experience of the previous prototyping process. It allows exploration with the same kind of navigation, analysis, and visualization tools as the second one, but with a more traditional user interface. The tools are available through icons that surround a large working area.

The different prototypes were developed using a hypermedia authoring tool for the Macintosh and were tested within the team, reviewed by peers, and subject to informal usability testing. Table 9.5 summarizes the main characteristics of the three prototypes.

FIGURE 9.5 Prototype 3 — EXPO '98 Environmental Spatial Multimedia CD-ROM.

A web version, available across platforms, programmed with Dynamic HTML in association with JavaScript, was developed later on, adapting the contents, interface, and visualization and exploration capabilities from the three prototypes [9].

9.3.2 PUBLIC PARTICIPATION WITHIN THE EIA PROCESS: A DISTRIBUTED APPLICATION

This case study involved the Portuguese agency responsible for the environmental impact assessment public consultation process. An application was developed that allows access via the Internet to the EIA Nontechnical Summaries and the visualization of GI associated to the location of the project using webmapping software. This application has been available on the Web since June 2001 (http://iambiente. pt/IPAMB_DPP/.

The development process had to deal with two distinct components. The first component is associated with the storage and management of organizational data on EIA public participation processes (e.g., the identification of the EIA projects currently under consultation). The development of this component implied the design and implementation of a database and corresponding web interface. The second component is intended to provide access to the EIA Nontechnical Summaries and explore the spatial and multimedia nature of the data. A template has been developed that allows the Portuguese agency responsible for the EIA public participation process to disseminate the studies on the Web in a more interactive and clear way.

TABLE 9.5
The EXPO '98 Environmental Exploratory System — Three Prototypes

	Description	Comments
Prototype 1: EXPO '98 Multimedia Spatial Information System for EIA	The prototype followed a structured approach, based on the different phases of the EIA process, intending to orient the user in a very objective way within the information and manipulation capabilities available for each EIA phase.	The system had the advantage of guiding the user through a very specialized thematic, but it was, as a result, restricted to a narrower group of users. It might, nevertheless be a better solution to adopt in meetings such as public audiences, where an assistant would use the application that could be seen and interactively used by that group of people.
Prototype 2: EXPO '98 Multimedia Environmental Exploratory System	The prototype follows a more flexible and exploratory approach that stimulates the user to explore the application without being attached to the EIA process structure. It is the one that gives the user more freedom to explore the information. The user can act as if he was exploring the EXPO '98 environmental story.	The system was to be used by a wider range of users, the ones that would visit the EXPO '98. Tests pointed out that it lacked guidance and consistency, either in terms of structure or in terms of user interface. Additionally the working area was considered too narrow.
Prototype 3: EXPO '98 Environmental Spatial Multimedia CD-ROM	The third prototype represents a compromise between the two previous approaches. Although very close to the second prototype, this final version allows the user to explore the same kind of navigation, analysis, and visualization tools as the previous one, but follows a more traditional kind of user interface design.	It intends to reach both the technical and the nontechnical public. The number of icons that gave access to the structural, associative and referential links was considered too high by the users.

This application took advantage of the GI available on the Portuguese Spatial Data Infrastructure (SNIG), as well as the technological developments in the fields of webmapping, and spatial visualization tools.

The web interface to the EIA processes database enables data search and retrieval by the general public and has a management module accessible only by professionals responsible for the updating tasks. The database structure accommodates information on three main stages of the EIA process: (1) scoping; (2) EIA studies; (3) postevaluation. For any of these categories it is possible to search information from the past, associated to finalized processes, and actual information associated to projects that are still under way.

For each EIA process that is under public consultation, a web page is obtained that integrates the main information concerning that specific consultation process, gives access to the nontechnical summary pdf file, and provides access to a web-mapping application that allows the user to browse through the GI that is available on the web for the municipalities involved in the project study area (Figure 9.6). The application also provides access to the metadata available at SNIG, the national geographic information infrastructure, concerning the study area.

Further work in this project aimed to evaluate the use of spatial multimedia and collaborative tools to improve the understanding of the environmental impacts and the proposed mitigation measures, using a case study that would not be restricted by data availability or scheduling obligations. These developments would promote the use of GI and multimedia data by the companies that develop EIA studies, through the availability of best practice examples.

9.3.3 THE USE OF ENVIRONMENTAL DATA COLLECTED BY CONCERNED CITIZENS: A MOBILE APPLICATION

Data collection has been one of the privileged areas of application of mobile spatial systems such as mobile GIS [33]. Within environmental management, Vivoni et al. [31], as mentioned previously, proposed an application of mobile GIS and wireless communications to support water quality data field measurements.

The popularity of mobile cellular phones makes them an attractive device to support citizenship activities such as voluntary environmental monitoring. According to Peng and Tsou [33], there is a high demand for mobile services within the GI community, both to support fieldworkers and mobile consumers. However, techno-logical and institutional barriers still exist. The major barriers are related to the large variety of mobile devices, the low bandwidth of the current wireless networks, and the uncertainty about the 3G mobile networks.

The display of maps on mobile phones presents specific interface problems due to the size of mobile phones' screens. As a result, display clutter usually occurs when displaying dense areas. Labeling is also limited. Color cannot be used to differentiate map features, since some screens are monochromatic. Interaction with mobile phones is further hampered by the multidimensionality of the keyboard.

Related to the creation of systems to support public participation, issues like users' privacy and the need to have an attractive business model (at least for the mobile operator) have to be considered. Nevertheless, mobile applications for data collection present advantages such as automatic data geo-referencing, the possibility to instanta-neously communicate and access the data, and anytime anywhere accessibility.

To illustrate some of the potential benefits of using mobile phones to support public participation within environmental monitoring, one prototype is under devel-opment. This prototype is integrated in a research project, named Senses@Watch (http://panda.igeo.pt/senses/sp/english/index.asp), which aims to define and evaluate strategies to promote the use of citizens´ collected data, including the information used in environmental complaints. The project explores developments in the area of information and communication technologies to promote a wider use of the data collected by citizens. Within the Senses@Watch project a prototype of a web-based

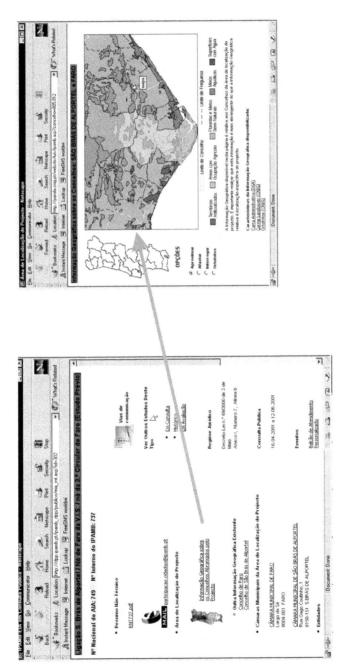

FIGURE 9.6 Webmapping application on the GI data concerning an EIA process under public consultation.

collaborative site has been developed. This prototype is based on the WWW, although an interface for data input and access through SMS and MMS, using mobile phones, is being developed.

The Senses@Watch collaborative site (Figure 9.7) is intended to support citizens to collect and manage the data collected by citizens within isolated environmental monitoring initiatives. It has been designed following the well-known metaphor of postcards. Citizens may use the site or mobile phones using MMS to create their postcards with photos, sounds, graphics, and text to describe an environmental problem. They can publish such e-cards on the WWW and send them to the authorities in charge. Three types of tools are available on the prototype, considering the major tasks performed by citizens when filling a complaint:

- Tools to support data collection and processing, which include geo-referencing tools, data annotation, and creation of metadata. The geo-referencing tools use a gazetteer and maps to help users to reference a complaint, minimizing the ambiguity of a reference to a place. Data annotation tools give the user the ability to underline specific issues within the images through the possibility of adding graphics, sounds, and text to the image. Metadata is used for data classification and user characterization and to enable the system to assess data fitness and facilitate data reuse. For example, data quality indicators can be built based on the history of each individual in what concerns his contribution to the system. The data input task was designed to be kept simple and minimize data entry.
- A clipart that intends to support the creation of multisensory messages in the context of environmental public participation. The clipart includes prerecorded images annotated with graphics, icons, texts and nonspoken sounds, as well as short textual descriptions that translate sensory data into environmental quality information. The user can explore this case library and select an annotated image to illustrate a specific environmental situation.
- Data access and visualization tools, namely thematic, temporal, and spatial searches. These tools are based on webmapping services. They allow users to access the data collected by other citizens and overlay it onto data from other sources such as orthophotos.

It is possible to access the Senses@Watch collaborative system using mobile phones. The mobile applications developed within the project are intended to explore research questions such as what type of tools should be available through mobile phones to support citizens' within their complaint process? How should data be presented according to the size and quality of most mobile phones? And what type of interface is more appropriate? Two major approaches have been followed:

- The development of an interface through mobile devices to access to the web-based collaborative site developed within the project
- The creation of an application that enables citizens to geo-reference the data collected based on the positioning system of the operator, save that information on a database, and make it accessible through the Web

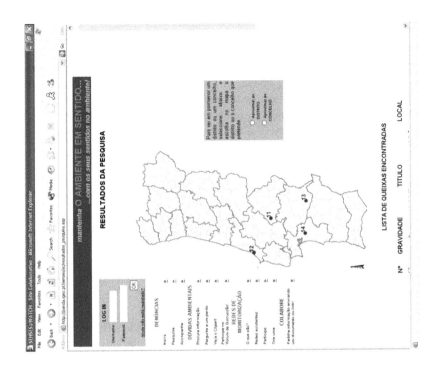

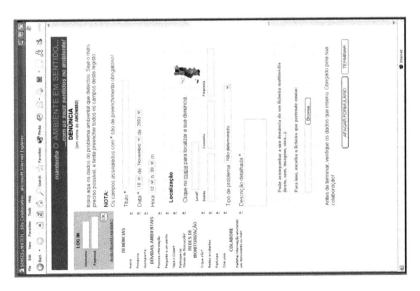

FIGURE 9.7 Senses@Watch collaborative site.

Access to the full contents of the Senses@Watch collaborative site is made by a WAP application. Additionally, citizens may send information through SMS or MMS to the site. The application to automatically geo-reference the data collected by citizens using their camera-equipped mobile phones is still under development, since it requires the operator agreement to provide such service and the consent of the user to release personal information such as the user's location. Since geo-referencing using the positioning system of the operator may be inaccurate (especially within rural areas) the application being developed includes the possibility to use a gazetteer together with maps to allow users to identify the data collection location and reduce the ambiguity of a reference to a place. Further developments include usability testing of the geo-referencing tools.

9.4 SUMMARY AND RESEARCH QUESTIONS

The technological developments observed within the ICT modified some characteristics of spatial multimedia systems. From stand-alone applications demanding sophisticated computer configurations, spatial multimedia systems have become distributed and so common that they are not always labeled as such. More recently, the emergence of mobile applications has made the use of multimedia spatial information even more pervasive. Nevertheless, the key concepts of spatial multimedia systems apply to any of these platforms.

Environmental management can benefit from some of the key features of spatial multimedia systems: (1) the availability of realistic representations of the environmental problems and (2) easy-to-use tools to access, capture, explore, analyze, and present the data. The spatial multimedia concepts developed early on (for example, the use of realistic representation of the world and easy-to-use interfaces) still apply within application design. These two ideas have made, since the beginning, the use of spatial multimedia suitable to environmental management. Accordingly, it is possible to find within the environmental management and planning field a large number of applications and studies about its advantages and disadvantages.

The case studies presented show different issues of spatial multimedia application. The EXPO '98 application illustrates the advantages of using spatial multimedia tools to enhance the comprehension of environmental problems and systems. The case study involving the Portuguese agency responsible for the environmental impact assessment public consultation process shows how the use of the Internet and webmapping tools may support the dissemination of EIA information, helping to improve public participation in EIA. New developments to the implemented work include the exploration of visualization and spatial multimedia interaction tools to create best practices guidelines to be followed both by environmental companies and by the public agencies responsible for managing the processes. The mobile application developed within Senses@Watch represents a first approach on how to disseminate environmental data through the use of mobile phones. On the other hand, the Senses@Watch mobile application is intended to explore the use of mobile phones to facilitate data collection activities by citizens within an environmental monitoring context.

Research questions such as what type of tools, what type of interfaces, and how to measure the success of such applications still have to be investigated. Additionally, in a context of almost ubiquitous computing, spatial multimedia may support the task of augmenting human senses. It may be used to increase human senses' utility range (e.g., provide traffic images of bottleneck route points) or to translate sensory information into quality data (e.g., indicate if a strange odor is a sign of air pollution problems). The use of spatial multimedia to augment human senses needs further research and may involve future developments such as geo-visualization advances and augmented reality.

On the other hand, the field of massive online games is an area that has registered significant advances in terms of interface and communication, namely the use of multiple representations of users and scenarios. Such developments should be analyzed to understand which developments should be brought to spatial multimedia and environmental modeling.

ACKNOWLEDGMENTS

This research was partially funded by the POCTI/MGS/35651/99 and by JNICT/DGA under research contract no. PEAM/C/TAI/267/93. The authors would like to thank to all members of these two projects' teams.

REFERENCES

1. Pickles, J., Arguments, debates, and dialogues: The GIS–social theory debate and concerns for alternatives, in *Geographical Information Systems: Principles, Techniques, Management, and Applications,* Longley, P., Goodchild, M., Maguire, D. and Rhind, D., Eds., John Wiley, New Jersey, 1999, pp. 49–60.
2. Carver, S., Evans, A., Kingston, R., and Turton, I., Public participation, GIS and cyberdemocracy: evaluating on-line spatial decision support systems, *Environ. Plann. B: Plann. Des.,* 28, 907, 2001.
3. Peng, Z-R., Internet GIS for public participation, *Environ. Plann. B: Plann. Des.,* 28, 889, 2001.
4. Weiner, D., Harris, T.M., and Craig, W.J., Community participation and geographic information systems, presented at Workshop on Access and Participatory Approaches in Using Geographic Information, Spoleto, Italy, December 5–9, 2001.
5. Fonseca, A., Gouveia, C., Câmara, A., and Silva, J., Environmental impact assessment with multimedia spatial information systems, *Environ. Plann. B: Plann. Des.,* 22, 637, 1995.
6. Raper, J., Progress towards spatial multimedia, in *Geographic Information Research: Bridging the Atlantic,* Craglia, M. and Couclelis, H., Eds., Taylor & Francis, London, 1997, 525.
7. Shiffer, M.J., Towards a Collaborative Planning System, *Environ. Plann. B: Plann. Des.,* 19, 709, 1992.
8. Raper, J., *Multidimensional Geographic Information Science*, Taylor & Francis, London, 2000.
9. Câmara, A., *Environmental Systems. A Multidimensional Approach*, Oxford University Press, Oxford, UK, 2002.

10. Steinmetz, R., Ruckert J., and Recke, W., Multimedia Systeme, *Informatik Specktrum,* 13, 280, 2002.
11. Câmara, A. and Raper, J.F., Eds., *Spatial Multimedia and Virtual Reality,* Taylor and Francis, London, 1999.
12. Fonseca, A., The Use of Multimedia Spatial Data Handling for Environmental Impact Assessment., Ph.D. Thesis, Universidade Nova de Lisboa, Monte da Caparica, 1998.
13. Gibbs, S. and Tsichritzis, D., *Multimedia Programming: Objects, Environments and Frameworks,* ACM Press, New York, 1995.
14. Shneiderman, B., *Designing the User Interface. Strategies for Effective Human-Computer Interaction,* Addison-Wesley, Reading, PA, 1998.
15. Goddard, J.B. and Armstrong, P., The 1986 Domesday Project, *Trans. Inst. Geogr.,* 11(3), 290, 1986.
16. Openshaw, S., Wymer, C., and Charlton, M., A geographical information and mapping system for the BBC Domesday optical disks, *Trans. Inst. Geogr.,* 11(3), 296, 1986.
17. Rhind, D.W., Armstrong, P., and Openshaw, S., The Domesday machine: a nationwide GIS, *Geogr. J.,* 154, 56, 1988.
18. Shiffer, M., Augmenting Geographical Information with Collaborative Multimedia Technologies, in *Proceedings of Auto Carto 11,* McMaster, R.B. and Armstrong, M.P., Eds., American Society for Photogrammetry and Remote Sensing, 1993, 367.
19. Câmara, A.S., Pinheiro, M., Antunes, P., and Seixas, M.J., A new method for qualitative simulation of water resources systems: 1. Theory, *Water Resour. Res.,* 23(11), 2015, 1987.
20. Câmara, A.S., Ferreira, F.C., Nobre, E., and Fialho, J.E., Pictorial modelling of dynamic systems, *Syst. Dyn. Rev.,* 10(4), 361, 1994.
21. Fedra, K. and Loucks, D.P., Interactive computer technology for planning and policy modelling, *Water Resour. Res.,* 21(2), 114, 1985.
22. Loucks, D.P., Taylor, M.R., and French P.N., Interactive data management for resource planning an analysis. *Water Resour. Res.,* 21(2), 131, 1985.
23. Fernandes, J.P., Fonseca, A., Pereira, L., Faria, A., Henriques, L., Garção, R., and Câmara, A., Visualization and interaction tools for aerial photograph mosaics, *Comput. Geosciences,* 24(4), 465, 1997.
24. Nobre, E. and Câmara, A., Spatial simulation by sketching, in *Spatial Multimedia and Virtual Reality, Research Monographs in GIS,* Câmara, A.S. and Raper J., Eds., Taylor & Francis., London, 1999, 47.
25. Dellavalle, R.P., Hester, E.J., Heilig, L.F., Drake, A.L. Kuntzman, J.W., Graber, M., and Schilling, L.M., Going, going, gone: lost internet references, *Science,* 302, 787, 2003.
26. Rojas, P., Now *Bloggers Can Hit the Road, Wired News,* retrieved from http://www.wired.com/news/wireless/0,1382,57431,00.html in December 2003.
27. Cunningham, M.D., The City Scan Project (WWW Document), retrieved from http://www.city-scan.com/section.php?section=moreinfo&page=resources in December 2003.
28. Jasnoch, U., GIS-based location services: a new service for the city of Darmstadt, *GeoInformatics,* 24, March 2003, pp. 24–25.
29. Krug, K., Mountain, D., and Phan, D., WebPark—location-based services for mobile users in protected areas. *GeoInformatics,* 26, March, 2003, pp. 26–29.
30. Rheingold, H., Moblogs seen as a crystal ball for a new era in online journalism. *Online Journalism Review,* University of Southern California. Retrieved from http://www. ojr.org/ojr/technology/1057780670.php in December 2003.

31. Vivoni, E.R., Camilli, R., Rodriguez, M.A.A., Sheehan, D.D., and Entekhabi, D., Development of mobile computing applications for hydraulics and water quality field studies, in *Hydraulic Engineering Software IX*. WIT Press, Montreal, CA, 2002.
32. Ydreams, LiveAnywhere Traffic. Let your phone be your pilot. [WWW Page] Retrieved from http://www.ydreams.com/solutions.php?sec=2&s_sec=1&s_s_sec=2 in May 2003.
33. Peng, Z.R. and Tsou, M.H., *Internet GIS: Distributed Geographic Information for the Internet and Wireless Networks*, John Wiley, Hoboken N.J., 2003.
34. Rosen, E.C., Haining, T.R., Long, D.D.E., Mantey, P.E., and Wittenbrink C.M., REINAS: A real-time system for managing environmental data, *Int. J. Software Eng. Knowledge Eng.*, 8(1), 35, 1998.
35. Hale, S., Bahner, L., and Paul, J., Finding common ground in managing data used for regional environmental assessments, *Environ. Monit. Assess,* 63(1), 143, 2000.
36. McLaughlin, R.L., Carl, L., Middel, T., Ross, M., Noakes, D.L.G. Hayes, D.B., and Baylis, J.R., Potentials and pitfalls of integrating data from diverse sources: lessons from a historical database for Great Lakes stream fishes. *Fisheries,* 26(7), 14, 2001.
37. Longley, P.A., Goodchild, M.F., Maguire, D.J., and Rhind, D.W., *Geographic Information Systems and Science*, John Wiley & Sons, Sussex, UK, 2001.
38. MacEachren, A.M., Cartography and GIS: extending collaborative tools to support virtual teams, *Prog. Hum. Geogr.,* 25, 431, 2001.
39. Al-Kodmany, K., Supporting imageability on the World Wide Web: Lynch's five elements of the city in community planning, *Environ. Plann. B: Plann. Des.,* 28(6), 805, 2001.
40. Kingston, R., Carver, S., Evans, A., and Turton, I., Web-based public participation geographical information systems: an aid to local environmental decision-making. *Comput. Environ. Urban Syst.,* 24(2), 109, 2000.
41. Wiggins, L. and Shiffer, M., Planning with Hypermedia. Combining text, graphics, sound and video, *J. Am. Plann. Assoc.,* Spring, 226, 1990.
42. Shiffer, M.J., Interactive multimedia planning support: moving from stand-alone systems to the World Wide Web, *Environ. Plann. B: Plann. Des.,* 22, 649, 1995.
43. Ferreira, F., Digital video applied to air pollution emission monitoring and modeling. In *Spatial Multimedia and Virtual Reality: Research Monographs in GIS,* Câmara, A. S. and Raper J., Eds., Taylor & Francis, London, 1999, 47.
44. Neves, J.N., Gouveia, C., and Bento, J., Portugal Digital: Flying over geographic information, *GIS PLANET 1998 Annual Conference Proceedings,* USIG, Lisbon, 1998.
45. Câmara, A., Neves, J.N., Muchaxo, J., Sousa, I., Costa, M., Nobre, E., Mil—Homens, J., and Rodrigues, A.C., Water quality management in virtual environments, *J. Infrastructure Syst.* ASCE, 116(3), 417, 1998.
46. Andrienko, G.L., Andrienko, N.V., Interactive maps for visual data exploration, *Int. J. Geogr. Inf. Sci.,* 13(4), 355, 1999.
47. Rinner, C., Argumentation maps: GIS-based discussion support for on-line planning. *Environ. Plann. B: Plann. Des.,* 28(6), 847, 2001.
48. Carver, S., Blake, M., Turton, I. and Duke-Williams, O., Open spatial decision-making: evaluating the potential of the World Wide Web, in *Innovations in GIS 4,* Kemp, Z., Ed., Taylor & Francis, London, 1997, 267.

10 Computer Support for Discussions in Spatial Planning

Claus Rinner

CONTENTS

10.1 INTRODUCTION

Spatial planning deals with the problem of distributing the limited resource "space" among different uses and users. It can be highly challenging to find a balanced land-use pattern, for example, in urban agglomerations. Different interest groups such as residents, industry, and ecologists will claim different desirable land uses for a given area. Spatial planning is also about locating unwanted land use such as waste facilities. In this case, interest groups (e.g., city councils, neighborhood organizations) and individuals will fight nearby locations. This situation is known as the NIMBY problem: "Not In My BackYard!"

In democratic societies, decisions such as those in spatial planning are made by political representatives in cooperation with public administration and residents. The final decision will usually be based on a number of consecutive prior decisions or choices, which are made by different groups of stakeholders. At any of these decision levels, there are two important methods to reach a conclusion: consensus finding, or voting. Both will be preceded by more or less intensive discussions and argumentation. The ultimate goal of discussions is to achieve sustainable development by integrating the objectives of diverse stakeholders. Thus, we argue that discussions are a crucial element of spatial planning procedures and are to be integrated with planning and decision support techniques.

Discussions will have diverse formats in different planning projects. For example, the number of participants may vary from only two to hundreds and more; participants may get together or stay separated in space and/or time; discussion may be unmoderated or moderated and structured. Nevertheless, discussion contributions (statements, messages, arguments, articles) in spatial planning will commonly contain a spatial reference. This does allow linking discussion support to spatially enabled decision support techniques, as argued in this chapter.

In Section 10.2, we will review general theories on argumentation and introduce major concepts of computer-supported cooperative work. Next, geographically referenced discourse will be analyzed in more detail, leading to the argumentation map model (Section 10.3). Section 10.4 develops use cases for GIS-based discussion support, and Section 10.5 presents some existing applications. Finally, we will speculate about future developments in computer support for discussions in spatial planning (Section 10.6).

10.2 ARGUMENTATION THEORY AND CSCW

Argumentation theorists analyze rational human discourse on a variety of levels. According to van Eemeren et al. [1], "Argumentation is a verbal and social activity of reason aimed at increasing (or decreasing) the acceptability of a controversial standpoint for the listener or reader, by putting forward a constellation of propositions intended to justify (or refute) the standpoint before a rational judge."

An important aim of argumentation analysis is structuring discourse. Formal models of argumentation were put forward by Toulmin [2] in *The Uses of Argument* and Kunz and Rittel [3] in "Issues as Elements of Information Systems." Both approaches suggest a limited set of types of basic argumentation elements, and a set of relations between these. Toulmin models argumentation elements as data, claims, or warrants. Warrants back claims, which in turn are based on data. Similarly, Kunz and Rittel use issues, positions, and arguments. Root issues are assumed to draw different positions that are supported or opposed by arguments.

While operated manually at the beginning, IBIS have quickly been computerized (e.g., in the gIBIS [graphical IBIS] research tool by Conklin and Begeman [4]), which was further developed into the commercial QuestMap product and more recently the open source tool compendium, by Jeff Conklin. In general, argumentation models can easily be visualized using graphs, if they are composed of elements of different types and relations between these elements. Nodes in an argumentation graph represent specific argumentation elements, while edges in a graph represent relations between the elements. For example, QuestMap uses icons at the nodes of a "dialog map": question marks, light bulbs, and plus and minus signs represent dialog elements of types such as question, idea, and pro and con, respectively (see http://www.cognexus.org/index.htm, link to "Dialog Mapping").

The common usenet newsgroups (e.g., comp.infosystems.gis) are another example of structured discourse support by (simple) computer visualization. The argumentation structure of newsgroups consists of threads that are initiated by a message with a new subject line. Additional messages within the thread reply to the initial message (at least they should). Visualization of this structure in most newsreader

TABLE 10.1
Examples of Discussion Settings in Spatial Planning, and Supporting Computer Tools

	Same Time	Different Time
Same Place	Community meeting: 2D, 3D, and animated project visualization; note keeping	Speaker series, shared Internet access: Video recording, argumentation recording and structuring
Different Place	Video conference, chat room: Shared text, graphics documents; virtual worlds	Internet newsgroups, forums, guestbooks: Argumentation recording and structuring, hyperlinking

software is achieved through indentation of the subject lines in the overview list of a discussion. The importance of visually representing the structure of discourse is underlined by the recent publication of Kirschner et al. [5].

On a more general level, computer support for structured discussions can be subsumed under the label of computer-supported cooperative work. CSCW examines the enabling techniques for collaboration in groups. Specific topics include groupware systems, network technologies, human–computer interaction, and the social implications of computer-supported distributed work environments.

Techniques for CSCW differ due to varying cooperation settings. A helpful distinction can be made between same place or different place, and same time or different time, cooperation (Ellis et al. [6]). This applies to discussions as well. Discussions between people meeting at the same time are called "synchronous" and include community meetings (same place) and video conferences or chat (different place). Discussions over a longer time period (different time, "asynchronous"), in which participants do not respond immediately to each other include speaker series (same place) and Internet newsgroups and forums (different place). Newsgroups can also be used in an asynchronous but same place setting, if common public Internet access points are used by participants.

Table 10.1 summarizes these examples of distinct discussion settings and adds useful computer tools for the four categories. The different place/different time setting is the focus of the remainder of this chapter, because it requires the most generic argumentation support, and generates a natural need for computer and network support for remote and asynchronous discussions.

10.3 MODELING GEOGRAPHICALLY REFERENCED DISCOURSE

In order to define the specific aspects involved with discussions in spatial decision-making, we will examine a sample of arguments that came up in a community meeting in a preliminary phase of planning a new office building in the German city of Münster. In the following citations, references to geographical objects identified on maps, plans, in a movie, or memorized, are emphasized by the author:

- Inhabitant (inquiring): "Our house on the north side of the adjacent road already suffered structural damage when the existing shopping mall next to the planned building was constructed."
- Inhabitant (enraged): "The layout of the parking entrance will significantly increase traffic in front of my living room window. When I moved here, this neighborhood was designated as a residential area, but it turned out that we already have traffic problems, due to the large electronics store in the mall."
- City planner (balancing, points to schematic road map and landscape photograph): "This building has been designed so that the historical silhouette of Münster, which is visible when approaching the city on this stretch of highway, will not be occluded. The silhouette will be modified though, because the new building will stick out between the towers of the cathedral and this church."
- Architect (matter-of-factly, displays 3D fly-by movie of building and surrounding streets): "This project is designed in a lightweight fashion similar to our much acclaimed building in the south of town. The grass-grown ramparts are designed to resemble the Omnisports center in Paris."

Obviously, geographic references do appear in the arguments in urban planning discussion. The most typical case probably is inhabitants referring to their own dwelling and to their neighborhood. Stakeholders involved in projects on a job level, instead of a personal level, may tend to refer to wider geographical areas such as to the city as a whole, or to comparable places in other cities. According to the previous section, Internet newsgroups are common instruments to facilitate remote, asynchronous discourse. Thus, it seems natural to seek ways of integrating newsgroups with digital maps and GIS. However, the geographical reference of argumentation elements poses some methodological problems that will be discussed in the sequel.

Figure 10.1 visualizes some of the geographical references used in the arguments cited above. Additional objects that were referred to, such as the cathedral and a comparable building, are difficult to include on a map in planning scale. Please note that some of the references are likely to be available as geographical objects on a digital planning map (buildings as a whole, road section), while others are part of objects or may best be thought of as coordinate locations (living room window). With respect to the situation sketched in Figure 10.1, we are going to assess different components involved in modeling geographically referenced argumentation.

Geographic information systems are designed to integrate spatial (geometric) data with thematic (attribute) data. For example, census maps visualize socioeconomic data values with reference to enumeration units such as census blocks. Handling arguments as just another attribute of geographic objects in a planning GIS would not accurately represent the "real world" — an argument is an entity on its own, which can be used independently of a geographical representation, and which can refer to more than one geographical object, thus being more than a "flat" attribute.

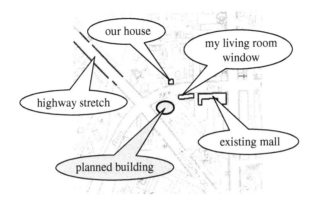

our house

my living room
window

highway stretch

existing mall

planned building

FIGURE 10.1 Sketch for excerpt of urban planning discussion. (Geographical references according to scenario in text. Background map modified from www.muenster.de. With permission.)

Argumentation Element
Message_ID
Author
Date
Argument_Type
Title
Text

Geographic Reference Object
Object_ID
Shape
Centroid
Area/Length
Land-Use

FIGURE 10.2 Object models for isolated argumentation elements and geographic reference objects.

Figure 10.2 suggests object models for both argumentation elements and geographic objects. This model reflects the individual object identities in both the argumentative and the geographical spaces. However, relationships between the objects have to be added. On the one hand, geographical objects have implicit spatial relations with each other, which may or may not be reflected in GIS data models. We will add a self-reference to the geographic object class to hint at topological relations. On the other hand, arguments in a structured discussion necessarily have (topo)logical relations with each other (e.g., a reply-to relationship). In this modeling approach, argumentation elements preserve a dependency structure according to the argumentation model chosen in a specific application, so that two sets of topological objects are combined.

Figure 10.3 includes these relations within each object class and, most importantly, the geo-argumentative relations between objects of the two classes. Argumentation elements can refer to one or more geographic objects, while a geographic object can be referenced by one or more arguments. In addition to direct references between arguments and geographical objects, the model in Figure 10.3 also includes

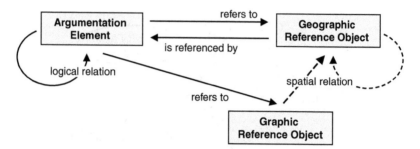

FIGURE 10.3 The argumentation map model links argumentation elements and geographic objects. (Extended from Rinner, 1999. With permission.)

TABLE 10.2
Optimal Functionality of Argumentation Map Implementations

Presentation of spatially referenced discussion
Input of arguments with geographical reference
Retrieval of argumentation elements or geographical objects or both
Analysis of the status of a spatially referenced debate

---

arguments referring to graphical helper objects. This means that discussants can specify an exact point location as a graphic reference for their contributions. Users also would be allowed to draw features on a map (e.g., a free-form ellipse around an area of concern) and link their argument to this graphic. If a direct reference to a geographic feature such as a street or planning area is missing, the spatial relation to planning features could be reconstructed using standard topological operations in GIS. This is an extension of the argumentation map model originally proposed by Rinner [7]. The following section outlines the potential uses of the model.

10.4   THE USES OF ARGUMENTATION MAPS

The term "argumentation map" describes a conceptual model that relates objects in a computer-supported discourse (argumentation elements) with objects in a geographical database (e.g., within a GIS). Implementations of argumentation maps thus can be used to support spatially referenced discussions. The functionality of argumentation maps can be classified according to typical information system functions such as data input, retrieval, analysis, and presentation. To follow an order of increasing complexity of these functions from a user point of view, we will shift the presentation function to the first place in this list (see Table 10.2).

The *presentation* of the current status of a spatially referenced discussion naturally involves maps. It is, however, not obvious what should be presented on an argumentation map (in the narrower sense) in addition to the geographic situation, which is at the center of discussion. In addition to the occurrence of arguments at specific locations or with predefined geographical objects such as roads, or land-use polygons, additional attributes of these arguments might be interesting to visualize. For example, to display the spatial pattern of approving versus objecting arguments, the type of argument (e.g., pro, contra) needs to be represented by color hue, or shape of a symbol. Also, the display may be limited to subsets of arguments (e.g., those arguments put forward by a single participant, or group of participants). Besides using cartographic maps, presentation of a debate will also require text-based displays of structured argument lists and contents of individual arguments. In the hypermedia setting of a Web-enabled argumentation map, the presentation function will allow users to navigate through discussion-related documents based on cartographic displays.

The *input* function of an argumentation map should support discussants in submitting a new argument together with its spatial reference. This process could be started from either a discussion forum or a mapping component showing the geographical area subject to discussion, but it will eventually involve both of these components. For example, in addition to writing the text of a discussion element, the participant would be asked to provide at least one spatial reference by clicking a location on the map or selecting a geographical object. Or, some kind of interaction with a map (e.g., double-clicking a location or geographical object) would open a dialog for the input of a discussion contribution, which would then be related to existing messages in a discussion forum. Text and geo-reference input are the primary functions to support active participation in a planning debate.

The *retrieval* functions include querying the discussion for arguments referring to selected geographical object(s), as well as querying the map for objects referenced by selected argument(s). These topological queries should be combinable with each other and with attribute queries. For example, it should be possible to search for messages of a certain author (attribute query) that refer to a selected object (topological query) and all its neighbors (another topological query). Some of these query types use the topology internal to the geographical domain or to the argumentative domain, while at least one part of the query crosses the two domains. The theoretical implications of combined topology are discussed further below.

As the most advanced class of functions of an argumentation map, the *analysis* of geo-referenced discussion uses existing data to generate additional information of use to participants or observers of a discussion. Summary statistics such as counts of arguments with certain characteristics (author, submission date, argument type) or dominant argument type per geographical area might be useful to understand the current state of a debate. Geo-argumentative analysis requires arguments to refer to geographical objects rather than coordinate locations, so that arguments referring to the same object can be identified. Alternatively, arguments referring to coordinate locations could be related using GIS analysis such as point-in-polygon to identify those reference locations falling in the same geographical object. The straightforward way of reporting the results of an analysis operation again is visualizing them using some regular cartographic method such as chart maps for counts, or area shading (choropleths) for averages.

The topological relations within, and between, the two domains involved in an argumentation map have interesting theoretical implications. In particular, if links

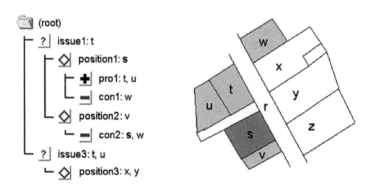

FIGURE 10.4 Geographically referenced IBIS discussion to demonstrate geo-argumentative relations. (From Rinner, 1999. With permission.)

cross the geographical domain and the argumentative domain twice, formal relations between objects in one domain appear that may be of practical use. For example, object *s* in Figure 10.4 is linked to argument *position1,* to which *con1* is responding in the discussion structure. *Con1* in turn is linked to object *w.* Although *w* is not a neighbor or otherwise directly spatially related to *s,* there seems to be a link between the two objects, at least according to some of the discussion participants. We term such links as "geo-argumentative relations," which allow for defining a distance relation between objects of the geographical domain via the argumentative domain, or vice versa. No other field is known to the author where such connections between objects via "parallel universes" including the geographical domain would be exploited.

10.5   EXISTING APPLICATIONS

In the past five years or so, several isolated attempts have been made to implement computer tools that would support geographically referenced discourse.

CrossDoc is a research prototype developed by Tweed [8] to visualize argumentation structures by modeling networks of documents. In the planning application described in the chapter, the document network includes a planning map and arguments, which are structured according to the issue-based information system (IBIS) framework (Kunz and Rittel [3]). CrossDoc thus provides a fully visual index to the structure and spatial reference of argumentation. The tool was conceived of as a stand-alone, integrated desktop tool for argumentation recording, a "decision journal." It is bound to the Apple Macintosh platform.

A cooperative hypermap is suggested by Rinner [9] to support the online link of planning maps with discussion contributions. A perspective view on an urban development plan is provided together with an input form. Users can create a 3D flag to represent an argument; user input includes the type of argument according to the IBIS model, a link to a message in a discussion forum that would contain the argument, and the spatial reference. The reference is achieved by placing the flag on top of the plan element to which the message was referring. The virtual reality modeling language (VRML) is used in conjunction with a Java applet to implement the user interface within a standard Web browser, as shown in Figure 10.5. This

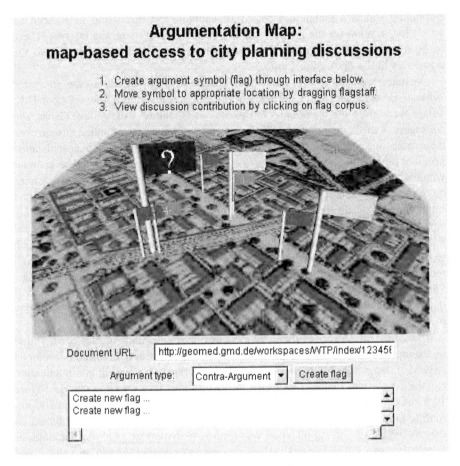

FIGURE 10.5 VRML-based 3D display of an argumentation map.

prototype version does not include a server component that would be required to store user input. With respect to the argumentation map model outlined above, this application is limited, in that it provides only for coordinate-based spatial references and allows only one-to-one relations between arguments and locations. In the original version, the third dimension was used to represent larger and smaller flags, depending on the level of the corresponding argument (e.g., largest for issues in an IBIS model), while the planning map was drawn as a texture on a flat surface. However, a virtual round table variant was implemented as a student project, with the map placed on a table in a virtual planning office showing a more "immersive" environment for discussion support.

VRMLView is presented by Lehmkühler [10] as an experimental combination of a three-dimensional mapping component and a newsgroup component. The 3D scene is developed using VRML as well, and represents a hypothetical planning scene. A mouse click in the scene opens a discussion forum in another Web browser window, which includes an input form for new contributions. This prototype suggests combining standard techniques for (3D) mapping and discussion to achieve an accessible solution for the greatest number of potential users on the Internet. There is, however, no link between individual elements of the planning "map" and the arguments in the discussion forum.

Virtual Slaithwaite is introduced by Kingston et al. [11] as a case study for "virtual" decision-making. In a real-world development effort, residents of a U.K. village were given the opportunity to discuss local planning issues. Steve Carver and colleagues at the School of Geography, University of Leeds, have studied this environment from different perspectives related to virtual society. The original application provides users with a village map and a comment frame combined in a Web browser window. In the initial version, comments would be placed at coordinate locations on the map and represented by point symbols. Subsequent users thus would get an impression of where there are comments. By clicking on symbols, corresponding comments would be displayed in the comment frame. The current, Java-based version of Virtual Slaithwaite allows the user to select a geographical object to which to link a comment. The resulting link structure, however, could not be verified in a recent visit of the site (http://www.ccg.leeds.ac.uk/slaithwaite/). In theory, Virtual Slaithwaite provides a map-based access to discussion contributions. On the side of the discussion however, argumentation structure is not supported, nor is any display of messages in a list or graph offered.

Based on the Descartes thematic mapper (Andrienko and Andrienko [12]) and on the argumentation map concept, Dialogis Software & Services GmbH developed ArguMap, later called NoteMap, a map-based forum for planning communication endorsed by urban planners in Bonn, Germany. The Java applet shown in Figure 10.6 could display an aerial image of the city, together with point symbols representing questions, answers, and comments to planning issues. When published on the city's Web site, the application was limited to displaying planners' annotations to selected issues. The reasons include the fear of overwhelming input volume and misuse of the system. No option for discussion of these issues by concerned citizens was included in this version.

Hans Voss and colleagues at the Spatial Decision Support team of Fraunhofer Institute for Autonomous Intelligent Systems (AIS) are working on the arguably most advanced approach to supporting spatial discourse. Coupling two existing software tools in AIS, the Zeno discussion forum, and the Descartes thematic mapper, was first suggested by Rinner [13]. Voss et al. [14] describe the most recent design for integrating structured discourse and spatial analysis and mapping. Their two systems, now called Dito and CommonGIS, will support many-to-many relations between user comments and geographical objects on maps. In addition to this and other conceptual requirements, the authors put forward recommendations to achieve a consistent graphical user interface. For example, geo-referenced comments should be represented on a special annotation layer. Further requirements refer to technical issues such as performance, synchronization, and security.

A typical feature of many argumentation map implementations is the storage of user input on a central server. Kolbe et al. [15] suggest a different approach. They introduce a map annotation tool for bicycle tourists in the Ruhr valley, Germany.

Computer Support for Discussions in Spatial Planning 177

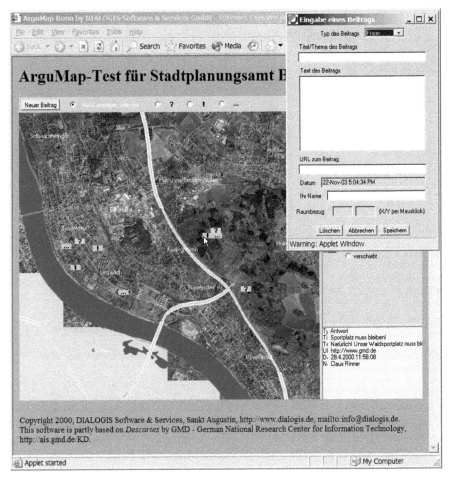

FIGURE 10.6 ArguMap — prototype of a map-based planning discussion forum. (From Dialogis GmbH. With permission.)

Their "cooperative Web map" prototype allows users to add comments, and link pictures, to a centrally stored bike route map. Instead of storing this input on the server, it is built into a complex URL, which can be sent to other users. Recipients of such a URL will view the central Web site augmented by the sender's annotations, while other visitors of the Web site will not know about the annotations. Using the URL as a container for Web page annotation is implemented such that it does not require any software installation on client computers as long as a modern Web browser is available. The central Web site is realized as a script that will handle annotations if they are provided in the URL. In this case, annotations will be placed in HTML layers on top of the default contents of the page (see example in Figure 10.7). If no annotations are provided in the URL, the bike map will appear as it is. This appears to be a highly innovative concept for Web-based, peer-to-peer cooperation, but Kolbe and colleagues also mention possible limitations of their approach.

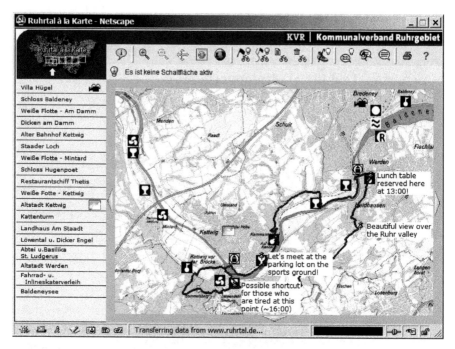

FIGURE 10.7 Cooperative Web map. (From Ruhrtal à la Karte, http://www.ruhrtal.de. With permission.)

Page annotations made by a user at one point in time may become invalid if the background page on the server changes, which is beyond the user's control. The amount of information contained in a URL is limited by Web browsers in a non-standardized way (e.g., 2048 bytes for Internet Explorer). Finally, the peer-to-peer approach lacks any option to get an overview of comments made on a Web map, which will disqualify it for certain applications that require a public forum rather than fragmented discussion groups.

10.6 THE ROAD AHEAD

Integrated computer support for both map display and participation in planning debates is an important milestone on the way to sustainable development, namely in urbanized democracies. But the conceptual peculiarities of geographical information, as well as human discourse, have so far prevented integrated tools from being developed, except for very specific applications. We anticipate that comprehensive computer support for discussions in spatial planning will only be established and widely accepted when popular mapping and discussion tools are combined. This might involve solutions as simple as MapQuest on the GIS side, and Usenet news-groups on the argumentation side. This might also require scaling down our expectations with respect to the optimal spatial discourse environment outlined above and by other researchers such as Voss et al. [14].

Nonetheless, investigating the character of geographically referenced argumentation has already contributed to geographic information science in that it poses specific demands on conceptual data models. Arguments as a type of media to be linked to maps are uncommon in GIS. The most similar to these are map hyperlinks (or hotlinks), which geo-reference HTML pages, photos, or movies. In the context of digital libraries, Goodchild [16] discusses the generalized concept of geographically referenced "information-bearing objects." In contrast to these, however, arguments typically refer to each other, in addition to having a geo-reference.

Helpful visualizations and analysis functions have been proposed in applications such as those described in the previous section. But computer support for spatial discussions needs to be founded on a theory of geographically referenced information objects. More examples from different application domains (planning, design, business, conservation, etc.) need to be collected, and prototypes be implemented, to test which user groups will benefit from this approach.

REFERENCES

1. Van Eemeren, F.H., Grootendorst, R., Snoeck Henkemans, F., Blair, J.A., Johnson, R.H., Krabbe, E.C.E., Plantin, C., Walton, D.N., Willard, C.A., and Woods, J., *Fundamentals of Argumentation Theory: A Handbook of Historical Backgrounds and Contemporary Developments,* Lawrence Erlbaum, Mahwah, NJ, 1996, 5.
2. Toulmin, S., *The Uses of Argument,* Cambridge University Press, Cambridge, UK, 1958.
3. Kunz, W. and Rittel, H.W.J., Issues as elements of information systems, Technical Report 0131, Institut für Grundlagen der Planung, University of Stuttgart, Stuttgart, Germany, 1970.
4. Conklin, J. and Begeman, M.L., gIBIS: a hypertext tool for exploratory policy discussion, in *Proceedings of the Conference on Computer-Supported Co-operative Work (CSCW'88),* Portland, OR, 1988, p. 140.
5. Kirschner, P.A., Buckingham Shum, S.J. and Carr, C.S., *Visualizing Argumentation: Software Tools for Collaborative and Educational Sense-Making,* Springer, London, 2003.
6. Ellis, C.A., Gibbs, S.J. and Reln, G.L., Groupware: Some issues and experiences, *Commun. ACM,* 34(1), 38, 1991.
7. Rinner, C., *Argumentation maps — GIS-Based Discussion Support for Online Planning,* GMD Research Series No. 22/1999, Sankt Augustin, 1999.
8. Tweed, C., An information system to support environmental decision making and debate, in *Evaluation of the Built Environment for Sustainability,* Brandon, P.S., Ed., Spon, London, 1997, p. 67.
9. Rinner, C., Using VRML for hypermapping in spatial planning, in *Proceedings of the Int. Conference on Urban, Regional, Environmental Planning, and Informatics to Planning in an Era of Transition,* Sellis, T. and Georgoulis, D., Eds., Athens, 22–24 October, 1997, p. 818.
10. Lehmkühler, S., Virtual Reality Modeling Language — 3D-Standard des World Wide Web/Chance für die Raumplanung [Virtual Reality Modeling Language — 3D-standard of the World Wide Web/Opportunity for spatial planning], in *Computergestützte Raumplanung. Beiträge zur CORP'98,* Schrenk, M., Ed., Selbstverlag des IEMAR, TU Wien, Vienna, 1998.

11. Kingston, R., Carver, S., Evans, A., and Turton, I., A GIS for the public: Enhancing participation in local decision making, in *Proceedings GIS Research United Kingdom (GIS-RUK)*, Southampton, April, 1999.

12. Andrienko, G. and Andrienko, N., Interactive maps for visual data exploration, *Int. J. Geogr. Inf. Sci.,* 13(4), 1999, 355.

13. Rinner, C., Argumaps for spatial planning, in *Proc. First Int. Workshop on Telegeoprocessing (TeleGeo'99)*, Laurini, R., Ed., Lyon, 6–7 May, 1999, p. 95.

14. Voss, A., Denisovich, I., Gatalsky, P., Gavouchidis, K., Klotz, A., Roeder, S., and Voss, H., Evolution of a participatory GIS, *Computers, Environment and Urban Systems,* 28(6), 2004, pp. 635–651.

15. Kolbe, T.H., Steinrücken, J. and Plümer, L., Cooperative Public Web Maps, in *Proc. Int. Cartographic Congress (ICC)*, Durban, 2003.

16. Goodchild, M.F., Towards a geography of geographic information in a digital world, *Comput. Environ. Urban Syst.* 21(6), 1997, 377.

11 Integration of GIS and Simulation Models

Andrea Giacomelli

CONTENTS

11.1 INTRODUCTION

An integrated and multidisciplinary approach is essential when addressing sustainable development issues. This comes from the need to consider environmental, economic, and social factors, in addition to introducing spatial and temporal perspective to ensure the exhaustive analysis of a problem.

Geographic information systems (GIS) provide a convenient way to analyze and represent information bearing a spatial component, and their data management architectures permit an integrated view of extremely diverse types of information, such as digital cartography, remotely sensed imagery, and survey data from a variety of sources. These capabilities propose GIS technologies as a natural framework to deal with much of the complexity embedded into the research and decision-making processes related to sustainable development.

Simulation models represent another category of tools, which may be used for planning or decision support. Several GIS packages offer simulation capability for simple processes, but specialized models will be required to refine the level of analysis, or may represent consolidated tools that are preexisting to GIS in a given context. For this reason, GIS and simulation models are often proposed to the decision-maker as an "integrated tool."

Integration also comes into play in GIS from a substantially different angle. An information system may be seen as an organized set of tools, data, and people, delivering a service, improving a business process, or providing decision support in cases where the spatial component of the system is prevailing.

The combination of these various needs for integration represents a substantial challenge in the implementation of GIS-based architectures. In the case of sustainable development issues, the acquisition of data and the deployment of tools will need to follow appropriate strategies, often over several years of time, in order to gain an appropriate configuration. Finally, the strategies related to the people involved in the GIS (technicians and end users) are in fact the key element in the whole integration process, involving needs assessment, training, and review of communication flow and organizational aspects.

11.2 WHAT IS A GEOGRAPHIC INFORMATION SYSTEM?

The trend of growing awareness and concern toward the environment and the need to make more rational choices to deal with environmental issues have gradually brought geographic information — generally intended as "information which can be related to a location on the Earth, particularly information on natural phenomena, cultural, and human resources" [1] — to be considered as the basis for an improved understanding of many of the environmental problems afflicting our planet.

In parallel, the complexity and heterogeneity of geographic information, as represented by maps, remote sensing imagery, monitoring network logs, socioeconomic data, etc., have required the development of adequate technologies for its representation, processing, and management.

Geographic information systems (GIS) have progressed, since the 1960s, in response to these needs and in tandem with related technologies and sciences, such as database management systems, scientific visualization, image processing, geostatistics, and remote sensing [2]. To give an idea of this growth, it may be interesting to note that efforts in assessing the number of GIS packages available reported circa 90 different GIS circa twenty years ago [3], to over double this number [4] fifteen years later. As an evolution of cartography-based applications, geographic information systems have assumed a role of primary relevance in domains such as natural resource management and land planning. At the same time, GI technologies have also flourished within other substantially different domains, such as facility management, tourism, telecommunications, transport, and archaeology. Corbin [5] provides an extremely intuitive exemplification of the role and importance of geographic information in various fields of interest.

To understand the role of GIS technologies in the context of sustainable development, a general definition of a geographic information system, highlighting its key characteristics, can provide a useful starting point.

In the various definitions of GIS given in the literature, common features that emerge are those of the interaction between groups of users (either represented by analysts or policy makers), the importance of data, and the availability of an array of analysis and decision-support tools. Given the intrinsic multidisciplinarity with which GIS has evolved, several definitions, stemming from different perspectives, have been proposed.

We may encounter definitions emphasizing a functional flow in the use of spatial information, such as "a system for capturing, storing, checking, manipulating, analyzing and displaying data which are spatially referenced to the Earth"[6]. Other definitions follow a focus on data and content: for example, GIS is proposed as "an information system that is designed to work with data referenced by spatial or geographic coordinates. In other words, a GIS is both a database system with specific capabilities for spatially-referenced data, as well as a set of operations for working with the data" [7].

Finally, we may find definitions adding to data and tools a "human resources" component. Dueker and Kjerne [8] identify a GIS as "a system of hardware, software, data, people, organizations, and institutional arrangements for collecting, storing, analyzing, and disseminating information about areas of the Earth." Chrisman [9], defines GIS as an "organized activity by which people measure and represent geographic phenomena then transform these representations into other forms while interacting with social structures."

On one side, so much yearning for a "primary definition" may be interpreted as a merely encyclopedic effort, with little relation to operational issues and practical actions. However, working in the GIS arena we find that such references provide an important minimum common denominator, which can often be the only shared reference between extremely diverse communities of users, concurring with their efforts within GIS initiatives or projects.

11.3 GEOGRAPHIC DATA

Geographic data used in relation to sustainable development applications and studies make it possible to describe the state of the environment and it inhabitants, as well as the distribution and trend of various indicators of pressure and impact on the system. Typically, these data sets are complex, voluminous, and characterized by heterogeneities and discrepancies due to data acquisition policies which, in time, have not always been harmonized across different countries or levels of administration, thus leading to significant resources being devoted primarily to data conversion or reprocessing tasks, rather than actual data analysis.

In the past few years, these issues have been clearly identified and recognized as a priority by the key subjects responsible for the acquisition and management of geographic information, such as national mapping agencies, major Earth-observation data providers, and communities of software developers and users [10].

The adoption of accepted standards to describe content, temporal and spatial reference, and quality specifications of a given dataset (i.e., metadata) should enable providers to produce data sets with known characteristics that may be more easily harmonized across boundaries, spatial scales, and geographic projection, allowing a seamless merging of the most diverse sources of geographic information.

The availability and diffusion of standards for data and metadata represent the basis for the simplification and enhancement of directory and data retrieval systems which are fundamental for the exchange of information between the numerous players involved in the analysis and definition of policies related to sustainable development problems.

The scenario that is envisioned in the evolution of GIS is strongly influenced by the development of increasingly powerful information and communication technologies [11], leading to the creation of spatial data infrastructures. These are expected to enable unprecedented levels of geographic data and knowledge sharing, with positive impact both on the degree of understanding of large-scale environmental issues, and on the economic resources required to manage these data sets and processes. The spatial data infrastructure concept, and the vision underlying its development, is in fact shared by the major institutional actors globally involved in the production of geographic information and in the adoption of policies involving spatial information for decision support [12–14].

11.4 TOOLS

The data acquired and maintained by various information providers, such as national mapping agencies, census bodies, imaging sensor data providers, or local administrations, need an adequate set of tools to enable processing, analysis, and ultimately, decision support based on the derived information.

The tools available in a geographic information system may be classified into three broad categories, covering distinct domains of functionality: editing, analysis, and visualization [15].

Editing functions allow various forms of data capture, coordinate transformation, topology corrections, and other basic operations related to data acquisition and maintenance.

Analysis functions represent the arena where the power of GIS really comes into play, allowing calculations and data-merging operations that cannot be undertaken otherwise. For example, if we take tabular data referring to county boundaries, with information on population and per capita production of waste, a simple spreadsheet allows us to derive significant statistics and perform ranking or charting operations (e.g., "which are the ten most populated counties?" or "which produce more waste?"). With the aid of a basic GIS package, the same information may be linked to a geographic layer with the corresponding administrative boundaries, making it immediately available to perform complex spatial queries. These can highlight not only the ranking among counties in relation to one of the above-mentioned variables, but also to identify clustering of population or waste production in specific parts of the region, or other spatial patterns.

Visualization functions, albeit implicit in GIS software packages — which are designed primarily to convey information by displaying map-based data — boast an impressive combination of tools, covering two- and three-dimensional visualization or animation. In addition, far from being an automated process, the choice of appropriate visualization parameters (such as graphic layout, thematic legends, or scale) is key to insuring that the information derived through complex analyses is not "filtered" or biased at the visualization stage.

The different levels of functionality described above are generally made available to the end user in the form of software packages that will be installed on the user's computers (be it in a corporate client-server environment or in a "home setting" with a single PC), and will be integrated with other software tools such as database management applications, image processing tools, and in general, any application that may be dealing with spatially related information. Compared to the picture outlined in broad GIS application surveys mentioned in the introduction, more recent trends, which may be observed by monitoring announcements on dedicated mailing lists, Web sites, and actual market offerings in GIS-related exhibitions and presentations, are showing a pattern which is consistent with the development of information technologies in other domains of application. Next to a consolidated diffusion of GIS application suites proposed by a few major software houses, capable of covering any GIS-related need, we have two lines of interest. On one side we see the persistent presence of packages that are minor in terms of market share, but turn out to be effective (and often economically more convenient) to address specific needs for technical users.

Another increasingly important way to deploy GIS functionality within software applications is represented by web services [16]. Such an approach removes the need for the end user to install and maintain complex software and data sets, which are instead accessed via a standard web browser. These applications, in some cases, are not even perceived by the end user as part of a geographic information system, as may be seen in examples like the numerous web sites offering the on-line determination of travel itineraries, applying GIS routing algorithms to road network models.

11.5 PEOPLE

The three categories of tools described above not only represent a functional classification in the complexity of a GIS platform, but may also be related to specific categories of users involved in a geographic information system. First of all, we may identify a limited set of users dedicated to data creation and editing. We have then a wider group of specialists and technicians who are responsible for the analysis and use of the geographic information provided by the "creators" in a wide array of application domains. An image generated by an earth-observation satellite (covering an area in the range of several tens or hundreds of square kilometers), following the first-level processing conducted by the provider of the data, will then be requested and utilized by different analysts; land planning experts, ecologists, and utility managers may easily be using the same starting image to derive information functional to their specific domain of interest.

Last but not least, we find the end users, either represented by decision-makers or simply "data viewers." These users do not need to have any specific knowledge of the geographic data processing, nor of the intricacies of spatial analysis, and will take advantage of the result of the efforts of the creators and analysts in the form of map-based results or other types of reporting.

In an environment where the technologic drive is extremely strong, as may be seen with the frequency of introduction on the market of new GIS applications, package updates, and proposals for IT innovation in business processes, people remain the key component to be considered in any GIS-related architecture.

This may seem a trivial consideration and will be shared by any end user, technician, or scientist involved in GIS efforts. However, practical experience demonstrates that a daily recall must be made during any project with a relevant GIS component. Furthermore, this "awareness maintenance" activity will often need to be structured and budgeted adequately within the activity, by means of meetings, seminars, training, and other forms of communication [17].

11.6 WHAT ARE MODELS?

Simulation models represent the main tool used by scientists and analysts to investigate the dynamics of a system [18]. For the purpose of a general classification, we may consider three broad and not necessarily distinct categories of model or modeling approach:

1. Empirical and regression models, which express some heuristic or statistical connection between observed phenomena, or between "inputs" and "outputs," without concern for the underlying physics, biology, and chemistry, or "inner workings" of the system
2. Conceptual and analytical approaches, which greatly simplify the underlying physics etc. for reasons such as tractability, simplicity, and convenience
3. Physically based or process-based models, which strive for as complete a description as possible of the underlying physics etc., within a deterministic or stochastic framework and within the limits of the processes and observations of interest

This hierarchy also encompasses to some degree the possible ways in which models can be used: the simplest models of the first category can serve as useful screening tools; conceptual models can generate what-if scenarios; the third class is ideally suited to research and can be used to conduct exploratory simulations to test new hypotheses and parameterizations, for example. In practice, there is a great deal of overlap between the ways different models can be used.

We can include, as a fourth category to the three above, that of combined approaches, in the sense of composite models which use one approach for a given subset of processes or subsystem and another approach for a second subsystem. Process models, based as they are on fundamental governing equations, are the most multipurpose, flexible, and extendable of the approaches, though these comprehensive models are not without their limitations and drawbacks. Chief among these are over-parameterization and uncertainty, in the sense that most models have not been validated in all their detail, owing, in part, to a mismatch between model complexity and the level of data which is available to test and calibrate the models.

Several GIS packages offer simulation capability for simple processes, but specialized models will be required to refine the level of analysis, or may represent consolidated tools, which are actually preexisting to GIS in a given context of application. For this reason, GIS and simulation models are often proposed to the decision-maker as an integrated tool [19].

11.7 HOW GIS AND MODELS INTEGRATE

Understanding the spatiotemporal behavior of environmental processes and state variables at large scales involves the use of many different types of data, obtained from field measurement, remote sensing, digital terrain models, and numerical simulation. In such a context, the support of GIS data processing functions can be particularly valuable for the modeler. For example, the design of a numerical grid representing an aquifer for application of a simulation model for groundwater flow can be automated and more directly linked to the mappable features in the study area. This makes the process more intuitive and relieves the user from tedious and error-prone processing tasks [20], while improving the accuracy in the description of the site under examination. At the same time, undertaking a modeling study in a GIS context provides a basis for the simplification of the interaction between the different players involved (data providers, modelers, and decision-makers), through the establishment of a common data structure, which can be visualized using the same GIS-based visualization tools.

With the complexity of models and the variety and volume of data that needs to be processed in environmental studies, pre- and postprocessing tasks related to modeling efforts rely not just on GIS, but on a host of other software tools such as scientific visualization systems, image processing software, and database management systems. Combining these data, models, and tools into a robust and user-friendly system is a research topic that has seen approaches ranging from so-called "loose" integration to "tight" integration [21–24].

The aim of loose integration is to facilitate the use of different tools within a single application by converting data to the correct data formats (which are tool dependent), and by providing interfaces to guide the user through the different steps that are involved in the overall process. Tight integration, on the other hand, aims at merging different tools in a single package, offering full functionality and interactivity between the utilities originally belonging to separate systems.

While the tight integration approach may deliver full functionality, interactivity, user friendliness, and speed, it also usually requires profound changes in the constituent tools. The key architectural modification concerns the use of a common data model, shared by the different tools embedded in the system, typically GIS-based visualization and data management, and model-based simulation. This is important, since it allows reducing the "overhead" associated with any operation by eliminating the need to convert data from one format to the other, reducing actual processing time and simplifying data management strategies. However, from a cost–benefit analysis perspective, the effort of attaining tight integration may result in excessive implementation time and/or costs.

Another approach to GIS-model integration is to move from modeling linked to GIS to modeling within GIS. Generally, this is achieved by implementing fundamental modeling primitives as intrinsic GIS functions, such as the advection–dispersion equation for groundwater transport, or by characterizing spatial response functions, via time–area diagrams [25]. The outcome of such an approach is clearly dependent on the type of model being considered, and several models have already been successfully integrated within different GIS packages. Some successful examples of integration in the field related to hydrologic and agricultural applications include the DRASTIC methodology for groundwater vulnerability mapping [26], the AGNPS and ANSWERS models for nonpoint source pollution modeling [27], and the BASINS application, an integrated watershed-based modeling system for water quality assessment and analysis of point and nonpoint sources of pollution [28].

Whichever technical solution is adopted to integrate geographic databases, GIS functionality, and ancillary tools with simulation models, the increased potential of the resulting system must be adequately supported in order to avoid improper or inexpert use. Given the greater ease of use insured by user-friendly interfaces and a higher "sense of intuitivity" concerning the information which is being treated, there is, paradoxically, an increasing risk as such tools become more sophisticated. This has led to the suggestion of yet another strategy for GIS-model integration, plausible in situations where the use of sophisticated analytical tools is not warranted by the amount or quality of data available, whereby reliance on quantitative estimates is replaced by a qualitative understanding of the pattern of hydrological response, and simple GIS-based reasoning is used to assist in the decision-making process [29].

11.8 HOW GIS, MODELS, AND PEOPLE INTEGRATE

The overall issue of integration between GIS and models does not *per se* solve the issue of making the resulting set usable to convey information to the public or to decision-makers.

The efforts in integration between the data management capabilities of GIS packages and the data analysis capabilities of models, discussed in the previous sections, represents an important step toward the improved treatment of the diverse sources of information on the state of the environment. At the same time, if we go back to the definition of GIS as a "data-tools-people" compound, we understand that the picture proposed above pertains mainly to the integration of tools (visualization, data management, simulation) and data used by these tools, but may not necessarily bind all of the people involved in the process.

Integration between GIS and models has been — and remains — a complex arena, where environmental modelers and computer scientists need to interact heavily in order to strike a difficult compromise. On one side, we have systems that propose robust and highly complex modeling tools, through maybe poor communication interfaces. On the other side, we have integrated modeling environments boasting extremely complex levels of interactivity, more and more often via web-based platforms, but with a more limited reliability in the results represented.

Independently of the specific integration strategy, the data from environmental observation and the results from simulation models must be presented in a form that is understandable and effective for the policy maker. While this may appear as a truism, the effort to actually convey information from the scientist to the policy maker is often far from a secondary issue in an operational context [30,31].

Simulation models will typically refer environmental parameters or indicators to grid units which are a convenient segmentation of space from a simulation standpoint. However, a policy maker's view of the same problem will normally be referred to administrative units (such as regions or counties) or to yet more complex spatial units, often indicated indirectly. For example, in the context of emergency management planning, it may be practical to require the estimation of availability of a given resource to travel time. In this context, geographic information systems can, in fact, make the difference. Spatial analysis primitives made available in GIS packages — such as overlay, buffering, and map algebra — paired by standard database query languages, allow "translating" data between nonhomogeneous spatial references with great simplicity. This opens the possibility of integrating the knowledge domain of the modeler and the knowledge domain of the policy maker by reprocessing data in the appropriate way, while insuring that these transitions are documented by adequately maintaining metadata information for the data sets that are generated.

Geographic information systems, notwithstanding the highly intuitive nature of map-based representation of information, should not be considered as the sole or dominant tools to be proposed in a decision support system for a policy maker. Rather, GIS should be placed at the same level as other information management tools, together with visualization (in general) and databases [32]. This is a point which appears to be, to a certain extent, neglected in the first approach to GIS in a given organization [33], where users tend to expect map-based representation of results as the natural output for analyses deriving from the system.

Identifying the priority issues raised by the need to transmit information from scientists and analysts to policy makers leads in turn to the definition of strictly technical problems, where GIS play an important role. However, since any tool applied in the definition of a policy acquires a political connotation in itself, we should also consider whether the application of GIS to the policy process differs in this respect from the application of other more "traditional" tools or models, such as those used for economic planning or welfare policy analysis. In this respect, King and Kraemer [34] suggest that geographic information has a greater possibility, compared to other modeling bases, to act as a "boundary object" with a greater acceptance by different parties, while at the same time the very breadth of GIS applicability to policy problems makes it likely that GIS will be drawn into many different kinds of policy debates.

11.9 DISCUSSION AND CONCLUSIONS

The technologies and know-how related to geographic information systems are proving to be an extremely effective combination of tools and skills to acquire, process, analyze, and manage geographic data. The integration of GI technologies with modeling makes available to the decision-maker an unprecedented combination of resources to improve the understanding and insight in relation to sustainable development problems.

To further increase the potential in the hands of researchers, technicians, and end users, communication technologies have substantially reduced the time, and often the costs, of transferring information between different subjects.

Still, significant challenges currently exist in the GIS and modeling arena. For example, we can mention the difficulty of deploying efficient data and metadata management practice within large-scale projects in operational contexts, or the issues encountered in the communication between scientists and end users in the appreciation of the results of a modeling simulation.

While these aspects tend to be seen as technicalities to many of the actors involved in sustainable development initiatives, they represent key factors to be considered for the success of long-term projects involving conspicuous amounts of information and large groups of users.

This issue is quite understandable in cases like studies on greenhouse gas emissions, or water resource management in large, transboundary catchments. However, it is also interesting to note how, in a globalized context, even apparently local-scale problems may warrant a complex approach in terms of combination of tools, information sources, and people required to address the issue. We may take, for example, the remediation of an industrial facility with contaminated soil; it is not infrequent to have situations where the facility, its owner, and the subjects called to analyze the site and propose a remediation strategy belong to different countries. In this case, the geographic location of the site may bring to confrontation (and require integration of) data standards, modeling approaches, regulatory settings, and policy practices deriving from different nations. To further enrich the picture, GIS and modeling tools will easily be at different stages of actual diffusion in the countries to which the different actors belong. For example, data used by the company owning the facility to describe key information on the site will be stored in an environmental information system which is compliant with data standards derived from the home country of the owning firm. This is perfectly consistent to insure an efficient view of information across sites in different countries from a corporate viewpoint, but may often pose data conversion or reprocessing issues in communicating the same information to local environmental authorities.

It should be noted that such issues will not by default translate into problems or potential for failure, if they are evaluated in the scoping phase of a project and are accounted for in the life cycle of the site remediation program. Insuring the resolution of these types of initiatives consistently, in cases of growing complexity, and in different regions of the world represents the challenge GIS and modeling specialists are facing, and will need to master, in order to demonstrate the potential of these scientific and technologic tools in support of sustainable development.

REFERENCES

1. Association for Geographic Information (AGI), GIS Dictionary, A Standards Committee Publication of the Association of Geographic Information, Version 1.1, STA/06/91. UK, 1991. http://www.geo.ed.ac.uk/agidict/welcome.html.
2. Ehlers, M., Edwards, G., and Bedard, Y., Integration of remote sensing with geographic information systems: a necessary evolution, *Photogrammetric Eng. Remote Sensing*, 55(11), 1619–1627, 1989.

3. Smith, D.R., Selecting a turn-key geographic information system, *Comput. Environ. Urban Syst.*, 7, 335–345, 1982.

4. United Nations Environment Programme, A survey of spatial data handling technologies 1997: Environment Information and Assessment Technical Report UNEP/DEIA/TR.97.13, 1998.

5. Corbin, C., Picture Book, D3.6.6, Ginie Project, European Umbrella Organisation for Geographic Information (EUROGI), 2004. Available online at http://www.lmu.jrc.it/ginie/doc/d366_picturebook_v1.pdf.

6. Department of the Environment, Handling Geographic Information, Technical report, Her Majesty's Stationary Office, 1987.

7. Star, J. and Estes, J., *Geographic Information Systems: An Introduction,* Prentice Hall, Englewood Cliffs, NJ, 1990.

8. Dueker, K. and Kjerne, D., Multipurpose Cadastre: Terms and Definitions, Bethesda, MD, *Technical Papers, 1989 ACSM-ASPRS Annual Convention*, Vol. 5, 1989, pp. 94–103.

9. Chrisman, N.R., What does "GIS" mean? *Trans. GIS*, 3(2), 175–186, 1999.

10. OGC Technical Committee, *The OpenGIS Guide,* The OpenGIS Consortium, Wayland, MA, 1998.

11. European Commission, A Strategic View of GIS Research and Technology Development for Europe, technical report, EUR 18126, 1998.

12. INSPIRE Environmental Thematic Coordination Group, Lillethun, A., Ed., Environmental Thematic User Needs Position Paper, Version 2, European Environmental Agency, 2002.

13. DeMulder, M.L., DeLoatch, I., Garie, H., Ryan, B.J., and Siderelis, K., A Clear Vision of the NSDI, http://www.geospatial-online.com/geospatialsolutions/article/articleDetail.jsp?id=89953, 2004.

14. Hall, M., Spatial Data Infrastructures in Australia, Canada and the United States, K.U. Leuven, 2003. Available online.

15. Longley, P.A., Goodchild, M.F., Maguire, D.J. and Rhind, D.W., *Geographic Information Systems and Science*, Wiley & Sons, New York, 2002.

16. Landgraf, G., Evolution of EO/GIS interoperability towards an integrated application infrastructure, in *Proceedings of the Second International Conference on Interoperating Geographic Information Systems*, Springer, Zurich, Switzerland, 1999, pp. 29–40.

17. Smith, D.A. and Tomlinson, R.F., Assessing costs and benefits of geographical information systems: methodological and implementation issues, *Int. J. Geogr. Inf. Syst.*, 6 (3), 247–256, 1992.

18. Oreskes, N., Schrader-Frechette, K. and Belitz, K., Verification, validation, and confirmation of numerical models in the Earth sciences, *Science*, 263, 641–646, 1994.

19. Goodchild et al., Eds., *GIS and Environmental Modeling: Progress and Research Issues*, GIS World Books, Fort Collins, CO, 1996.

20. Kuniansky, E.L. and Lowther, R.A., Finite-element mesh generation from mappable features, *Int. J. Geogr. Inf. Syst.*, 7(5), 395–405, 1993.

21. Batty, M. and Xie, Y., Modelling inside GIS: Part 1. Model structures, exploratory spatial data analysis and aggregation, *Int. J. Geogr. Inf. Syst.*, 8(3), 291–307, 1994.

22. Livingstone, D. and Raper, J., Modelling environmental systems with GIS: Theoretical barriers to progress, in *Innovations in GIS*, Worboys, M. F., Ed., Taylor & Francis, London, 1994, pp. 229–240.

23. Nyerges, T.L., Understanding the scope of GIS: its relationship to environmental modelling, in *Innovations in GIS*, Worboys, M.F., Ed., Taylor & Francis, London, 1994, pp. 75–93.

24. Paniconi, C., Kleinfeldt, S., Deckmyn, J. and Giacomelli, A., Integrating GIS and data visualization tools for distributed hydrologic modelling, *Trans. GIS*, 3(2), 97–118, 1999.

25. Maidment, D.R., Environmental modelling within GIS, in *GIS and Environmental Modelling: Progress and Research Issues*, Goodchild, M.F. et al., Eds., GIS World Books, Fort Collins, CO, 1996, pp. 315–323.

26. Merchant, J.W., GIS-based groundwater pollution hazard assessment: A critical review of the DRASTIC model, *Photogrammetric Eng. Remote Sensing*, 60(9), 1117–1127, 1994.

27. Wilson, J.P., GIS-based land surface/subsurface modelling: new potential for new models? in *Proceedings Third International NCGIA Conference on Integrating GIS and Environmental Modelling*, CD-ROM, NCGIA, Santa Barbara, CA, 1996.

28. EPA, BASINS Version 2.0 User's Manual: Better Assessment Science Integrating Point and Nonpoint Sources. U.S. Environmental Protection Agency, Washington, DC, 1998.

29. Grayson, R.B., Blöschl, G., Barling, R.D. and Moore, I.D., Process, scale and constraints to hydrological modelling in GIS, in *HydroGIS 93: Application of Geographic Information Systems in Hydrology and Water Resources*, IAHS, Wien, 1993, pp. 83–92.

30. Davies, C., and Medyckyj-Scott, D., GIS usability: recommendations based on the user's view, *Int. J. Geogr. Inf. Syst.*, 8 (2), 175–189, 1994.

31. Reeve, D. and Petch, J., *GIS Organisations and People: A Socio-Technical Approach*, Taylor & Francis, London, 1999.

32. Peirce, M., Computer-based models in integrated environmental assessment. Technical Report AEAT-1987, AEA Technology, 1998.

33. Hendricks, P.H.J., Information strategies for geographical information systems, *Int. J. Geogr. Inf. Sci.*, 12, 6, 621–639, 1998.

34. King, J.L. and Kraemer, K.L., Models, facts, and the policy process: the political ecology of estimated truth, in *Environmental Modelling with GIS*, Goodchild, M.F., Parks, B.O. and Steyaert, L.T., Eds., Oxford University Press, Oxford, UK, 1993, pp. 353–360.

12 Microsimulation and GIS for Spatial Decision-Making

Dimitris Ballas

CONTENTS

12.1 INTRODUCTION

Simulation-based spatial modeling is an expanding area of research that has a lot of potential for the evaluation of the socioeconomic and spatial effects of major developments in the regional or local economy. Among the most crucial issues that concern policy makers is the prediction of the effect of alternative policies in the short term, as well as in the long term. There is a long history of modeling work in geography and regional science that focuses on the assessment of the various short- and long-term effects of major socioeconomic regional or local developments. As Wilson [1] points out, cities and regions are extremely complicated and can be seen as complex spatial systems. A regional socioeconomic system comprises several subsystems that interact with each other through socioeconomic and spatial mechanisms [2,3]. In addition, components of cities and regions are systems that involve a large number of interacting components [1]. It can be argued that the main subsystems that make up an urban economy are as follows [2,3]:

- Housing market
- Labor market
- Service sector
- Land market
- Transport

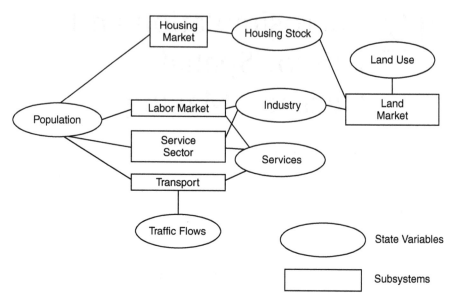

FIGURE 12.1 Main subsystems and variables of an urban system. (From Bertuglia et al., 1987. With permission.)

Further, Tadei and Williams [3] point out that the main variables for describing the structure of an urban system are as follows:

- Population
- Housing stock
- Industry (economic base)
- Services
- Land use
- Traffic flows

Similarly, Wilson [1] argues that the basic entities with which urban and regional analysts deal can be identified broadly as follows:

- People
- Organizations
- Commodities, goods and services
- Land
- Physical structures and facilities

Figure 12.1 depicts schematically the main subsystems and variables of an urban system and their interrelations.

Geographers and regional scientists have traditionally been involved in building models that investigate and analyze the role that these entities play in urban and regional systems. Recently, these modeling efforts have been aided by an accelerating growth in the volume, variety, power, and sophistication of the computer-based tools and methods available to support urban and regional analysis and policymaking. New developments in hardware and software systems have enabled significant advances to be made in the storage, retrieval, processing, and presentation of spatially referenced data [4]. There has also been significant progress in the development of geographical information systems (GIS) for socioeconomic applications [4–6]. Further, there has been an increasing availability of a wide range of new data sources in both the public and private sectors and an increased power and portability of personal computers bringing high-quality computer graphics and presentation facilities [7,8].

In this new environment, there have been many spatial models that have shed new light on patterns and flows within cities and regions. These models, when combined with relevant performance indicators, have been very useful in measuring the quality of life for residents in different localities [8,9]. However, relatively little is known about the interdependencies between household structure or type and their lifestyles, including the events they routinely participate in and hence their ability to raise and spend various types of income and wealth. The modeling of interdependencies requires a different level of urban and regional system representation. As Batty [10] points out: "To move to better dynamic representations of urban processes suggests that individuals rather than groups or aggregates must form the elemental basis of these simulations" [10, p. 26].

In this context, spatial microsimulation offers a potentially powerful framework for the representation of the urban and regional systems and entities outlined above. It can be argued that there is an increasing need for such a framework, given the current trends in socioeconomic polarization and inequalities between and within cities. It has been argued elsewhere (see, for instance, Hutton [11]) that a socioeconomic system that degrades the environment, excludes large sections of the population from secure employment, and causes social disintegration is not sustainable. Spatial microsimulation combined with GIS technologies can be employed to monitor trends in socioeconomic polarization and inequalities and evaluate the sustainability of local socioeconomic systems if past trends were to continue. Further, the major advantages of the spatial microsimulation method concern the ability to address a series of important policy questions, which could aim to ameliorate socioeconomic polarization. The remainder of this chapter introduces spatial microsimulation modeling methodologies and discusses their potential to be combined with GIS frameworks to aid local decision-making.

12.2 SPATIAL MICROSIMULATION METHODOLOGIES

Microsimulation models aim at building large-scale data sets on the attributes of individuals or households (and/or on the attributes of individual firms or organizations) and at analyzing policy impacts on these micro-units [12–14]. Further, by permitting analyses at the level of the individual, family, or household they provide the means of assessing variations in the distributional effects of different policies [15,16]. In addition, microsimulation modeling frameworks provide the possibility of defining the goals of economic and social policy, the instruments employed, and also the structural changes of those affected by socioeconomic policy measures [17]. Microsimulation methodologies have become accepted tools in the evaluation of economic and social policy and in the analysis of tax-benefit options and in other areas of public policy [15].

Various types of microsimulation models can be distinguished. For instance, Mertz [16] classifies them between static models, which are based on simple snapshots of the current circumstances of a sample of the population at any one time, and dynamic models that vary or age the attributes of each micro-unit in a sample to build up a synthetic longitudinal database describing the sample members' lifetimes. Further, in a population analysis context, Falkingham and Lessof [18] distinguish between static models, dynamic population models, and dynamic cohort models. Dynamic population models take a sample of the population as an initial micro database and project the micro-unit attributes forward into time. For instance, each micro-unit of the sample is aged individually by an empirically based survivorship probability [16,18]. Moreover, other demographic events may be simulated on the basis of the existing individual attributes. On the other hand, dynamic cohort models employ the same dynamic procedures as dynamic population models [18]. However, these models are not underpinned by the characteristics of a real sample unit, but rather, the simulation process creates or synthesizes micro-units and forecasts the whole process from birth to death [18].

The first geographical application of micro-analytic simulation or microsimulation was developed by Hägerstrand [19], who employed micro-analytical techniques for the study of spatial diffusion of innovation. Nevertheless, it can be argued that the basis for spatial microsimulation of households and individuals was founded in the 1970s. In particular, Wilson and Pownall [20] were among the first to address the aggregation difficulties that were associated with traditional comprehensive spatial models of urban systems.

Birkin and Clarke [13] built on the original framework that was formulated by Wilson and Pownall [20] and developed one of the first applied spatial microsimulation models, which they called SYNTHESIS. The latter was used to create small-area microdata and generate incomes for individuals. Further, Williamson [21] has also included gross income as a variable in his OLDCARE microsimulation model. The latter was used for the spatial analysis of community care policies for the elderly. In particular, one of the main aims of OLDCARE was to estimate the prevalence and severity of disability amongst the elderly. The model aimed at synthetically reconstructing a micropopulation, and it applied the iterative proportional fitting (IPF) technique on data from the 1981 Census of U.K. population and other sources.

Moreover, Williamson and Voas [22] worked toward providing more robust and reliable estimates of income at the small-area level. They argue that income estimation at the small-area level may be seen as a multilevel analysis problem where variables at individual and area levels may interact.

Nevertheless, it should be noted that the microsimulation modeling effort in the United Kingdom has been focused on modeling earned income rather than income from wealth. In contrast, in the United States there have been efforts to use spatial microsimulation models for the estimation and analysis of the distribution of wealth [23]. In particular, Caldwell and Keister [23] present CORSIM, which is a dynamic microsimulation model that has been under development at Cornell University since

1986. CORSIM has been used to model wealth distribution in the United States over the historical period 1960 to 1995 and to forecast wealth distribution over the future [23]. It is noteworthy that over 17 different national microdata files have been used to build the model, which incorporated 50 economic, demographic and social processes by means of approximately 900 stochastic equations and rule-based algorithms [23]. Furthermore, Caldwell et al. [24] review the geography of wealth and show how CORSIM has included many variables relating to assets and debts, and stress the importance and policy-relevance of detailed spatial estimates of household wealth. They also point out that spatial microsimulation frameworks can be used to shed new light on the local impacts of major national policy changes.

Wegener and Spiekermann [25] explore the potential of microsimulation for urban models, focusing on land-use and travel models. They argue that a new generation of travel models has emerged which requires more detailed information on household demographics and employment characteristics at the small-area level. They also point out that there are new neighborhood-scale transport policies aimed at promoting public transport, walking, and cycling. These policies require detailed information on the precise location of the population and its activities. Wegener and Spiekermann [25] also stress the need for urban models to predict not only the economic but also the environmental impacts of land-use transport policies. In order to model the environmental impacts, there is a need for small-area forecasts of emissions from stationary and mobile sources as well as of emissions in terms of the affected population. After outlining the main characteristics of a microanalytic theory of urban change, Wegener and Spiekermann [25] report on modeling efforts carried out at the University of Dortmund to integrate microsimulation into a comprehensive urban land-use transport model.

Another spatial microsimulation example in a transport policy context is the work of Veldhuisen et al. [26]. They present RAMBLAS, which is a regional planning model for the Eindhoven region in the Netherlands and is based on the microsimulation of daily activity patterns. In the context of this model, daily activity patterns are used as a basis for predicting the spatial distribution of the demand for various transport services in the urban system. Microsimulation techniques are employed to predict traffic flows and the demand for transport services. The main aim of RAMBLAS is to estimate the intended and unintended consequences of planning decisions related to land use, building programs, and road construction for households and firms [26]. Microsimulation methodologies were used to predict which activities will be conducted where, by whom, when, and for how long, the transport mode involved, and which route is chosen to implement the activities.

A recent example of comprehensive spatial microsimulation modeling is the work of the Spatial Modeling Centre in Sweden [27]. Vencatasawmy et al. [27] built on previous microsimulation modeling efforts to construct SVERIGE (System for Visualizing Economic and Regional Influences Governing the Environment), which is a national spatial microsimulation model for Sweden. SVERIGE is aimed at studying the spatial consequences of various national, regional, and local-level public policies. The database used for this model comprises longitudinal socioeconomic information on every resident of Sweden for the years 1985 to 1995.

Bearing in mind the various classifications, an integrated definition to micro-simulation modeling may be given. In particular, it can be argued that the micro-simulation method typically involves four major procedures:

1. The construction of a microdata set (when this is not available)
2. Monte Carlo sampling from this data-set to "create" a micro-level population
3. What-if simulations, in which the impacts of alternative policy scenarios on the population are estimated
4. Dynamic modeling to update a basic microdata set

Although the number of published microdata sets is increasing around the world, there are still many cases in which this kind of data is not available at the desired spatial scale [13,14,28]. In the case of the United Kingdom, a very important source of microdata is the Samples of Anonymized Records (SARs). These are samples of individual census records that are anonymized in various ways ensuring that there is no breach of the confidentiality of the census and that no individual can be identified from the data [29,30]. However, the spatial scale at which these data sets are released is, at best, the regional or district scale. In cases when official microdata are not available at the desired spatial scale, this kind of data can be estimated with the use of existing data sets and a variety of techniques ranging from iterative proportional fitting methods to linear programming and complex combinatorial optimization methods [28,31]. These techniques make up stage 1 of the microsimulation procedure outlined above.

The process of recreating a spatially disaggregated microdata set using existing data that are available at more aggregate spatial scales can be clarified with use of an illustrative example. Let us assume that there is a need to investigate the relationships between sex (S), age (A), number of dependent children (DC), tenure (T), educational qualifications (Q), economic position (EP), and socioeconomic group (SEG) for a given population group X in location i. It is possible to obtain from data sources, such as the 1991 Census of the U.K. population for the residents of a specified area (e.g., at the enumeration district level), separate tabulations of:

- Sex by age by economic position (Small Area Statistics table 08)
- Tenure by number of dependent children (Small Area Statistics table 46)
- Level of qualifications by sex (Small Area Statistics table 84)
- Socioeconomic group by economic position (Small Area Statistics table 92)

From these tabulations, the respective conditional probabilities can be calculated, and then the problem would be to estimate the probability:

- $p(xi,S,A,T,DC,Q,EP,SEG)$

given a set of constraints or known probabilities:

- $p(xi,S,A,EP)$
- $p(xi, DC, T)$

- p(xi,Q,S)
- p(xi,SEG,EP)

There are a number of ways to solve this problem, such as linear programming models, discrete choice models, balancing factor methods in spatial interaction models, and iterative proportional fitting techniques [14,31,32].

Spatial microsimulation models are becoming increasingly policy relevant and are being used for local and regional policy analysis. The next section demonstrates the policy relevance of spatial microsimulation by showing how SimLeeds, which is a spatial microsimulation model for the city of Leeds, has been used for socio-economic impact assessment and social policy analysis at the small-area level.

12.3 HOW GIS AND SPATIAL MICROSIMULATION CAN BE USED FOR DECISION SUPPORT

It has long been argued that it is very important for national and local governments to be able to predict the effects of policy changes upon local incomes and employment. Nevertheless, traditional modeling tools used by economic geographers and regional scientists have focused on answering questions relating to policy impacts upon regions, rather than cities and smaller areas. Clearly, the regional or city-wide impact analysis of regional or budget policy changes is an extremely useful task with very important political implications, and it is vital for the formation of appropriate regional or urban policies aimed at tackling or ameliorating regional employment and growth problems. Nevertheless, it should be noted that cities and regions are composed of smaller areas, which differ considerably in population size and demographic structure, labor force skills, income and spending power, consumption patterns, accessibility to workplaces, and many other important characteristics [33,34]. Furthermore, there are considerable variations in the household mix of smaller city areas, such as census wards, because these contain even smaller areas such as census enumeration districts. It would therefore be particularly useful if policy makers could predict the intraurban and intraward impacts of regional and urban policy changes. In other words, it would be important for policymaking purposes if the following questions could be answered:

- Which neighborhoods will suffer/benefit most from a policy change?
- Which households will be most affected?
- What will be the intraregion, intraurban and intraward impact of a possible plant closure/development?

Spatial microsimulation models can be employed to estimate such impacts. For instance, SimLeeds was a spatial microsimulation model developed in the Java programming language in the context of doctoral research [33], which implemented different approaches to conditional probability analysis for microsimulation modeling in order to create a small-area microdata population for the city of Leeds. The SimLeeds model outputs were then imported, analyzed and mapped with the use of

TABLE 12.1
A Selection of *SimLeeds* Variables

Micro-Unit Attributes

Location (place of residence) at the ED level
Location (workplace) at the ward level
Age
Sex
Marital status
Tenure
Employment status
Industry (SIC)
Socioeconomic group
Earned income
Job seekers allowance (JSA)

proprietary GIS software. The SimLeeds model initially focused on a relatively small number of individual attributes such as those described in Table 12.1.

It should be noted that the variables described in Table 12.1 were available at the smallest area level for which census data were available in the United Kingdom (enumeration district level). This spatially detailed information can be extremely useful for local planners and policy decision-makers, because it can be employed to perform spatial policy analysis. The potential of spatial microsimulation models for such policy analysis has been illustrated by the estimation of the spatial impacts of an engineering plant in Leeds [33,34]. It was assumed that the hypothetical plant was located at the area of Seacroft in East Leeds (see Figure 12.2). Figure 12.3 depicts the observed travel-to-work flows to Seacroft. As can be seen, Seacroft, Whinmoor, and Halton are the wards with the largest numbers of individuals who work in Seacroft. In particular, 20% of the people who work in Seacroft live in Seacroft. In addition, 17.8% of the people who commute to Seacroft come from Whinmoor, and 8.7% come from Halton. Therefore, it can be reasonably expected that these wards will be the most affected in the event of a plant closure in Seacroft.

Nevertheless, it should be noted that the extent of the impact on these wards depends on their socioeconomic and demographic structure. Further, there is considerable variation in the socioeconomic mixture of the population within wards. In particular, wards are made up of enumeration districts (EDs), which are small areas with an average of 200 inhabitant households. SimLeeds has been used to provide useful insights into the analysis of the intraward impacts of the plant closure. For instance, Figure 12.4 depicts the estimated spatial distribution for the whole Leeds Metropolitan District of male employees, aged between 16 and 29, who work in the Metal Goods, Engineering, Vehicles industry (SIC3) and are skilled manual workers, live in owner-occupied housing, and whose marital status is single, widowed, or divorced. Respectively, Figure 12.5 depicts the same spatial distribution, focusing on the wards of Seacroft, Whinmoor and Halton, which, as seen above, have the highest numbers of economically active population that works in Seacroft.

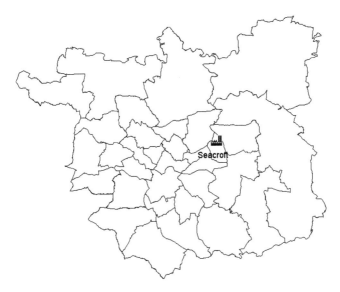

FIGURE 12.2 Location of the hypothetical plant. (From Ballas and Clarke, 2001, p. 302. With permission.)

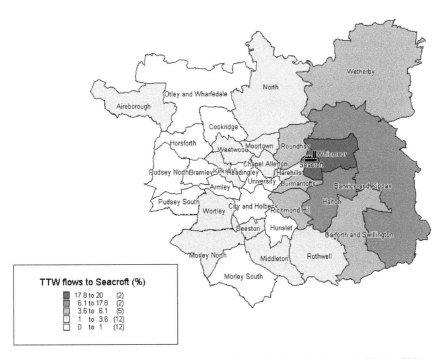

FIGURE 12.3 Travel to Work (TTW) flows to Seacroft. (From Ballas and Clarke, 2001, p. 303. With permission.)

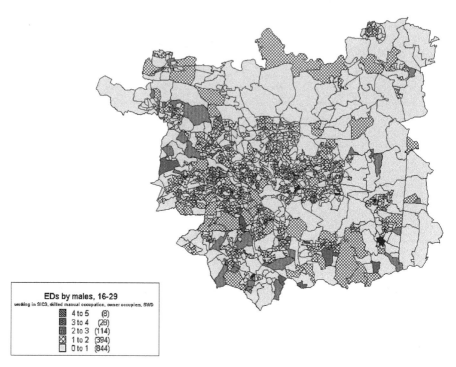

FIGURE 12.4 Estimated spatial distribution of estimated spatial distribution of SWD male employees, aged between 16 and 29, who work in the Metal Goods, Engineering, Vehicles industry (SIC3) and are skilled manual workers and owner-occupiers

These maps are typical products of a spatial microsimulation model: the ability to produce estimates of population groups, which are not available directly from published sources.

In the context of the impact analysis presented here, it was assumed that the hypothetical plant had 3000 employees and that the plant's workforce structure is as depicted in Figure 12.6.

As can be seen, most of the plant's workforce belongs to the Skilled Manual category (45% of the total workforce) and to the Managerial and Technical category (25% of the total workforce). Figure 12.7 shows the estimated spatial distribution of the plant's varied workforce at the ward level. For instance, 5% of the 3000 employees of the plants (i.e., 150 employees) belong to the Professionals etc. group. These 150 employees were distributed first to different wards and then to different EDs on the basis of the actual travel-to-work flows at the ward level and of the estimated spatial distribution of Professionals at the ED level.

In the context of this example, it is possible to estimate the overall initial income loss from the plant closure on the basis of the estimated earned income of the plant employees. Ballas et al. [35] built on this work to show how it is possible to estimate the spatial distribution of estimated income loss in various localities around Leeds (see Figure 12.7).

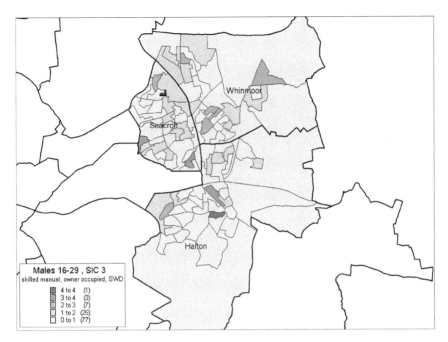

FIGURE 12.5 Spatial distribution of spatially disaggregated microgroup around Seacroft. (From Ballas and Clarke, 2001, p. 304. With permission.)

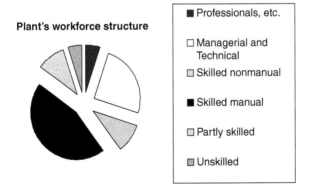

FIGURE 12.6 The plant's workforce. (From Ballas and Clarke, 2001, p. 304. With permission.)

The information presented above can be very useful for planners and local policy makers, as it can enable them to foresee potential problems that could be caused by major changes in the local economies (such as plant closures, redundancies, etc.). Spatial microsimulation and GIS can also be used to estimate the local impacts of social policies that could target groups affected from events such as plant closures. For instance, if policy makers wanted to assist the families of unskilled manual

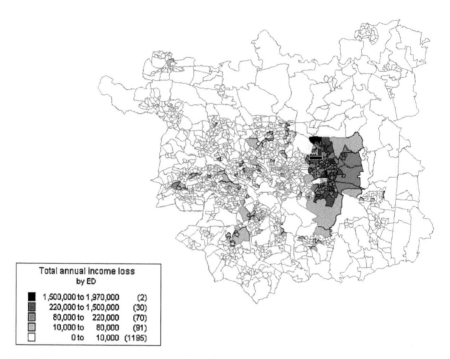

Total annual income loss
by ED

■	1,500,000 to 1,970,000	(2)
▨	220,000 to 1,500,000	(30)
▦	80,000 to 220,000	(70)
▢	10,000 to 80,000	(91)
□	0 to 10,000	(1195)

FIGURE 12.7 Estimated spatial distribution of annual income loss localities around Leeds. (From Ballas et al., 2002, p. 20. With permission.)

workers who lost their jobs at the plant they could formulate a policy aimed at increasing their income in some way. This could, for instance, be done by altering existing social policies, such as the child benefit. For example, if it were assumed that the child benefit for unskilled workers for the eldest eligible child was increased to £20 per week, then it would be possible to predict the likely spatial impact of this change. Figure 12.8 depicts the estimated spatial distribution of the increase in benefits that would result from such a policy change. This kind of modeling can provide insights into the design of social policies so that they can take the spatial implications into account (for a more detailed discussion of this kind of policy analysis see Ballas and Clarke [36]). In this context, social policies can be the alternative to traditional area-based policies. In particular, if the national government or local authority decides that it wants to help specific localities within the city, this can be done with spatially oriented social policies and programs rather that with traditional urban regeneration schemes and other area-based policies.

One of the major advantages of spatial microsimulation frameworks is that they can also be used to analyze social policy in a geographically oriented proactive fashion. For instance, spatial microsimulation can be employed to identify deprived localities in which poor individuals and households are overrepresented. Spatial microsimulation modeling can then be used to answer questions such as: "What social policy should be applied in order to improve the quality of life of residents

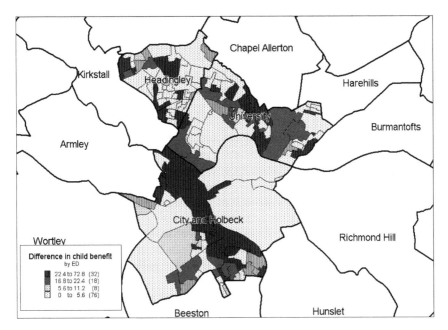

FIGURE 12.8 Spatial distribution of the Child Benefit change under a hypothetical policy change. (From Ballas and Clarke, 2001, p. 602. With permission.)

in the inner-city localities of Burmantofts in Leeds?" In other words, new social policies can be formulated on the basis of spatial microsimulation modeling outputs. These spatially oriented social policies can be seen as a substitute or an alternative to traditional area-based policies, and direct comparisons of their efficiency and effectiveness can be made.

GIS and spatial microsimulation models can also play a very important role in the ongoing debates on the role of potential of new technologies to promote local democracy and electronic decision-making. It can be argued that a model such as SimLeeds, developed in JAVA, which is a platform-independent programming language, can be put on the World Wide Web and linked to virtual decision-making environments (VDMEs). The latter are Internet World Wide Web–based systems that allow the general public to explore "real world" problems and become more involved in the public participation processes of the planning system [37–39]. Models such as SimLeeds can be used not only to provide information on the possible consequences and the local multiplier effects of major policy changes but also to inform the general public about these and to enhance, in this way, the public participation in policymaking procedures. For instance, Ballas et al. [40] present work funded by the Leeds City Council, which aims at extending the SimLeeds model by adding stand-alone mapping capabilities (using GeoTools technologies) and making it more user-friendly with the incorporation of graphical user interfaces (GUIs). The spatial decision support system that will result from this work will enable Leeds local policy makers to utilize the SimLeeds model, which was briefly described in this chapter.

12.4 CONCLUSION

This chapter argues that GIS, when combined with spatial microsimulation models, has a great potential for applied policy analysis. Spatial microsimulation is an important methodology in geography and regional science for examining the impacts of new or revised area- or social-based policies. This chapter reviewed work that illustrated how spatial microsimulation can be used for socioeconomic impact assessment at the microscale. It should be noted that the impact assessment of a hypothetical plant closure that was presented can be repeated for real case studies. New economic developments or disinvestments in the local labor market can be evaluated. Further, the spatial microsimulation approach to social policy evaluation that has been demonstrated in this chapter provides new possibilities for policy makers, who can evaluate the socioeconomic as well as the spatial impacts of proposed policy changes. Clearly, there is further potential to estimate a wide range of policies at the individual and household level and at different scales. Spatial microsimulation frameworks offer the possibility of analyzing systematically the impact that future government budgets and national local labor market policies at any geographical scale deem desirable.

However, it can be argued that just a fraction of the possible uses of spatial microsimulation in a labor market context has been shown here. As Birkin et al. [11] point out, the ongoing computational advances offer the enabling environment for large-scale microsimulations of all the subsystems that make up the economies of cities and regions. New developments in hardware and software, as well as increased data availability, offer the potential for modeling the interactions between all the entities and urban subsystems that have been outlined in the introduction of this chapter. In particular, models such as SimLeeds can be further developed in order to include more household variables and improved estimates of income and wealth and incorporate all the urban subsystems that make up the local socioeconomic structure (i.e., education, health provision etc.). Moreover, dynamic modeling procedures can be incorporated in order to render models such as SimLeeds capable of performing sophisticated what-if local multiplier analysis of different social policy initiatives.

Nevertheless, it should be noted that caution is necessary when using spatial microsimulation methodologies to perform what-if policy analysis and evaluation. The output of all microsimulation models, no matter how good, is always simulated and not actual data. The validity of the simulated data will always depend on the quality of the original data that are used and on the assumptions upon which the microsimulation model is based. Moreover, it will depend on the specific microsimulation methodology that is employed. It should also be noted that methodological elements depend on factors such as time constraints and available hardware. In addition, spatial microsimulation outputs generally depend on subjective judgments associated with the ordering of the conditional probability tables that are used as inputs and/or with the selection of the data sets that are used as small-area constraints. As Birkin and Clarke [8] point out, the modeler's art in microsimulation is to generate population characteristics in an appropriate order so that potential errors are minimized. These aspects should always be taken into account when using spatial microsimulation models for policy impact assessment.

It should be noted that further research is required in order to improve the performance of spatial microsimulation models such as SimLeeds and to highlight the sources of error. For instance, as Williamson et al. [41] point out, there are many ways in which combinatorial optimization methodologies can be fine-tuned, through the evaluation of the use of more or different census small area statistics (SAS) tables, or by changing the model parameters. Further, there is a need to build on existing work on the validity and reliability of microsimulation models.

Despite the data problems and the difficulties with calibration, spatial microsimulation is a methodology that is becoming increasingly popular. Williamson [41] examined the question of whether microsimulation is an idea whose time has come. It is hoped that the work presented here adds more to the argument of those who believe that the answer to this question is positive! As Clarke [14] points out:

> The era of a computational urban or regional geography based on the behaviour of individual households or firms not only provides a stimulating and exciting prospect for the years ahead but it is one which is now realistic and achievable. Clarke [14, p. 202]

ACKNOWLEDGMENTS

The research that led to the development of the SimLeeds model was funded by the Greek State Scholarship Foundation (SSF). The Census Small Area Statistics used are provided through the Census Dissemination Unit of the University of Manchester, with the support of the ESRC / JISC / DENI 1991 Census of Population Programme. The Census Sample of Anonymised Records are provided through the Census Microdata Unit of the University of Manchester, with the support of the ESRC / JISC / DENI. All Census data are Crown Copyright. All digitized boundary data are Crown and ED-LINE copyright.

REFERENCES

1. Wilson, A.G., *Complex Spatial Systems: the Modelling Foundations of Urban and Regional Analysis,* Prentice Hall, London, 2000.
2. Bertuglia, C.S., Leonardi, G., Occelli, S., Rabino, G.A., Tadei, R., and Wilson, A.G., Eds., *Urban Systems: Contemporary Approaches to Modelling,* Croom Helm, London, 1987.
3. Tadei, R. and Williams, H.C.W.L., Performance indicators for evaluation with a dynamic urban model, in *Modelling the City,* Bertunglia, C.S., Clarke, G.P. and Wilson, A.G., Eds., Routledge, London, 1994, p. 82.
4. Scholten, H. and Stillwell, J.C.H., Eds., *Geographic Information Systems for Urban and Regional Planning,* Kluwer, Dordrecht, 1990.
5. Longley, P. A., Goodchild, M.F., Maguire, D.J. and Rhind, D.W., Eds., *Geographical Information Systems: Management Issues and Applications,* Wiley, New York, 1999.
6. Martin, D., *Geographic Information Systems: Socioeconomic Applications,* Routledge, London, 1996.

7. Bertuglia, C.S., Clarke, G.P., and Wilson, A.G., Models and performance indicators in urban planning: the changing policy context, in *Modelling the City*, Bertuglia, C.S., Clarke, G.P., and Wilson, A.G., Eds., Routledge, London, 1994, p. 20.

8. Birkin, M., Clarke G.P., and Clarke, M., Urban and regional modelling at the microscale, in *Microsimulation for Urban and Regional Policy Analysis*, Clarke G.P., Ed., Pion, London, 1996, p. 10.

9. Clarke, G.P. and Wilson, A.G., A new geography of performance indicators for urban planning, in *Modelling the City*, Bertunglia, C.S., Clarke, G.P., and Wilson, A.G., Eds., London, 1994, pp. 55.

10. Batty, M., Review of *Modelling the City*, Bertunglia, C.S., Clarke, G.P., and Wilson, A.G., Eds., Routledge, London, *Progr. Hum. Geogr.*, 20, 260, 1996.

11. Hutton, W., *The State We're In*, Vintage, London, 1996.

12. Orcutt, G.H., Mertz, J., and Quinke, H., Eds., *Microanalytic Simulation Models to Support Social and Financial Policy*, North-Holland, Amsterdam, 1986.

13. Birkin, M. and Clarke, G.P., Using microsimulation methods to synthesize census data, Census Users' Handbook, in *GeoInformation International*, Openshaw, S., Ed., London, 1995, p. 363.

14. Clarke, G.P., Ed., *Microsimulation for Urban and Regional Policy Analysis*, Pion, London, 1996.

15. Hancock, R. and Sutherland, H., Eds., *Microsimulation Models for Public Policy Analysis: New Frontiers*, Suntory-Toyota International Centre for Economics and Related Disciplines — LSE, London, 1992.

16. Mertz, J., Microsimulation — A survey of principles developments and applications, *Int. J. Forecasting*, 7, 77, 1991.

17. Krupp, H., Potential and limitations of microsimulation models, in *Microanalytic Simulation Models to Support Social and Financial Policy*, Orcutt, G.H., Mertz, J., and Quinke, H., Eds., North-Holland, Amsterdam, 1986, p. 31.

18. Falkingham, J. and Lessof, C., Playing God or LIFEMOD — The construction of a dynamic microsimulation model, in *Microsimulation Models for Public Policy Analysis: New Frontiers*, Hancock, R. and Sutherland, H., Eds., Suntory-Toyota International Centre for Economics and Related Disciplines — LSE, London, 1992, p. 5.

19. Hägerstrand, T., *Innovation Diffusion As a Spatial Process*, University of Chicago Press, Chicago, 1967.

20. Wilson, A. and Pownall, C.E., A new representation of the urban system for modelling and for the study of micro-level interdependence, *Area*, 8, 246, 1976.

21. Williamson, P., Community care policies for the elderly: a microsimulation approach, unpublished Ph.D. thesis, School of Geography, University of Leeds, Leeds, UK, 1992.

22. Williamson, P. and Voas, D., Income imputation for small areas: Interim progress report, paper presented at the conference Census of Population: 2000 and beyond, Manchester, UK, 22–23 June 2000.

23. Caldwell, S.B. and Keister, L.A., Wealth in America: family stock ownership and accumulation, 1960–1995, in *Microsimulation for Urban and Regional Policy Analysis*, Clarke G.P., Ed., Pion, London, 1996, p. 88.

24. Caldwell, S.B., Clarke, G.P. and Keister, L.A., Modelling regional changes in US household income and wealth: a research agenda, *Environ. Plann. C: Gov. Policy*, 16, 707, 1998.

25. Wegener, M. and Spiekermann, K., The potential of microsimulation for urban models, in *Microsimulation for Urban and Regional Policy Analysis*, Clarke, G.P., Ed., Pion, London, 1996, p. 149.

26. Veldhuisen, K.J., Kapoen, L.L., and Timmermans, H.J.P., RAMBLAS: A regional planning model based on the micro-simulation of daily activity patterns, *Environ. Plann.* A, 31, 427, 2000.
27. Vencatasawmy, C.P., Holm, E., and Rephann, T. (1999), Building a spatial microsimulation model, paper presented at the 11th Theoretical and Quantitative Geography European colloquium, Durham Castle, Durham, 3–7 September 1999.
28. Ballas, D. and Clarke, G.P., GIS and microsimulation for local labour market policy analysis, *Comput. Environ. Urban Syst.* 24, 305, 2000.
29. Marsh, C., The sample of anonymised records, in *The 1991 Census Users's Guide,* Dale, A. and Marsh, C., Eds., HMSO, London, 1993, p. 295.
30. Middleton, E., Samples of anonymized records, in *Census Users' Handbook,* Openshaw, S., Ed., GeoInformation International, London, 1995, 337–362.
31. Williamson, P., Birkin, M., and Rees, P., The estimation of population microdata by using data from small area statistics and samples of anonymised records, *Environ. Plann.* A, 30, 785, 1998.
32. Fienberg, S.E., An iterative procedure for estimation in contingency tables, *Ann. Math. Stat.,* 41, 907, 1970.
33. Ballas, D., A spatial microsimulation approach to local labour market policy analysis, unpublished Ph.D. thesis, School of Geography, University of Leeds, UK, 2001.
34. Ballas, D. and Clarke, G.P., Towards local implications of major job transformations in the city: a spatial microsimulation approach, *Geogr. Anal.* 33, 291, 2001.
35. Ballas, D., Clarke, G.P., and Dewhurst, J., A spatial microsimulation approach to the analysis of local multiplier effects, paper presented at the 32nd Regional Science Association, RSAI — British and Irish Section conference, The Dudley Hotel, Brighton and Hove, UK, 21–23 August 2002.
36. Ballas, D. and Clarke, G.P., Modelling the local impacts of national social policies: a spatial microsimulation approach, *Environ. Plann. C: Gov. Policy,* 19, 587, 2001.
37. Carver, S., Turton, I., Kingston, R., and Evans, A., Virtual Decision-Making in Spatial Planning [on-line document], School of Geography, University of Leeds, UK. Available at http://www.ccg.leeds.ac.uk/vdmisp/vdmisp.htm. Accessed 13 August 1999.
38. Evans, A., Kingston, R., Carver, S., and Turton, I., Web-based GIS to enhance public democratic involvement, paper presented at the 4th International Conference on GeoComputation, Fredericksburg, VA, 25–28 July 1999.
39. Kingston, R., Carver, S., Evans, A., and Turton, I., Web-based public participation geographical information systems: an aid to local environmental decision-making, *Comput. Environ. Urban Syst.,* 24, 109, 2000.
40. Ballas, D, Kingston, R., and Stillwell, J., Using a spatial microsimulation decision support system for policy scenario analysis, in *Recent Advances in Design and Decision Support Systems in Architecture and Urban Planning,* van Leeuwen, J. and Timmermans, H., Eds., Kluwer, Dordrecht, 2004, p. 150.
41. Williamson, P., Microsimulation: An idea whose time has come?, paper presented at the 39th European Regional Science Association Congress, University College Dublin, Dublin, Ireland, 23–27 August 1999.

13 Using Geodemographics and GIS for Sustainable Development

Linda See and Phil Gibson

CONTENTS

13.1 INTRODUCTION

This chapter discusses geodemographic and lifestyle data and how these systems are built and integrated into GIS and spatial decision-support systems. Most of the applications of geodemographics are commercial, yet there are also many public sector and research examples. For sustainable development, geodemographics and lifestyle data have barely been utilized, yet they represent a valuable source of input that could be used in a range of different planning and resource management applications, particularly in support of Local Agenda 21 initiatives.

Geodemographics is the classification of spatially referenced small-area demographic data to create area typologies or descriptions of areas [1]. It was originally developed for research and planning purposes and can be traced back to the U.S. census tract studies undertaken in the early 1900s as well as the methods of social area analysis and factorial ecology. One of the main reasons for undertaking these types of studies was to investigate the social structure of cities [2]. In the 1980s, the emphasis of geodemographic analyses shifted toward the commercial sector. The roots of this shift can be traced back to the availability of national data from the first U.K. census in the 1960s and the use of mainframe computing for processing what was a large data set at the time. This finally made it possible to undertake a national analysis of spatially referenced demographic data. Three national classifications were commissioned by the U.K. Office for Populations, Censuses and Surveys, from which the first commercial product eventually developed [3]. The original classification, referred to as the ACORN® system (A Classification of Residential Neighbourhoods; CACI Limited, London), used 1981 census data. The procedure involved segmenting 41 census variables to produce two residential classifications: a higher-level classification with 11 residential types and a more detailed subdivision into 38 types. Each class or type was then assigned a short description or pen portrait designed to capture the type of people living in that area; for example, ACORN type B in the higher level classification was given the description "modern family housing, higher incomes" [4].

Geodemographics has proved to be a major geographical marketing research tool of enormous commercial value, in particular for the targeting of segments or subgroups of the population that display higher than average response rates when presented with the opportunity to purchase an existing or new type of product or service. By linking a customer address list to a geodemographic classification, it is possible to build a geodemographic profile of the customer base. Beaumont [5] organized the types of commercial applications of geodemographics into four main areas: branch location analysis, marketing management information systems, credit scoring, and direct or target marketing.

In addition to commercial use, geodemographics is still employed in many public sector applications (e.g., the use of geodemographics for health screening documented in Openshaw and Blake [6]). The descriptive summary and the differences between the area typologies in a spatial context can provide general information needed for policy formulation, decision-making, or resource allocation, especially when linked to a GIS. Many commercial geodemographic and lifestyle systems are either integrated within an existing proprietary GIS, or a customized spatial decision support system is built around a commercial geodemographic system with user-driven analysis and functionality.

This chapter will outline the methodology used to build geodemographic systems, from a database to a fully integrated mapping and analysis system. The use of lifestyle data and the development of customized geodemographic systems are also discussed. Finally, the role of geodemographic and lifestyle data in current and potential applications of sustainable development is outlined.

13.2 CURRENT GEODEMOGRAPHIC AND LIFESTYLE SYSTEMS

Geodemographics represents the largest business sector use of census data and applied spatial analysis. There are several commercial geodemographics packages available such as ACORN, mentioned previously, Mosaic® (Experian Corporation,

Nottingham, UK), DemoGraf and Microvision [7]. However, this is a fast-changing area, and new products are developing on a continuous basis. The breadth of products from each company is also increasing as more and more customized solutions are now being offered.

However, these commercial products can be expensive, so Stan Openshaw developed a system for use by academics and the public sector called GBProfiles91 [8], which was built entirely using census data. A web-based version of GBProfiles91, renamed LGAS (Leeds Geodemographic Analysis System), is available at http://www.ccg.leeds.ac.uk/linda/geodem/.

The system is described in more detail in [9]. It currently contains a series of geodemographic classifications for the United Kingdom for 1991, but will be upgraded to include U.K. data from the 2001 census.

Although geodemographic systems can provide an excellent overview of the population, there are also a number of weaknesses [10]. A census is only taken at certain points in time such as every five to ten years, depending on the country, so the data can soon become outdated. However, it has been argued by Openshaw [11] that residential areas change slowly over time, so perhaps this problem is not as serious as it seems. Another weakness is the problem of ecological fallacy, which occurs because geodemographic systems are built using small-area data. The ecological fallacy is that the area typology is assumed to apply to all the individuals living in a particular area, which is not necessarily true [12]. It is important to be aware of this problem when applying any geodemographic analysis. Another weakness is that census data do not take into account people's values or attitudes. This is particularly important if you are trying to take consumer behavior into account, but equally, attitudes toward environmental issues for purposes of sustainable development are absent. Finally, geodemographic systems only exist for certain countries, so their application to sustainable development is limited to these areas, unless customized systems are built for a particular application such as environmental management in Kerala, India [13].

In addition to census data, another source of information is lifestyle databases, which are built from customer responses to questionnaires [10]. They are a relatively recent phenomenon, having originated in the United States and spread to the United Kingdom in the mid-1980s. The key point is that these databases deal with the individual as the unit of interest, rather than an area. There are usually questions on basic demographics such as age, occupation, etc., followed by a section on hobbies and interests, newspapers read, TV programs watched, sports and gardening, etc. There are even sometimes questions on the environment, such as to which groups donations are made.

The first main player in the U.K. lifestyle database market was the U.S. company National Demographics and Lifestyle (NDL). The methodology that NDL specialized in and deployed to generate lifestyle data was in the form of questionnaires attached to the extended guarantee cards of consumer durables. Answers to these questions could be used to provide a comprehensive insight into that individual and their household, which could then be used for marketing purposes by clients of NDL. Three other major players are Computerized Marketing Technologies (CMT), ICD (International Communications and Data), and CSL (Consumer Surveys Limited). They operate in a different way to NDL, in that they distribute regular waves of questionnaires via magazines and door-to-door distribution. The majority of questions asked concerns the regularity of purchases of products and the particular brands purchased. Customers are incentivized to respond through store coupons and other offers [10].

In 1998, NDL and CMT were brought into the same holding company as Claritas® (Claritas, San Diego, CA), which was the first organization to create a lifestyle classification, which they called PRIZM® [14]. Both lifestyle data from questionnaire banks and nonlifestyle data were used in the creation of the system, but it was unique in that it did not use census data in the early versions. The latest version called PRIZM NE, however, now combines information from both census and lifestyle data.

Other commercial geodemographics firms have now added lifestyle data to their classifications to produce either a more comprehensive description of areas or for input into specialized or customized systems.

13.3 BUILDING A GEODEMOGRAPHIC SYSTEM

The development of the first national geodemographic systems was a mammoth task and required extensive computing power at the time. It is now possible to create large-scale classifications using a desktop PC and proprietary statistical software packages such as SPSS® (SPSS Inc., Chicago, IL) or Minitab® (Minitab Inc., State College, PA). In fact, it is not necessary to create a classification at the national level. It is also possible to create more regional or local area classifications, because the methodology is the same.

A geodemographic system is built in an iterative process, which is outlined in Figure 13.1. The steps are discussed in more detail in the sections that follow.

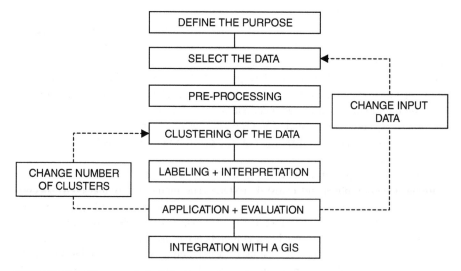

FIGURE 13.1 The steps in building a geodemographic system.

13.3.1 IDENTIFY THE PURPOSE OF THE CLASSIFICATION

The starting point for the development of any system is to identify the purpose of the classification, because this will drive the selection of the variables that will make up the system. Many geodemographic classifications are general purpose (i.e., they cover a range of data inputs that will build a comprehensive picture of an area). Age structure, family composition, and type of housing are examples of variables used in many geodemographic classifications. Depending on the country and the individual questions asked on the census forms, there can be many other variables used in a general purpose classification. In the U.K. Census, for example, one could also include indicators of ethnicity, birthplace, migration, illness, economic status, occupation, education, industry, workplace, transport mode to work, car ownership, and use of Welsh and Gaelic languages (in Wales and Scotland). Together, these variables can be used to develop a range of diverse area typologies. Many general-purpose classifications may be sufficient if area description is the main reason for development. The ability to show decision-makers a map describing the local area may be useful in itself for formulating policy and allocating resources [15]. Alternatively, it might be necessary to build a more customized system, in which case the variables selected may be a subset of the census data, or alternative data could be incorporated such as lifestyle or other survey data.

13.3.2 SELECT THE DATA

The main source of geodemographic data is the census, because it provides almost comprehensive coverage for a wide range of variables and is available for small areas. Commercial geodemographic systems also use other sources of data such as share ownership, unemployment, county court judgments (CCJs), and registers of company directors, which tend to provide more financial information to the classification, but it also reflects the commercial nature of these products [16].

The variables will normally be converted to percentages to allow different areas to be compared. Openshaw and Wymer [15] argue that one should try to use the smallest number of variables for the purpose required and that one should try to avoid variables that are highly correlated. The data should also be examined for outliers or suspect data, which should be removed prior to classification. However, transforming the data is not recommended, as the variables will lose their meaning when classified. This stage may need revisiting after the classification is created (Figure 13.1), because there is no unique or correct single set of variables to use. This step is part of an iterative and subjective process that will vary according to the experience of the system developer(s).

13.3.3 PREPROCESS THE DATA

There are many different types of preprocessing that could be applied to the data. The most important is a simple normalization so that different variables will be scaled on the same range, usually between 0 and 1. This ensures that variables with large ranges are equally weighted with variables that have much smaller ranges.

The input variables may be correlated, which should be avoided as much as possible. To deal with this, another type of common preprocessing operation can be used called principal components analysis (PCA). PCA replaces the original variables with a set of orthogonal factors or principal components that are completely uncorrelated. The clustering is then carried out on the factors, choosing sufficient factors so as to keep as much variance as possible from the original data. Openshaw and Wymer [15] suggest 95% variance as the cutoff point for factor inclusion. They also argue the opposite case, i.e., avoid the use of PCA because it is highly sensitive to skewed variables, which include many census variables. Also, there should be no reason to assume that the data are linearly related, which is an assumption of PCA. Therefore, normalization is the only preprocessing operation recommended, although with cheap computing power it is, of course, possible to try out different preprocessing techniques and compare the effects.

13.3.4 CLASSIFY THE DATA USING CLUSTER ANALYSIS

Classification or clustering is a process of data reduction. It is a way of moving from a large data set to a descriptive and manageable number of area typologies. There are many different clustering algorithms available, ranging from a simple k-means approach to algorithms developed in the field of artificial intelligence such as self-organizing maps [17] and machine learning techniques [17]. There are also fuzzy clustering algorithms [19] that can help characterize areas more accurately in situations where they would happily belong to more than one cluster type. However, an advantage of simple algorithms such as k-means is that they are commonly available in proprietary statistical packages such as SPSS and MINITAB. The task in this step is to select a clustering algorithm (or a sample of algorithms) and then apply the algorithm(s) to the data. The reader is referred to Krzanowski and Marriott [20] for a comprehensive overview of multivariate clustering techniques.

One important step when clustering is deciding on the number of clusters, which is not a straightforward task and is usually subjective. There are methods available to help choose the optimal number, but they generally involve creating many classifications of varying numbers and then looking for rapid changes in the slope of the scree or information loss plot [20]. This is another necessary iterative part of the development of the system (Figure 13.1), but experience gained over time will also play a part in helping to choose the right number.

13.3.5 CLUSTER LABELING AND INTERPRETATION

Once the clustering algorithm has divided the data into groups, the next task is to examine the cluster centers or the average behavior of each cluster in relation to the overall or global average. Here the global average could be the national, regional, or a more local average, depending on the size of the area chosen for classification. A common procedure is to calculate z-scores, which tells you the deviations from the global average for each variable. This then allows you to label the clusters. How you label is related back to the purpose of building the geodemographic system. Many proprietary systems use heavy marketing speak to summarize their area typologies in short, sharp phrases that instantly invoke an impression of the area. For example, the Microvision classification has a type called "Upper Crust," which clearly invokes a picture of a wealthy area [14]. Others are more descriptive and therefore paint a more comprehensive picture.

Once the cluster labels are applied, it is a useful exercise to undertake some groundtruthing to verify the label descriptions. There will always be an element of variation within a given area typology, but the point is to capture areas as best as possible, which can then be used for their intended purpose. One method of groundtruthing involves visiting an area and observing how well the labels fit. This is obviously not possible for very small-scale national classifications but a sample across different area types is a useful exercise. Expert knowledge is another method of groundtruthing; local experts will be able to verify the descriptions or, in some cases, help to fine-tune them. Finally, there is a growing database of properties on the Internet, either advertised privately or through estate agents. These databases can be used in combination with other groundtruthing methods to help verify area descriptions.

13.3.6 APPLICATION AND EVALUATION

Once the system has been developed, it can then be used in different types of applications. As mentioned previously, the most common application by far is commercial (i.e., for retail site analysis and store location, credit ratings, target marketing, segmentation of customer databases, advertising, and setting sales targets). Evaluation, on the other hand, is a much more difficult process. None of the commercial systems has ever been evaluated by independent experts, and it is rare to find comparisons between them. Moreover, a system can only be evaluated against the specific purpose for which it was originally required. For example, if used in a financial application, evaluation will be measured in profit. Evaluation of the system for public planning is more difficult. The ultimate criterion of plausibility of the results is a weak measure of evaluation [15], but one that requires further research.

13.3.7 INTEGRATION WITH A GIS

Once a geodemographic system is integrated with a GIS, it is then possible to begin mapping the area typologies that are geo-referenced with census boundaries. The mapping alone will provide a greater understanding of an area. However, the real power will become apparent when integration moves to decision support. Many proprietary geodemographics firms have either developed their own spatial decision support system as part of the geodemographics product, with specialized functions for commercial use, or teamed up with an existing proprietary package such as MapInfo® (MapInfo Corporation, Troy, NY) to offer an integrated solution. Thus, there is already a strong link between these two powerful tools.

13.4 BUILDING CUSTOMIZED SYSTEMS

Mainstream geodemographic classifications are regularly applied to diverse subjects of population analysis and market research. More recently, however, businesses have identified the potential for a new, more targeted approach through the development of customized geodemographic systems. Naturally, this advancement is driven by specific project aims or a client's output requirements. The process can also be limited, or indeed encouraged, by the existence and availability of suitable data to address those needs. Additional customer information, surveys, and lifestyle data can be used to develop systems with specific classification types (e.g., classifications of areas of financial risk, experience with computers and the Internet, propensity to take holidays, areas with different crime types and susceptibility to crime). The different combinations of data in customized systems opens up a whole new set of opportunities. As with general-purpose geodemographic systems, the vast majority of current applications appear in the commercial sector. However, these experiences will undoubtedly contribute to a smooth progression in the application of geodemographics to alternative areas, such as sustainable development, in the future.

A customized approach was adopted by the U.K. Post Office in 2002, with the intention of multiple application to the fields of market analysis, network planning, and product placement. The company was able to overcome an initial perceived lack of detailed customer information through manipulation of sales data to better understand its current consumer base. An emphasis was placed on the collection and integration of transactional information as an indicator of product popularity at the point of sale with a bridge to a proprietary geodemographic segmentation to describe the clusters of outlets. Post Office products are not restricted to traditional core mail and benefit payment functions, but now include banking, travel, and personal financial services. Access to data on product uptake by outlet enabled the analysis company, GeoBusiness Solutions Ltd., to construct a network classification based on enhanced customer information. The Post Office segmentation is currently being reviewed and enhanced, in collaboration with CACI® (CACI Limited, London, UK) and researchers at the University of Leeds, and is certain to evolve as new and updated information becomes available and improved analysis techniques are developed, such as the application of fuzzy techniques and genetic algorithms. These types of customized systems will become increasingly important as more specialized needs are addressed, both in the commercial and public sectors.

13.5 APPLYING GEODEMOGRAPHICS TO SUSTAINABLE DEVELOPMENT

A search of the literature and the Internet revealed very few currently reported applications of geodemographics and sustainable development, although applications of GIS and sustainable development are numerous. For example, the FAO (Food and Agriculture Organization of the United Nations) has utilized GIS technologies in a wide variety of projects, including studies of forestry, land use, mountain environments, water resources, ecosystem monitoring, and crop forecasting [21]. One successful investigation involved proximity analysis using Voronoi polygons to assess water availability in South Africa. The study region was partitioned to a tessellation of polygonal zones, each with a meteorological station at its center. Water level data could then be mapped for geocomputational analysis and applied to other related studies in the field of sustainable development. Another relevant application is the joint research of food insecurity, poverty, and vulnerability by the FAO, the CGIAR (Consultative Group on International Agriculture Research [22]), and the UNEP-GRID Arendal (UN Environment Program [23]). These groups are currently developing a GIS database, integrated with data of standardized formats, to support the mapping of food insecurity and poverty for application at national, regional, and global scales. The intention is that improved GIS methods and tools will be made available for general use in the search for solutions to these development issues. Project aims recognize the potential to "contribute to the accelerated accomplishment" of the WFS (World Food Summit [24]) goal to halve the number of undernourished by 2015, through provision of more complete, more accessible, and better-formatted decision-making information, all of which relate to goals of sustainable development. Given that GIS and geodemographics are becoming further integrated into spatial decision-support systems, we anticipate that geodemographics will become a vital part of the planning process in sustainable-development applications in the future. Indeed, the FAO web site provides the following comments to accompany the example described above, namely that, "demographic and administrative data can be added [into GIS systems] to provide projections for future supply-and-demand scenarios" when considering land suitability for types and intensities of use. Thus, there is already a move toward integrating basic demographic information in sustainable development work.

The World Bank is also financing a project in which geodemographics is being applied to the planning of environmentally sustainable activities in a forest reserve located in Kerala, India [13]. GIS is being used to determine environmentally sensitive areas, which are then fed into a model that tries to find optimal locations of new economic activity and the sources of labor for that new area. These new economic centers will be located in areas that are environmentally less sensitive, while at the same time trying to encourage the movement of people away from environmentally vulnerable areas. Geodemographics provides input to the model by supplying parameters such as attractiveness of an area. The modeling exercise is currently at the stage of gathering of data for the geodemographic model. This clearly illustrates the less than global coverage of geodemographic systems at the present time and the usefulness that these systems might possess for these types of planning activities in the developing world.

Another application of geodemographic information to sustainable development issues has been partially published by PROCIG, a geographic information project from Central America. This is an ongoing two-year promotion of the integration of census data with digital administrative boundary maps. Data have been integrated by GIS specialists from Panama, Costa Rica, Nicaragua, Honduras, El Salvador, Guatemala, and Belize to create an "unprecedented opportunity for progress in applying information to development in Central America" [25]. Costa Rican collaborators have analyzed statistical data on human population concentration (and factors of age/sex demography) in relation to the proportion of cultivated land by district, in a model of existing and future pressures on their protected areas. Accessibility to protected areas was also studied using additional information from satellite imagery (road networks, population centers, and digital elevation models). This study could potentially be advanced by analysis of further geodemographic information. For example, variables such as income, availability of private transport (car ownership), propensity to travel, personal interests, etc. could factor strongly when considering accessibility to protected zones or conservation areas.

All these examples clearly illustrate that GIS represents an important approach to supporting decisions in development, but the lack of socioeconomic data in developing countries severely restricts the use of geodemographic and lifestyle data. In those countries, however, where geodemographic and lifestyle systems already exist or where the data for building customized systems are available, geodemographics can be used to support Local Agenda 21 initiatives. Yet once again, there is little evidence from either the literature or the Internet of the uptake of these systems. For both planning and resource allocation, geodemographics provides a comprehensive picture of a population and how it is geographically distributed. Geodemographic systems also provide a picture of deprivation, which can be used as a proxy for a map of poverty. As poverty alleviation is one of the main goals of Agenda 21, these systems can be used to target more deprived areas with the appropriate policy.

These types of systems could also be used in more of a commercial way to exploit the power of direct marketing and advertising. For example, Local Agenda 21 groups are often voluntary, and targeting individuals for potential membership would be one extremely useful application of geodemographics in its more traditional role. Lifestyle data specifically addressing contributions to environmental organizations or the importance of environmental issues could be used to target individuals that might be interested in participation. Raising awareness and education is another area in which the power of both geodemographic and lifestyle systems could be harnessed. Surveys of environmental awareness could be used in combination with these systems to determine if certain area typologies are less "environmentally aware," allowing educational programs to be planned in those areas that need awareness raised.

As highlighted in the earlier section on customized systems, geodemographic classifications are increasingly being developed as solutions in response to specific needs or issues. Although the construction of a customized system may initially seem more complex (and possibly more expensive if data must be collected), it will allow the developers and users to understand essential geographical data more readily, thus enabling the most efficient process of policy development and resource management in the future.

13.6 CONCLUSIONS

Geodemographics and lifestyle data are currently being used extensively in the commercial world, yet there is a great deal of potential for applications in the public sector, in particular for sustainable development. There are many commercial packages available, although it is also possible to build your own system or a customized version using proprietary statistical software and a desktop PC. It is argued that geodemographics and lifestyle data can be used to support a range of Local Agenda 21 initiatives, as well as for targeting individuals for participation in environmental issues and raising awareness.

At present, the situation is more difficult with respect to the developing world. The potential applications are also numerous, but there are a set of barriers that must first be overcome before the widespread use of geodemographics and lifestyle data will be possible. These barriers apply equally to the use of GIS [26]. A major problem is simply the lack of socioeconomic data, because collection is expensive and time consuming. Governments must be convinced of the importance of collecting comprehensive socioeconomic information for planning purposes before these systems can even be built, and the key decision-makers must also be aware of the benefits and potential applications of the technology. A culture of information exchange must be nurtured, and the spatial data infrastructures must be put into place. Finally, the software and expertise must be available. Although these obstacles may seem difficult, they are certainly not insurmountable. Some of these barriers also exist in the developed world and may be one explanation for why there is still little evidence of the use of geodemographic and lifestyle data in applications of sustainable development. However, it is anticipated that this situation will change in the future as GIS and spatial decision support systems become an integral part of planning and resource allocation, especially with the increasing pressure to address environmental issues and alleviate poverty.

ACKNOWLEDGMENTS

The authors would like to thank the British Council in Italy for the funding provided through the British Council–MIUR/CRUI Agreement 2003.

REFERENCES

1. Birkin, M., Customer targeting, geodemographics and lifestyle approaches, in *GIS for Business and Service Planning,* Longley, P. and Clarke, G., Eds., GeoInformation International, Cambridge, UK, 1995, p. 104.
2. Brown, P.J.B., Exploring geodemographics, in *Handling Geographical Information,* Masser, I. and Blakemore, M., Eds., Longman Group UK Limited, Avon, 1991, p. 221.
3. Webber, R.J., The use of census-derived classifications in the marketing of consumer products in the United Kingdom, *J. Econ. Soc. Meas.,* 13, 113, 1985.
4. Beaumont, J.R. and Inglis, K., Geodemographics in practice: developments in Britain and Europe, *Environ. Plann. A,* 21, 587, 1989.
5. Beaumont, J.R., GIS and market analysis, in *Geographical Information System: Principles and Applications,* vol. 2, Maguire, D.J., Goodchild, M.F. and Rhind, D.W., Eds., Longman Scientific & Technical, London, 1991, p. 139.
6. Openshaw, S. and Blake, M., Geodemographic segmentation systems for screening health data, *J. Epidemiol. Comm. Health,* 49, S34, 1995.
7. Hooks, A., The Geodemographics Knowledge Base, 2002, http://www.geodemo-graphics.org.uk.
8. Openshaw, S., Blake, M., and Wymer, C., Using neurocomputing methods to classify Britain's residential areas, in *Innovations in GIS,* Fisher, P., Ed., Taylor & Francis, London, 1995, p. 97.

9. Rees, P., Denham, C., Charlton, J., Openshaw, S., Blake, M., and See, L., ONS classifications and GB profiles: census typologies for researchers, in *The Census Data System*, Rees, P., Martin, D. and Williamson, P., Eds., John Wiley & Sons, Chichester, UK, 2002, p. 149.

10. Birkin, M., Clarke, G.P., and Clarke, M., *Retail Geography and Intelligent Retail Planning*, John Wiley & Sons, Chichester, UK, 2002.

11. Openshaw, S., Making geodemographics more sophisticated, *J. Mark. Res. Soc.*, 31, 111, 1989.

12. Openshaw, S., Ecological fallacies and the analysis of area census data, *Environ. Plann. A*, 16, 17, 1984.

13. Varma, A.D.K., GIS for planning environmentally sustainable activities in Kulathu-puzha reserve forest, Kerala, India, no date, http://www.gisdevelopment.net/application/environment/overview/frpf0003pf.htm.

14. Claritas, Claritas: Adding Intelligence to Information, 2003. http://www.claritas.com

15. Openshaw, S. and Wymer, C., Classifying and regionalizing census data, in *Census Users Handbook*, Openshaw, S., Ed., GeoInformation International, Cambridge, UK, 1995, 239.

16. Sleight, P., *Targeting Customers: How to Use Geodemographics and Lifestyle Data in Your Business*, 2nd ed., NTC Publications Limited, Henley-on-Thames, UK, 1997.

17. Kohonen, T., *Self-Organization and Associative Memory.* Springer Verlag, 1984.

18. Quinlan, J.R., *C4.5 Programs for Machine Learning*, Morgan Kauffman Publishers, San Mateo, California, 1993.

19. See, L., Openshaw, S., Fuzzy Geodemographic Targeting, in *Regional Science in Business*, Clarke, G. and Madden, M., Eds., Springer, Berlin, 2001, p. 269.

20. Krzanowski, W.J. and Marriott, F.H.C., *Multivariate Analysis Part 2: Classification, Covariance Structures and Repeated Measurements*, Edward Arnold, London, 1995.

21. FAO, Food and Agriculture Organization of the United Nations, 2003, http://www.fao.org.

22. CGIAR, Consultative Group on International Agriculture Research, 2003, http://www.cgiar.org.

23. UNEP-GRID Arendal, UN Environment Programme Poverty Mapping, 2003, http://www.povertymap.net.

24. WFS, World Food Summit, 1996, http://www.fao.org/wfs.

25. PROCIG, A GIS Analysis of Population and Agricultural Pressures on the Protected Areas and Conservation Areas System, 2003, http://www.procig.org.

26. Hall, P.A.V., Use of GIS Based DSS for Sustainable Development: Experience and Potential, 1996, http://www.qub.ac.uk/mgt/papers/devel/hall.html.

14 Multivariate Spatial Analysis in Epidemiology: An Integrated Approach to Human Health and the Environment

Stefania Bertazzon and Marina Gavrilova

CONTENTS

"PEOPLE DIE each year because no one *BOTHERS to properly analyze DISEASE and DEATH* data for unusual localized concentrations."

Stan Openshaw

14.1 INTRODUCTION

The recent outbreak of SARS (severe acute respiratory syndrome) and its dramatic diffusion from a remote region of China to a narrow segment of population in Toronto (Canada); the spread of West Nile virus throughout regions of the United States, carried by wild birds and transmitted by mosquitoes; the geographical dimension of the international beef market in relation to bovine spongiform encephelopathy (BSE) (mad cow disease) are just a few examples from the current headlines. These examples confirm the compelling need for advanced spatial analyses of human health in its spatiotemporal dynamics and environmental determinants. They imply an urgent need for efficient and accurate models to assist in management and prevention of health-related emergencies. Our chapter addresses applications of the multivariate spatial analysis as the solution to the problems stated above, proposing the use of geometry-based methods for building efficient model estimates and predictions.

Defined in modern terms as the study of the distribution and determinants of disease and injuries in human populations, epidemiology deals with series of spatial processes, namely, the disease occurrence and its various determinants. The scope of this chapter is to analyze the spatial variation of disease (i.e., *where* the cases occur) and to investigate the relationship between disease incidence and socioeconomic factors (*why* the clusters of disease incidence occur at those locations). We discuss the use of a multivariate spatial regression model as a tool to address these questions. The goal of the model is to analyze current patterns and to simulate and predict future trends of disease incidence, based on the spatial variation of a set of socioeconomic indicators. The model represents an effective tool for policy, planning, service providing, and facility location. The novelty of the model is its spatial thrust. To guarantee its efficiency, we propose an innovative approach to the measurement of distance, based on the use of alternative metrics. This is an original approach that expands the scope of our discussion to fundamental questions in GIScience. The use of alternative metrics can provide a uniform criterion to estimate spatial autocorrelation, aiding in the definition of spatial contiguity and multivariate spatial correlation.

The application of regression analysis to spatial data raises computational as well as conceptual issues. Computational issues arise from statistical inefficiency, caused by spatial dependence in the data. The rigidity of current distance evaluation methods induces difficulties in the estimation of spatial autocorrelation. Conceptual issues range from the pressing need for standardization and interoperability in GIS (geographical information systems) to the conceptualization of space in the context of diverse spatial processes. The scope of this chapter is to discuss effective ways to address such issues by means of advanced computational geometry methods, based on the use of alternative distance functions. Within this approach, we concentrate on incorporation of L_p metrics in spatial regression analysis. The method draws on the relationship between the algebraic (attribute related) and topological (location related) data, coupled with advanced algorithms for spatial data analysis. This provides an innovative environment for spatial regression analysis, as well as the solid base for the development of the new and efficient methods for representation and visualization of geographical data based on topological as well as algebraic properties of the data.

14.2 CURRENT RESEARCH IN HEALTH AND ENVIRONMENT

14.2.1 GIS AND SPATIAL ANALYSIS IN HUMAN HEALTH AND ENVIRONMENT

In her paper from the 4th conference on GIS and Environmental Modeling, Couclelis [1] notes that "while environmental models integrating natural processes have already achieved a certain maturity, those seeking to combine human and natural processes are still in their infancy." Several reasons may have contributed to such a delay, including a persistent dichotomy within the discipline of geography (human versus physical) and, more generally, in the academic community (social sciences versus natural sciences). As a result of such dichotomy, the "human" and the "environmental" researchers have developed independent and often diverging tools and methodologies: one of the most evident discrepancies within GIS is the dichotomy between a dominant raster data model in the environmental sciences as opposed to a vector data model in the social sciences. Yet the vector/raster debate masks a more profound difference in the conceptualization of human versus natural spatial processes, with significant impacts on the analytical tools applied in either field, resulting in different analytical and predictive methodologies utilized in the two domains. As the discipline of GIScience matures, signals that unifying solutions may be achieved in the near future are apparent, even though their effective feasibility may not be imminent. The lack of integration between studies on natural and human processes may be viewed as a general framework in which some of the shortcomings of GIS/spatial epidemiology research can be seen.

A special issue of *Transactions in GIS* (2000, vol. 4) featured an editorial by Rushton [2] on the potential role of GIS in improving public health. Rushton's analysis on the lack of interaction between the GIS and health research community highlights at least five aspects. These include: the lack of a core group of researchers that recognize themselves as a community; the lack of recognition of the need for basic research in health and GIS, resulting in the lack of appropriate funding; the deficiencies of GIS health records, specifically the lack of temporal and historical information that could form the basis for identifying correlations between patients and the environment; the need for privacy in the records, which generally results in the release of health data at some spatially aggregated level; and finally the intrinsic characteristics of human disease as a spatial process. The last aspect, in particular, points out how relative rarity of incidence affects the meaningfulness of traditional geographical analytical methods, requiring novel techniques for pattern and cluster analysis.

Specific solutions to the joint analysis of human health and the environment are proposed for particular cases, such as Crabbe et al. [3] in their spatiotemporal analysis of asthma response in relationship to pollution exposure. Their methodological proposal focuses on the data collection and management phases, while the GIS application focuses more on the descriptive than on the spatial analytical stage.

An edited book on spatial epidemiology by Elliott et al. [4] contributes to the current debate. While certainly not exhaustive, this selective summary testifies to the rising interest, particularly in the GIS community, in human health applications and their integration with environmental and other socioeconomic factors. We would certainly consider this book as a remarkable contribution in the field of spatial analysis, more than GIS, featuring a number of high-profile contributions from both the medical and the spatial analysis community. The level of sophistication in spatial analysis applied to epidemiology is certainly of the highest quality, presenting some cutting-edge statistical techniques. The general framework for epidemiological analysis, set by the authors, consists of a probabilistic framework, which outlines the number of applicable analytical tools (logit, Bayesian models). Specific spatial analysis models (e.g., spatial autoregressive specifications) are explicitly considered, yet their particularity tends to result in *ad hoc* solutions.

In the context of the present debate, our work seeks to address the ever-pressing problem of an appropriate specification of a spatial autocorrelation model. This problem lies at the foundation of most spatial analytical techniques. Specifically, we propose a geometrically based model, aimed at obtaining a more flexible measurement of distance, which, in turn, is used in spatial autoregressive specifications. The flexibility induced by the method renders it particularly suitable for multivariate analysis, as the method can deal with data collected (or released) at varying spatial units. The multivariate framework with the topology-based methodology constitutes the basis for accurate and efficient analysis of the dynamic interaction between human health and the environment.

The application of advanced GIScience and geometric tools further allows us to produce effective representations and maps, which will aid the communication of the scientific results to the general public. The further use of state-of-the-art 3-dimensional visualization methods provides a novel environment for representation and analysis of the disperse data, facilitates communication among stakeholders and interest groups, and helps to increase public participation and achieve a more effective planning of health care provision.

14.2.2 ALTERNATIVE DISTANCE FUNCTIONS

The phenomena of two disciplines being close to each other but lacking active exchange of knowledge/methodology is true in respect to geometry and GIS. There has been some previous communication between the geometry and GIS communities, however on a very informal level. Delivering his invited lecture at the 11th ACM Symposium on Computational Geometry, Michael Goodchild [5] stated that much more could be done to bring these two disciplines together, and more benefits could be obtained, by further integration of the methodologies developed in both disciplines. Specifically, the study of spatial autocorrelation may benefit from the use of computational science approaches to spatial objects representation, distance measurement, contiguity and connectivity analysis, and geometric visualization [6–9]. Other interesting geometric issues include handling of approximate and inconsistent data and matching similar features from different databases [6,10]. On the other hand, geographic problems already visible in the geometric community include interpolation of surfaces from scattered data, hierarchical representations of terrain information, and boundary simplification [5].

The research presented in this chapter is concerned with providing a flexible method for the computation of spatial autocorrelation as a tool for specifying efficient spatial regression models. Development of advanced computational methods for a variety of diverse instances, specifically with respect to distance evaluation and spatial contiguity, is the focus of our investigation. The feasibility of utilization of metrics alternative to the traditionally used Euclidean, such as Manhattan (L_1), supremum (L_∞), Minkowski (L_p), Hausdorff (min-max), and Mahalanobis metrics, serves a dual purpose of providing flexible ways for computing distance, as well as measuring spatial attribute correlation. While Euclidean space has been used as a meaningful framework for representing and analyzing geographical phenomena due to the traditional association of geographical entities in Euclidean geometry, a need for an innovative alternative has been pressing [11]. Namely, in the case of spatial regression analysis, geographic space attributes and their relationship cannot be fully captured by the Euclidean model.

In order to connect geometric representation with the spatial analysis on a deeper level, application of new tools based on universal and generalized methodologies should be utilized. For instance, the L_p type metrics, such as Manhattan or supremum, have been successfully used in cartography for analysis and navigation purposes [6]. Also, the increased complexity of analyzed data and a wide spectrum of applications have resulted in richer, more complex geographical models, requiring wider-spectrum measurements of geometric relationship and analysis of geographic form (see Longley [12]). Investigation of alternative methods for data representation and analysis, including non-Euclidean metrics, emerged as one of the pressing necessities of today.

14.2.3 SPATIAL REGRESSION MODELS

The number of spatial regression techniques discussed in the academic literature has undergone impressive growth in recent years [13–15]. While the increased availability of spatial data and the increased user-friendliness of specialized software have certainly played a role, the main reason for these recent developments should be traced in an increased awareness of the inadequacy of traditional analytical techniques to deal with spatial data, according to Openshaw and Alvanides [16]. Some of the most recent techniques, including geographically weighted regression, aim at enhancing the flexibility of spatial models, by allowing the regression coefficients to vary over subregions in a study area (consult Fotheringham et al. [15]). Such approaches represent important developments, bearing great potential for advanced applications in the current data rich environment. But the bulk of spatially regressive, or autoregressive models aims at increasing the efficiency of the regression estimates, affected by the spatial dependence typically present in spatial data. Stemming from the field of spatial econometrics [17,18], such techniques emerge as a response to the severe limitations induced in the models by spatial dependence. Paraphrasing Waldo Tobler [19], spatial dependence can be described as the relatedness of things as a function of their distance; in more formal terms, the correlation between attributes of spatial units can be linked functionally to the distance among those units.

The property of spatial dependence became the root that was recognized by early quantitative geography (see Haggett et al. [20]). Later it became a cornerstone for the development of the theory of spatial processes, as described by Gatrell [21]. It was not until several decades later that the devious effects of spatial dependence began to surface, as they were identified by spatial econometricians Paelinck and Klaassen [18], who started to unveil the suboptimality of regression estimates obtained in the presence of spatially autocorrelated data, and to propose operational solutions. The analytical dilemma can be summarized as follows: there cannot be a spatial process without spatial dependence, and there cannot be an efficient regression model with spatial dependence; *ergo,* spatial processes cannot be efficiently modeled by regression analysis. The dilemma is even more dramatic than this; statistical efficiency is defined in terms of minimum variance of the parameter estimates. In applied regression analysis, spatial dependence violates the assumption of independence among sample units; such dependence results in a reduced provision of information, which implies redundancy of the sample and, ultimately, an effective loss of degrees of freedom, which is not apparent, nor immediately quantifiable. The result is an increased variance of the estimates. In turn, parameter estimates characterized by inflated variance are less reliable than parameters obtained under standard assumptions. There are at least two important dangers hiding in inefficient parameter estimates.

The first risk is to rely on an unreliable parameter and use the model for description, or simulation and prediction; considering the low confidence associated with the parameter estimate, the "true" value of the parameter might be so distant that the estimated parameter might be entirely misleading. Trusting inefficient parameter estimates can thus result simply in wrong predictions. The second, more subtle, danger is associated with the inference procedures typically used in model selection. The parameter variance is a crucial element in most of the traditional inferential tests; inflated variance results in a bias toward type II error (i.e., rejection of a null hypothesis when it is true). In model selection procedures, this bias results in the specification of models that assume as significant a nonsignificant relationship between dependent and independent variables. These two risks together result in the specification of models that produce nonreliable parameter estimates of nonsignificant functional relationships!

The operational solution to this problem is the specification of a *spatial* regression model. Even though several specifications have been proposed within spatial statistics, the general framework is a generalized model (GLS — generalized least squares model), which includes a model of the spatial autocorrelation present in the data. If a spatial autocorrelation model (SAM) is correctly specified, an inverse of SAM in the GLS specification rids the model of the redundancy, restoring the optimality of the variance, and producing efficient parameter estimates and allowing for correct inferential procedures.

The GLS approach is not free from criticism. In many respects, it can be viewed as a "patch" solution, where the spatial component of the spatial process is treated in a separate model and only introduced as an inverse component in the main model. More sound solutions can be foreseen in the near future, as alternative data models are developed, which will allow for overcoming some of the rigidity of the current representation of spatial processes. Currently, despite its conceptual limitations, the GLS approach remains an operationally practicable solution and an effective method to attain efficient models. One of the main conditions for the efficacy of GLS models is the specification of a correct spatial autocorrelation model, and one of the main factors determining the spatial autocorrelation function is the measurement of distance.

14.3 METHODOLOGY

The following sections describe the proposed methodology and its application to the case study of a spatial regression model for heart disease instances in Alberta, Canada. The methodology rests on two cornerstones: a spatial regression analysis approach using non-Euclidean distance functions, and finding the extent of spatial dependence using computational methods.

14.3.1 SPATIAL REGRESSION ANALYSIS

The concept behind the spatial regression analysis is based on the notion of proximity among studies' objects (instances, events, parameters, etc.) traditionally measured using the Euclidean distance. Waldo Tobler [19] referred, as the first law of geography, to his proposition that "Everything is related to everything else, but near things are more related than distant things." What the first law of geography expresses (rather informally) is the property known as spatial dependence. Spatial regression analysis is one of the tools commonly used to study spatial dependence.

14.3.1.1 Spatial Regression Expression

Formally, the spatial regression equation does not differ from the classical case: the only difference is the nature of the variables, in that both dependent (y) and explanatory variables (x) are *spatial*, because they refer to objects or events located in space

The classical regression model can be written in matrix form: $Y = X \beta + \varepsilon$.

The classical method for the estimation of the unknown parameters, β_k and σ^2, is known as OLS (ordinary least squares) and consists in minimizing the squared error vector ε. Calculating the first- and second-order conditions for a minimum for $\varepsilon\varepsilon$, the OLS estimator, is a combination of the observed X and Y vectors:

$$\beta_{OLS} = (XX)^{-1}\,(XY)$$

The OLS method relies on a number of assumptions, which guarantee the optimality of the estimates. Particularly, the error vector ε should be *identically* and *independently distributed* (i.i.d.) or:

$$\varepsilon \sim N\ (0,\ \mathbf{\Sigma}^2),\ \text{where}\ \mathbf{\Sigma}^2 = \sigma^2\ \mathbf{I}\ \text{(the variance matrix expression).}$$

14.3.1.2 Spatial Autocorrelation

Unfortunately, the property of independent distribution is unlikely to be met in spatial data, which are known to be generally affected by spatial dependence, which Anselin [17] defines as "the existence of a functional relationship between what happens at one point in space and what happens elsewhere," or:

$$y_i = f\ (y_1,\ y_2,\ ...,\ y_N),$$

which is the formal expression of a *spatial process.*

Spatial dependence implies, also intuitively, some redundancy of information; this renders standard regression methods inefficient. Formally, the presence of spatial *dependence* in a data set violates the hypothesis of *independence* in the error distribution; hence, the OLS estimator is no longer the best estimator, the one with minimum variance, and the property of efficiency is no longer guaranteed. In the presence of spatial dependence, the OLS estimator is still unbiased, but the violation of the efficiency property renders it not only unreliable, but also potentially misleading. The dependence among observations introduces non-null values of the cross products, or covariances, in the variance matrix, which in this case takes the form:

$$\mathbf{\Sigma}^2 = \sigma^2\ \mathbf{\Omega},$$

where the diagonal identity matrix $\mathbf{I}$ is replaced by the full matrix $\mathbf{\Omega}$.

The correlations ω_{ij} in the matrix $\mathbf{\Omega}$ result from the dependence among observations, and can be interpreted as the effect of spatial dependence, or the result of some spatial process for each pair of observations. The entire $\mathbf{\Omega}$ matrix is thus the expression of the spatial process(es) at work. It should thus be clear that, while disconcerting from the statistical standpoint, this is the crucial, and most distinctive, component of a *spatial regression* model.

In less formal, but more intuitive terms, the distribution of the OLS estimator can be represented by the "flatter" curve in Figure 14.1.

14.3.1.3 GLS Model

Within regression analysis the standard solution to an autocorrelated error is known as GLS, or generalized least squares method, and is based on defining an inverse of the matrix $\mathbf{\Omega}$, $\mathbf{\Omega}^{-1}$. Applying this inverse matrix, the GLS estimator is computed, as:

$$\beta_{GLS} = (\mathbf{X}\ \mathbf{\Omega}^{-1}\ \mathbf{X})^{-1}\ (\mathbf{X}\ \mathbf{\Omega}^{-1}\ \mathbf{Y})$$

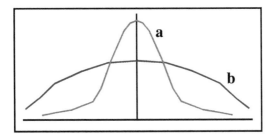

FIGURE 14.1 Variance of the OLS estimator in the case of independent distribution (line a) versus autocorrelated errors (line b).

where the inverse matrix restores the variance matrix to its optimal properties. The GLS estimator is optimal, or B.L.U.E. (best linear unbiased estimator).

14.3.1.4 Contiguity Matrix

Statistically, the GLS solution is acceptable; the cause of inefficiency is removed, and the optimality of the estimates is restored. In the case of *spatial* regression, the cause of inefficiency is spatial dependence, which is the most fundamental property of any spatial process; its removal *per se* is problematic from the *geographical* point of view, since ridding the model of its inefficiency corresponds to ridding it of its spatial component. The focus of the analysis should then be shifted to the process of constructing the matrix Ω^{-1}. Constructing an inverse of the matrix Ω means, in fact, specifying a spatial autocorrelation model (SAM).

Spatial dependence is composed of two elements: location and attribute. Location information refers to spatial objects, or locational units, such as census districts or weather stations, typically represented by points, areas, or lattices. Attribute information refers to the observed features, or variables, of those spatial units, such as population or rainfall. Consequently, the spatial autocorrelation model must include both components. Location and attribute are explicitly accounted for in the Ω^{-1} matrix:

$$\Omega^{-1} = W * C$$

where W (weight) represents the locational feature, and C the attribute feature. While C expresses the sign and value of the dependency (correlation), the locational feature expresses its spatial extent, or range. The most significant feature in this context is clearly the locational feature, which is also the novelty in the spatial regression approach. W is known, technically, as a *contiguity matrix*. In its simplest specification it is a binary structure, while some more complex specifications include various types of weights to decrease the effect of distance. Despite its crucial role in spatial regression analysis, the contiguity matrix has received proportionally very little attention in the literature (Griffith and Layne [22]), yet its importance is widely acknowledged. Any application of spatial regression analysis relies on some ambiguous, largely artificial, often application-specific definition of contiguity. Due to its

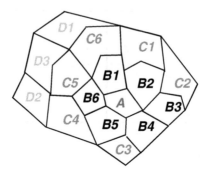

FIGURE 14.2 Contiguity based on **shared** borders.

FIGURE 14.3 Contiguity based on a **threshold** distance.

ambiguity, highlighted by Anselin [17], there are several practical possibilities of defining contiguity: shared borders and threshold distance are among the most common.

For area units, contiguity is typically based on shared borders, as shown in Figure 14.2. The criterion is, in general, applicable to area data; a good topological database is required. The criterion also allows for the definition of several orders of contiguity; the choice of the number of orders to be introduced in a model, or the weight to be assigned to each, does introduce an external element of choice. The most crucial issue is probably that in most applied cases, the spatial objects, or areas, used for such calculation are *artificial* (sampling points, census districts, or lattice cells). The definition of contiguity, as well as any subsequent measurement, depends largely on the location, size and shape of artificial objects, often built for the sake of statistical analysis (see Bertazzon [23]).

The classical alternative is the definition of a threshold distance, as exemplified in Figure 14.3.

This criterion is applicable to any kind of spatial data (area, point, line, or lattice data). It is not unambiguous, since the threshold must be determined from some external knowledge, and there are, generally, no means of establishing such threshold from within the model.

14.3.1.5   Spatial Autocorrelation Model

For all of the above reasons, the question arises whether it would be possible to follow a standard criterion for the definition of contiguity. In other words, is it possible to obtain an explicit, endogenous model of spatial autocorrelation? An interesting solution comes from the field of geostatistics: the empirical variogram provides a description of how the data are related (correlated) with distance. The semivariogram function was originally defined by Matheron [24] in 1971 as half the average squared difference between points separated by the distance:

$$\gamma(h) = \frac{1}{(2|N(h)|)} \sum_{N(h)} (z_i - z_j)^2$$

where N(h) is the set of all pair-wise Euclidean distances $i - j = h$, and z_i, z_j are the values of the variable at each pair of locations. The crucial parameter in the present analysis is the *range* value: the distance at which data are no longer autocorrelated. The range value is not infrequently used to determine *endogenously* the threshold for distance-based contiguity.

The most radical, conceptual difference between this approach and the topological one described above is the role taken by *relatedness* and *nearness* in each case. Although, in the topologically based contiguity (be it border or distance based), it was the locational feature that drove the treatment of nearness and set the divide between related and unrelated things, in this approach, *both* locational and attribute features determine *jointly* the autocorrelation model and, thus, the contiguity matrix. The distance element, or the h interval, enters as a measurement unit, but is evaluated by the γ function in conjunction with the attribute correlation, and based on the pattern of γ, a function of z *and* h, the range is determined. In the topological model, things are related because they are near, and only near things can be related. The semivariogram model considers both relatedness and nearness, location and attribute.

As shown in Figure 14.4, attribute and locational data are entered in a spatial autocorrelation model (SAM, i.e., the semivariogram). The model produces a range R, which is used to create a binary contiguity matrix CM, and the matrix assigns null or non-null values W to the attribute values C in the Ω^{-1} matrix, which models the spatial dependence present in the data. The Ω^{-1} matrix is finally input in the calculation of the GLS estimator to restore its efficiency.

14.3.2   COMPUTATIONAL GEOMETRY METHODS IN SPATIAL REGRESSION ANALYSIS

In the context of the spatial autocorrelation model, the semivariogram can provide a range at which the spatial autocorrelation ceases to exist. It combines the attribute

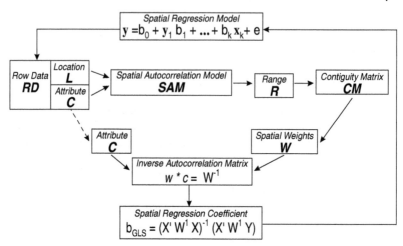

FIGURE 14.4 Endogenous approach to contiguity.

feature with the locational feature, to determine the range of autocorrelation and to measure spatial attribute correlation. In general, the semivariogram shows the variance of the difference between attributes at a given proximity to each other. A number of parameters in the application of the semivariogram approach dictate the general applicability of the method and influence the analysis, such as measuring similarity, defining geographical adjacency, computing attribute weights, and selecting sample pairs. This assists in creating a model with meaningful analytical indices of spatial autocorrelation, as well as in creating a background for development of an effective method of representation and visualization of geographical data. Further application of computational geometry-based methods for proximity representation and nearest-neighbor computation assists in the determination of the spatial extent of the spatial process, making it less arbitrary and less application-specific (see Gavrilova [25]).

14.3.2.1 Benefits of the Geometry-Based Approach

Using the geometry-based approach, different measures and metrics for autocorrelation evolve. Conceptually, the proposed method exists in the four-dimensional research space, encompassing algebraic, geometrical, applied, and conceptual issues. An algebraic factor is at the core of the semivariogram model itself, which is defined as an algebraic expression depending on the values of the attribute feature and a set of distinct pairs at which the feature is sampled. A geometric factor is introduced by the relationship of the geographical data to its geometrical representation and by the proposed approach to employ computational geometry methods for spatial data analysis and data processing. Applied aspects are brought forward by the analysis of applied data samples, which represent environmental, socioeconomic, and health-related spatial processes. Finally, the conceptual factor originates from the unique topology-attribute-based method for model visualization. Conceptual properties of the model, such as time, contiguity, and spatial dependence, will be visually represented and qualitatively analyzed. To the best of our knowledge, there is no analogous research currently conducted at other research establishments.

14.3.2.2 Non-Euclidean Metrics in Spatial Regression Analysis

We have already discussed the benefits of the use of a semivariogram-based spatial autocorrelation model. Our next methodological step is the replacement of the traditional Euclidean metric with alternative approaches for distance computation in the model.

Specifically, Manhattan (L_1), supremum(L_∞), power (Laguerre geometry), Hausdorff (max-min), Mahalanobis, and general L_p metrics are considered both for measuring the distance between the pair of sample points and as a parameter in the autocorrelation model. In order to proceed further, some geometric concepts must be introduced more formally.

Consider the d-dimensional real Cartesian space R^d containing d-tuples of real numbers $(x_1, x_2, ..., x_d)$ called *points*. Euclidean space E^d can be considered as an instance of R^d, where the Euclidean distance between two points $\mathbf{x} = (x_1, x_2, ..., x_d)$ and $\mathbf{y} = (y_1, y_2, ..., y_d)$ from E^d is computed as:

$$d(\mathbf{x}, \mathbf{y}) = \sqrt{\sum_{i=1}^{d}(x_i - y_i)^2}$$

In the sequel, 2D is used to denote a 2-dimensional space, and 3D is used to describe a 3-dimensional space. Geometrical objects in R^d can be characterized in terms of properties of sets of points in R^d. Concept of L_p metrics arises when the rules to compute the distance $d(x,y)$ between two points $x, y \in R^d$ is modified. In particular, in the *Minkowski* metric L_p the distance $d(\mathbf{x}, \mathbf{p})$ between points $\mathbf{x}(x_1, x_2, ..., x_d)$ and $\mathbf{p}(p_1, p_2, ..., p_d)$ is computed as:

$$d(\mathbf{x}, \mathbf{p}) = \left(\sum_{i=1}^{d} |x_i - p_i|^p \right)^{1/p}$$

The particular specification is obtained when parameter p is assigned some specific value. Thus, when $p = 1$, the *Manhattan* (L_1) distance is computed as:

$$d(\mathbf{x}, \mathbf{p}) = \sum_{i=1}^{d} |x_i - p_i|$$

The traditional Euclidean metric is obtained by assigning $p = 2$; thus, the Euclidean distance between two points is computed as:

$$d(x, y) = \sqrt{\sum_{i=1}^{d}(x_i - y_i)^2}$$

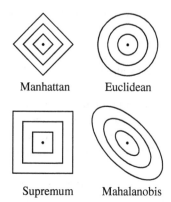

FIGURE 14.5 Distances in the Manhattan, supremum, Euclidean and Mahalanobis metrics.

When p is approaching infinity, the *supremum* or *Chebychev* (L_∞) metric arises, with the distance computed as:

$$d(\mathbf{x}, \mathbf{p}) = \max_{i=1..d} |x_i - p_i|$$

The *Mahalanobis metric* is a modification of the L_2 metric used in GIS. It is defined as:

$$d(x, p) = \sqrt{(x - p)' Cov(D)^{-1}(x - p)}$$

If the features of the vectors x and p are independent of each other, then the covariance matrix $Cov(D)$ will be the identity matrix, and the Mahalanobis distance would be the same as the Euclidean. The above concepts can be illustrated by Figure 14.5.

It is interesting to note that the *sphere* in the Manhattan metric is thus represented by the diamond, and the *sphere* in the supremum metric is a cube (see Figure 14.6).

The distance between a point $\mathbf{x}$ and a sphere $P = \{\mathbf{p}, r_p\}$ in the Manhattan or the supremum metric is defined as $d(\mathbf{x}, P) = d(\mathbf{x}, \mathbf{p}) - r_p$, which is illustrated by Figure 14.7.

Some of the most common applications of Manhattan and supremum metrics can be found in cartography and city planning. The special data structure corresponding to some geographical entities such as cities, regions, specific locations, etc. and representing the topological information on a set of those entities is built. This data structure is usually referred to as the Voronoi diagram. Thus, the Voronoi diagram represents the proximity information on the set of entities and splits the space into regions. Each of the regions is a locus of points that are closer to the given geographical entity than to any other entity. The Voronoi diagram can be used for performing various tasks, such as finding a nearest-neighbor for a given geographical object, finding all neighbors and tracing boundaries between regions,

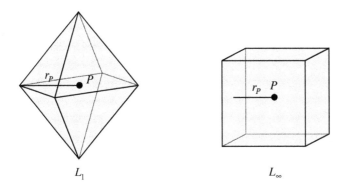

FIGURE 14.6 Spheres in the Manhattan and the supremum metrics in R^3.

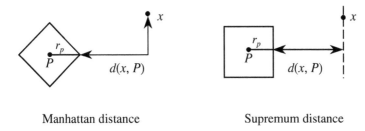

Manhattan distance   Supremum distance

FIGURE 14.7 Distance in the Manhattan and the supremum metric.

estimating how many other elements are located within each specified region, predicting the distribution of the geographical entities over the long term in the presence
of changes, and measuring some specific parameters. Each region, in addition, can
be attributed with some specific values, such as population, environmental, or health-
related statistics, etc. These attributes can be easily manipulated and studied using
the above data structure.

More formally, a *recilinear Voronoi region* $RVor(P)$ of the sphere $P \in S$ in the
L_1 (L_∞) metric in R^d is defined as the set of points that are closer to P than to any
other sphere $Q \in S$: $RVor(P) = \left\{ \mathbf{x} \in R^d \mid d(\mathbf{x}, P) \le d(\mathbf{x}, Q), \forall Q \in S - \{P\} \right\}$, where
the distance $d(\cdot, \cdot)$ is defined in the corresponding metric, according to Okabe [6].
The Voronoi diagram is defined as a collection of rectangular Voronoi regions,
corresponding to some geographical objects such as cities, rivers, mountains, etc.
Figure 14.8 shows the Manhattan Voronoi diagram for a set of sites (a) and the
supremum Voronoi diagram for a set of sites (b) in the plane.

While the Manhattan and supremum metrics are often used in city planning and
cartography due to the ease of distance calculation and the closeness of the obtained
regions to the city blocks, another modification of the Euclidean metric, namely, the
Mahalanobis metric, is specifically introduced to reflect the covariance between
studied entities. In GIS, it is used to sharpen the quality of habitat maps and to
improve classification.

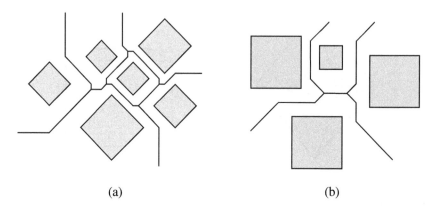

(a)   (b)

FIGURE 14.8 The rectilinear Voronoi diagram used in cartography in the Manhattan (a) and the supremum (b) metrics in the plane.

The particular features of the metrics need to be aligned closely with the properties of the analyzed data set, with the purpose of extracting the set of criteria for the appropriateness of the metric. The comparison of the rates at which the specific metric will reduce the variance in the statistics is another topic for in-depth investigation.

14.3.2.3   Algorithm for Valuating the Extent of Spatial Dependence

In the context of the spatial autocorrelation model, a combined attribute/topology-based criterion for *measuring the similarity* of locations is presented in this chapter. Such parameters as the boundary adjacency, common boundary length, or common boundary number of segments will be considered in building the matrix of locational similarities (the adjacency matrix). In the simplest form, the adjacency matrix represents the binary matrix of adjacent geographical regions, as in Geary's index [26]. Firstly, a *k-order adjacency* is defined between the areas that are not directly adjacent, but separated by *k* intervening areas. The Warshall's algorithm is applied to find the path matrix, P, of directed graph from its adjacency matrix. This methodology is used to analyze the locational similarities between remote as well as adjacent areas, and to study the range of the autocorrelation in the framework given above. The method is then combined with the use of the attribute feature as a weight to the locational feature to determine the critical distance and estimate the parameters of an autocorrelation model. Conceptually, this can be viewed similar to the semivariogram approach, where both attribute and location are used to determine the range (which was the good part of it), but is a more comprehensive approach, based on units and free from sampling biases. It also bears a very important conceptual feature, in that it somehow reverses the traditional approach to measuring autocorrelation, where the spatial, or locational feature is considered a weight to attribute feature. This assists in creating a model with meaningful analytical indices of spatial autocorrelation, as well as in creating a background for development of an effective method of representation and visualization of geographical data.

The neighbor pair selection of sample points within the semivariogram model can be performed by using a specified space partitioning method, such as a range–trees-based approach. The nearest-neighbor algorithms based on the Voronoi diagram methodology can also be used. The approach constitutes partitioning of the space onto Voronoi polygons, and then tracking neighbors of any specified sample point by following the Voronoi edges. The method is based on the well-known *nearest-neighbor property* [6], which allows finding neighbors of each Voronoi site simply by traversing the edges. The Voronoi-based method is deemed to require more preprocessing time than a tree-based method, but shows a very good running time and reasonable memory requirements during execution [6,27]. The developed method assists in the determination of the threshold distance that marks the spatial extent of the spatial process, making it less arbitrary and application-specific.

The multivariate spatial regression model estimates the parameters of the linear function that links the explanatory variables with the dependent variable. Spatial autocorrelation indices, such as Moran's I, provide a quantitative indication of the spatial autocorrelation present in the data. We utilize the different metrics approach for computation of Moran's index, and in the assigning weights based on attribute values correspondingly.

14.4 ANALYSIS AND APPLICATIONS

The methodology described above is illustrated on the example of the study of heart disease distribution in Alberta taking into account environmental and socioeconomic variables.

14.4.1 HEART DISEASE AND ITS ENVIRONMENTAL AND SOCIOECONOMIC DETERMINANTS

Heart disease (myocardial infarction) has become one of the leading causes of death in the developed world. "It is not obvious, however, what the relative importance is of such factors as stress, limited physical activity, smoking, high intake of calories, and high proportion of saturated fats, or what the relation is between these characteristics and elevated blood pressure, serum cholesterol, and triglycerides (blood fat)" [28]. All these factors are in turn related to a complex variable usually referred to as *lifestyle,* which can be represented by demographic indicators (e.g., age, sex), socioeconomic indicators (i.e., income, job type), and environmental indicators (e.g., recreational sports facilities, pollution). Preliminary analyses have shown evidence of a clear spatial pattern of the variables within the city of Calgary, indicating that the disease incidence should be analyzed as a spatial process, using appropriate analytical tools.

The proposed multivariate regression model is a vital tool in the hands of policy makers and planners, to predict *where* and *when* the disease is more likely to occur, and to effectively monitor zones of particular concern. The goal of the model is to analyze current patterns to simulate and predict future trends of disease incidence, based on the spatial variation of a set of lifestyle indicators. The novelty of the model is its spatial thrust. To guarantee its efficiency and flexibility, we propose to measure distance based on the use of L_p norms in evaluation of spatial autocorrelation. The method is an essential methodological tool, since it allows for a uniform treatment of diverse data (medical, socioeconomic, and environmental) and also for the selection of the most appropriate distance measure for the computation of autocorrelation. Our analysis can be used to locate clinics and health care services effectively. Within this study we propose some innovative measurements of distance, which can be used for a more effective route design that will lead to a more efficient provision of emergency care services.

The spatial multivariate specification of the model, based on L_p metrics, allows us to analyze, model, and predict the disease incidence based on demographic, socioeconomic, and environmental factors. This provides a broader base for the disease prevention and mitigation and enhances the predictive capacity of the model. Based on its ability to perform predictions and simulations, the model can be used to address *"what if"* questions. This is a useful feature in planning, as it can address, for example, the question, "What will happen if a new clinic is opened in a neighborhood?" The model can be used to evaluate the response of the disease incidence to a change in each lifestyle factor (elasticity) independently or jointly, allowing for accurate and specific simulations. Due to the flexibility of the method, a dynamic model can be specified, for the detection of spatiotemporal trends. The main feature of the proposed model is the use of alternative distance metrics for an accurate measurement of the spatial autocorrelation in the data. This ultimately ensures the *efficiency* of the model. We regard this as an extremely important outcome of the proposed method, in that our model can thus provide efficient (i.e., reliable) parameter estimates, a much needed property, that greatly enhances the usefulness and applicability of the estimated models.

14.4.2 The Data

The APPROACH Project is an ongoing data collection initiative, begun in 1995, containing information on all patients undergoing cardiac catheterization in Alberta (see Ghali et al. [29]). Preliminary analyses have shown evidence of a clear spatial pattern of the variable within the city of Calgary, indicating that the disease incidence should be analyzed as a spatial process, using appropriate analytical tools. The spatial distribution of cardiac catheterization cases in Calgary (Alberta) is represented in Figure 14.9.

In the multivariate spatial regression model, the spatial distribution of cardiac catheterization cases is the dependent variable, while the explanatory variables are census and other environmental variables. Example of the census variables are the population distribution and average age (Figure 14.10 and Figure 14.11).

14.4.3 Model Specification

The multivariate spatial regression model estimates the parameters of the linear function that links the explanatory variables with the dependent variable. Spatial autocorrelation indices, such as Moran's I, provide a quantitative indication of the

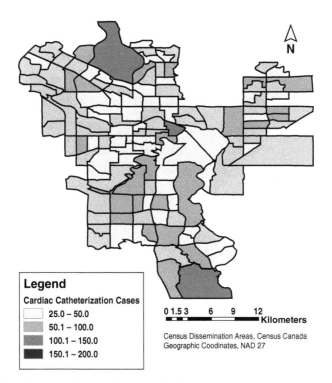

FIGURE 14.9 Cardiac catheterization cases in Calgary (Alberta).

spatial autocorrelation present in the data. Table 14.1 shows the different results in the calculation of Moran's I, obtained by use of different L_p norms. As highlighted in the table, using a different distance metric leads to different values of the spatial autocorrelation index; finding a "best" norm is thus a way to best represent the spatial dependence present in the data and, consequently, deal with it using the most appropriate specification of the GLS model.

One of the important considerations in the spatial regression model is the spatial resolution of the dependent and explanatory variables. The census variables are represented at the census dissemination scale, while the catheterization cases, recorded at the city block level, are represented at the census tract aggregated level in Figure 14.1. Moreover, environmental variables will have yet different spatial specifications (i.e., isolated polygons represent parks and sports facilities, while diffused variables, such as air pollution are best represented by an overlaying raster layer). Our proposed multivariate γ function will make use of the L_p norms to evaluate the cross-correlation among all these independent variables, and their correlation with the dependent variable, in a multivariate spatial autocorrelation model.

14.5 DISCUSSION

The use of L_p norms for the evaluation of distance in the spatial autocorrelation model is a novel tool that brings new insights into applied human health research

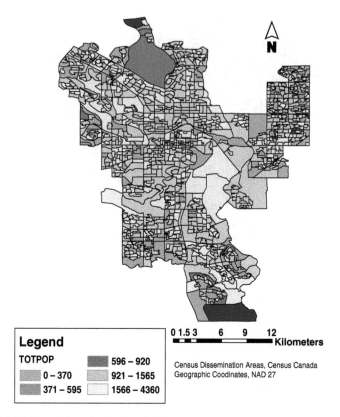

FIGURE 14.10 Population distribution in Calgary (2001 census).

using GIS. Our method achieves two primary goals that represent effective advancements in GIS research, objectives that are met through the use of L_p norms. The first one is the provision of a flexible tool for measuring distance. As it is widely acknowledged in the spatial analysis literature, one of the crucial, unresolved issues in spatial analysis is the definition of a uniform criterion for the assessment of spatial autocorrelation, a standard criterion to overcome the present plethora of ad hoc solutions for each application. The provision of such a solution, which can serve as a guiding criterion in medical research, is particularly important in medical research, in that it can assure the efficiency of parameter estimates. As noted above, current research does not always provide a method for efficient estimates, and the unreliability of estimates in the presence of inflated variance is a serious threat to the applicability of regression models. Moreover, the increased efficiency achieved by our method overcomes the bias generally induced by the inflated variance in the inferential procedures. This is a crucial feature of a multivariate model, where the significance of each explanatory variable should be tested with the highest degree of confidence. The above considerations introduce the importance of our method in the specification of multivariate models; the use of a flexible measure of distance is key to assessing the cross-correlation among a number of related variables. Using this feature, we can efficiently specify models that consider a number of explanatory

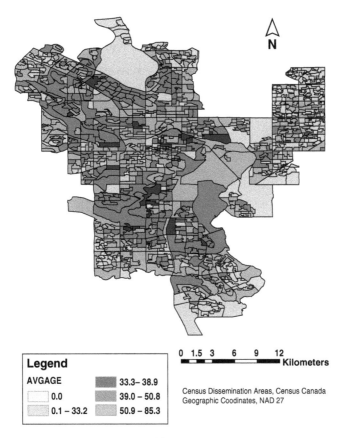

FIGURE 14.11 Average age in Calgary (2001 census).

variables. The latter point is also crucial for the specification of an integrated model, including human as well as environmental variables. As pointed out in literature within the current debate, there is an urgent need to effectively consider these factors in GIS research, and particularly in spatial analysis.

There are several aspects of the application of GIS and spatial analysis research to human health that need to be addressed urgently. Our work addresses specifically one of these aspects, providing a promising tool for efficient modeling of spatial processes involving human health. Efficiency is a crucial aspect of such models. We feel that our contribution can be instrumental in promoting the use of advanced spatial analysis techniques in human health research and its integration with the analysis of environmental processes.

14.6 FURTHER RESEARCH DIRECTIONS

Most of the methods developed in spatial autocorrelation are intended for use with static data [30]. Spatial movements are seldom of interest in this case. In the few examples where time is included explicitly in the methods, the meaning of patterns

TABLE 14.1
Moran's I Value Computed Using Different L_p Norms

Spatial Correlation Estimate	Euclidean (L_2)	Manhattan (L_1)	Supremum (L_∞)
Correlation	0.273	0.1684	0.2934
Variance	0.03444	0.03599	0.03262
Std. error	0.1856	0.1897	0.1806
Normal statistic	1.577	0.9909	1.733
Normal p-value (2-sided)	0.1149	0.3217	0.08311
Null hypothesis	No spatial autocorr.	No spatial autocorr.	No spatial autocorr.
Permutation p-value	0.06	0.135	0.052
Summary of	Min: −0.8133	Min: −0.6957	Min: −0.6467
permutation-correlations	1st Qu: −0.1248	1st Qu: −0.1204	1st Qu: −0.1399
	Median: −0.0215	Median: −0.0157	Median: −0.0284
Summary of	Mean: −0.0173	Mean: −0.0180	Mean: −0.0229
permutation-correlations	3rd Qu: 0.0963	3rd Qu: 0.0845	3rd Qu: 0.0894
	Max: 0.6887	Max: 0.6680	Max: 0.7255

to be searched for is reduced to comparing the clustering of points to complete spatiotemporal randomness [30]. The only approach available for handling time with spatial data from animal locations is the aggregation of observations into time slices, followed by the application of static analysis methods. Secondly, almost all of these methods including the point process theory [31] are only of marginal interest here, as they are not dealing with mobile objects.

Thus, the extension of the above ideas is aligned with the further generalization toward dynamic GIS models. Under *dynamic* GIS models, we mean the spatial data possessing the three major qualifiers: the time, the location and attributes. Creating the model of a multidimensional space, capable of handling time as a third dimension within the spatial autocorrelation model by the means of a 3D geometric interpretation is one of the promising approaches.

Another interesting topic is an investigation of the concept of relative space for contiguity study in the autocorrelation model. It is only natural to extend the concept to merge the geometrical and geographical concepts even further, applying the computational geometry methods for representing the attributed proximity information in the context of spatial contiguity. The use of additively weighted and multiplicatively weighted distance assignments in the framework of the Voronoi tessellation-based approach will allow for diverse conceptualization of spatial separation, much more general and flexible than traditional methods.

14.7 CONCLUSIONS

The purpose of this chapter was a critical discussion of the application of multivariate spatial regression models in spatial epidemiology, to specifically address the issue of inefficient parameter estimates in multivariate spatial modeling. The approach is innovative in that the spatial dimension is added to the traditional multivariate modeling; the approach to the measurement of distance using L_p norms and the specification of a spatial autocorrelation function based on such norms are the most original contributions of the work discussed.

The main benefits of the presented research are in the creation of an efficient multivariate spatial regression model for the analysis and prediction of the spatial pattern as a function of localized socioeconomic and environmental variables and in the development of methods based on non-Euclidean metrics more suitable for diverse applications and aiming at reducing the variance.

Finally, the described methodology can serve as a catalyst for further integration of computational science and geography disciplines, with the goal of finding an innovative solution to other pressing problems.

ACKNOWLEDGMENTS

We would like to acknowledge the GEOIDE network for supporting our research project "Multivariate Spatial Regression in the Social Sciences: Alternative Computational Approaches for Estimating Spatial Dependence": this chapter has been developed in the context of that project. We would also like to thank APPROACH project researchers for providing us with data and support for our work.

REFERENCES

1. Couclelis, H., Modeling frameworks, paradigms and approaches, in *Geographic Information Systems and Environmental Modeling,* Clarke, K.C., Parks, B.O,. and Cranes, M.P., Eds., Prentice Hall, Upper Saddle River, NJ, 2002, pp. 36–50.
2. Rushton, G., GIS to improve public health: guest editorial, *Trans. GIS.* 4: 1–4, 2000.
3. Crabbe et al., Using GIS and dispersion modelling tools to assess the effect of the environment on health. *Trans. GIS,* 4(3), 235–244, 2000.
4. Elliott, P., Wakefield, J.C., Best, N.G., and Briggs, D.J., *Spatial Epidemiology: Methods and Applications,* Oxford University Press, Oxford, 2000.
5. Goodchild, M., Computational Geography, invited lecture, *11th Annual ACM Symposium on Computational Geometry,* June 5–7, Vancouver, British Columbia, Canada, 1995.
6. Okabe, A., Boots, B., and Sugihara, K., *Spatial Tessellations — Concepts and Applications of Voronoi Diagrams,* 2nd ed, John Wiley and Sons, Chichester, 1999.
7. Bespamyatnikh, S. and Kelarev, A., An algorithm for analysis of data in geographic information systems, *13th Australasian Workshop on Combinatorial Algorithms* AWOCA, 2002.
8. Bertazzon S., 2002, Metaspace: from a model of spatial contiguity to the conceptualization of space in geo-analyses, *Joint International Symposium on Geospatial Theory, Processes and Applications,* Ottawa, July 8–12, 2001.
9. Bertazzon, S., Carlon, C., Critto, A., Marcomini, A., and Zanetto G., Integration of spatial analysis and ecological risk assessment in a GIS environment. The case study of the Venetian lagoon contaminated sediments, *GIS/EM4, 4th International Conference on Integrating GIS and Environmental Modeling,* Banff, A.B., Canada, September 2–8, 2000.

10. Gavrilova, M., Ratschek, H., and Rokne, J., Exact computation of Voronoi diagram and Delaunay triangulation, *J. Reliable Comput.,* 6(1), 39–60, 2000.

11. Miller, H. and Wentz, E., Geographic representation in GIS and spatial analysis: a common ground for Integration, *J. Geogr. Syst.,* 2, 55–60, 2000.

12. Longley, P., Foundations, in *Geocomputation: A Primer,* Longley, P., Brooks, S., McDonnell, R., and MacMillan, B., Eds., John Wiley, New York, 1998, pp. 3–15.

13. Anselin, L., Under the hood. Issues in the specification and interpretation of spatial regression models, *Agric. Econ.,* 27(3), 247–267, 2002.

14. Cressie, N.A.C., *Statistics for Spatial Data,* Wiley, New York, 1993.

15. Fotheringham, A.S., Brundson, C. and Charlton, M., *Quantitative Geography. Perspectives on Spatial Data Analysis.* Sage, London, 2000.

16. Openshaw, S. and Alvanides, S., Applying geocomputation to the analysis of spatial distributions, in *Geographical Information Systems: Principles and Technical Issues, vol. 1,* Longley, P.A., Goodchild, M.F., Maguire, D.J., and Rhind, D.W., Eds., 1999, 267–282.

17. Anselin, L., *Spatial Econometrics: Methods and Models,* Kluwer Academic Publisher, New York, 1988.

18. Paelinck, J.H.P. and Klaassen, L.H., *Spatial Econometrics,* Saxon House, Westmead, UK, 1979.

19. Tobler, W.R., Cellular geography, in *Philosophy in Geography,* Gale, S. and Olsson, G., D. Reidel Publishing Company, Dordrecht, Holland, 1979, pp. 379-386.

20. Haggett, P., Cliff, A.D., and Frey, A., *Locational Analysis in Human Geography,* Edward Arnold, London, 1977.

21. Gatrell, A.C., *Distance and Space: A Geographical Perspective,* Clarendon Press, Oxford, UK, 1983.

22. Griffith, D.A. and Layne, L.J., *A Casebook for Spatial Statistical Data Analysis,* Oxford University Press, Oxford, UK, 1999.

23. Bertazzon, S., A definition of contiguity for spatial regression analysis, in GISc: conceptual and computational aspects of spatial dependence, *Rivista Geografica Italiana,* Vol. 2, No. CX, June 2003, pp. 247–280.

24. Matheron, G., The Theory of Regionalised Random Variables and its Applications. Report nr 5. Centre de Morphologie Mathematique de Fontainebleau, 1971.

25. Gavrilova, M., On a nearest-neighbor problem under Minkowski and power metrics for large data sets, *J. Supercomputing, Special Issue on Computational Issues in Fluid Dynamics, Optimization and Simulation,* Kluwer, 22(1), 87–98, 2002.

26. Geary, R.C., The contiguity ratio and statistical mapping. *Incorporated Statistician* 5, 115–145, 1954.

27. Gavrilova, M. and Alsuwaiyel, M., Two algorithms for computing the Euclidean distance transform, *Int. J. Image Graphics,* Special Issue on Image Processing, World Scientific, 1(4), 635–646, 2001.

28. Ahlbom, A. and Norell, S., *Introduction to Modern Epidemiology,* Epidemiology Resources Incorporated, Newton Lower Falls, MA, 1984.

29. Ghali, M.D., William A., Knudtson, M.D., and Merril, L., Overview of the Alberta Provincial Project for Outcome Assessment in Coronary Heart Disease, *Can. J. Cardiol.* 16(10), 1225–1230, 2000.

30. Imfeld, S., Time, Points and Space — Toward a Better Analysis of Wildlife Data in GIS, Dissertation, Universität Zürich, Zurich, 2000.

31. Cressie, N. and Kornak, J., 2003, Spatial statistics in the presence of location error with an application to remote sensing of the environment, *Stat. Sci.,* 18(4), 436–456, 2003.

15 Zone Design in Public Health Policy

Konstantinos Daras and Seraphim Alvanides

CONTENTS

15.1 INTRODUCTION

One of the most advanced tools for the spatial aggregation of areal units is *zone design* [1,2]. Zone design can be used to aggregate A areal units through specific criteria such as contiguity, homogeneity, or similarity into Z zones, where $Z < A$. However, it is important to have in mind that when data for such small areas are grouped into larger social units, the modifiable areal unit problem is present [3]. Zone design has so far been concerned with optimizing the performance of a function, producing valuable results for planning and policy in various research contexts, but not in health administration. This chapter seeks to address the research gap by developing a zone design tool specifically for public health policy. The chapter consists of three main sections. The first section describes the structure of the U.K. National Health System, focusing on England and Wales. The second section introduces the reader to basic aggregation issues and presents the principles of the zone design tool. The final section applies zone design tools in order to produce geographies with populations likely to experience similar health care needs. A comparison of Limiting Long Term Illness (LLTI) for zones in England and Wales with the abolished District Health Authorities by the Health Authorities Act 1995 is presented.

In this approach, roughly 10,000 wards in England and Wales have been aggregated into 195 zones (following the 195 District Health Authorities) to create homogeneous zones in terms of LLTI. Two different objective functions have been developed in order to target two important geographical aspects. The first function builds compact zones using the LLTI variable, while the second function produces homogeneous zones, also using the LLTI variable. The comparison of these two zonations with the abolished Health Authorities provides useful insights and supports the case for a zone design approach to health administration policy in the concluding section.

15.2 THE NATIONAL HEALTH SERVICE
ORGANIZATION STRUCTURE

For the last 30 years the National Health Service (NHS) consisted of three strategic levels, and even today the changes to each level's name are equivalent to different allocation of roles and responsibilities. The District Health Authorities, as the middle level of the NHS, were valid from 1974 to 1996, and they were abolished by the Health Authorities Act 1995. They were replaced by Health Authorities, which have also been abolished and replaced by Primary Care Trusts. All these changes (including small and large changes to the boundaries of those levels) are aiming to identify regional health needs, plan appropriate programs and services, and ensure programs and services are properly funded and managed.

The National Health Service of the United Kingdom was set up on 5 July 1948 under an extremely weak economy. The wounds of the Second World War were still fresh, and many side effects such as population undernutrition, limited building materials, shortage of fuel, and housing crisis, together with the economic crisis in the United States, predetermined a rough environment for the establishment of the new health system. The main NHS aims since 1948 have been to increase the physical and mental health for all citizens by promoting health, preventing ill-health, diagnosing injury/disease, treating injury/disease, and caring for those with a long-term illness and disability, based on the need of the citizen and not the ability to pay. The NHS is accountable to Parliament, and it is managed by the Department of Health.

The Department of Health supports the government to improve the health picture of the population according to the Health Authorities consultancy. The administrative boundaries of the NHS organization in different levels have been changed numerous times since they were created in 1948. In the last decade, the NHS administrative organization consisted of two or three strategic levels, and every change made was equivalent to different allocation of roles and responsibilities. From 1982 to 1996 a structure of Regional and District Health Authorities (RHAs and DHAs, respectively) existed in England, and they were abolished by the Health Authorities Act in 1995 [4]. They were replaced by Regional Offices and Health Authorities, which have also been abolished and replaced by a new threefold structure of 8 Regional Offices (ROs), roughly 100 Health Authorities, and 481 Primary Care Groups (see Table 15.1).

Twenty years after the establishment of Regional and District Health Authorities in 1982 the government, in consultation with the Department of Health, decided to restructure the NHS organization. The reformed NHS, according to the "NHS Plan:

TABLE 15.1
Changes of the English and Wales NHS Organization Structure since 1982

	1982 to 1996	1996 to April 1999	April 1999 to March 2002	April 2002 to June 2003	July 2003 Onwards	
Level I	Regional Health Authorities (14)	Regional Offices (8)	Regional Offices (8)	Directorates of Health and Social Care (4)	—	England
	NHS Wales Department (NHSWD)	NHS Wales Department (NHSWD)	NHS Wales Department (NHSWD)	NHS Wales Department (NHSWD)	—	Wales
Level II	District Health Authorities (192)	Health Authorities (~100)	Health Authorities (~100)	Strategic Health Authorities (~100)	Strategic Health Authorities (28)	England
	Health Authorities (5)	Health Authorities (5)	Health Authorities (5)	Health Authorities (5)	NHSWD Regional Offices (3)	Wales
Level III	—	—	Primary Care Organizations (481)	Primary Care Organizations (~481)	Primary Care Organizations (300 + 3)	England
	—	—	Local Health Groups (22)	Local Health Groups (22)	Local Health Boards (22)	Wales

A plan for investment, a plan for reform" [5], was implemented to simplify the communication between the NHS and patient, as highlighted by the Department of Health (2000): "In the future, care and treatment will be redesigned around their [patients'] needs."

The new structure for health administration on 1 July 2003 was formed by 28 Strategic Health Authorities (SHAs) for England and 3 NHSWD Regional Offices for Wales, which are constituted from 303 Primary Care Organizations (300 Primary Care Trusts and 3 Care Trusts) and 22 Local Health Boards, respectively. The PCTs and LHBs work with local authorities and other agencies that provide health and social care locally, and are responsible to understand the needs of their local community. According to the NHS Plan, the PCTs and LHBs are now at the center of the NHS and will get 80% of the NHS budget. Because they are closer organizations to the citizen, the target of this plan is effectively a health and social care system with less bureaucracy and rapid financial support for the needs of the local community.

But how can this plan tackle the unacceptable socioeconomic or geographical variations and inequalities in health care? A positive view of these issues has been noted by Macara [6], where he argued that the processes of decision-taking are becoming more overt and more widely shared than in the past, because the system's paternalistic side has been replaced by partnership in health care such as the 26 Health Action Zones (HAZs), which are partnerships between the NHS, local authorities, the voluntary and private sectors, and local communities [7]. Moreover, he suggested that geographical and other inequalities are being tackled by the new NHS plan. Regarding the geographical dimension, national surveys (in 1991 and 2001) are referencing increase of uptake and use of geographical information systems (GIS) within all Health Authorities/Health Boards and Primary Care Organizations [8,9]. Also, the uptake and use of GIS within NHS organizations is concentrated at the Health Authority level, and few PCOs had personnel that were using GIS [10,11]. The growing awareness of the value of GIS [12,13] and the recognition of the spatial analysis importance in health care [14] are introducing a new NHS information era [15]. The use of spatial analysis and GIS to identify local areas for primary health care planning has been discussed by Bullen et al. [16]. In this document a clear presentation of three solutions in local areas' health planning has been proposed, and both geographical concepts and specific tools have been discussed. Certainly, the study by Bullen et al. [16] is not the only example that we can illustrate. A valuable literature review on many health-based studies is presented in Higgs and Gould [17], targeting the importance of using GIS in strategic health planning contexts. However, the majority of epidemiological studies analyze health data based on administrative areas, such as county districts [18] and census wards [19], rather than health-specific zones as suggested in this chapter.

15.3 MODIFIABLE AREAL UNIT PROBLEM AND ZONE DESIGN SYSTEMS

15.3.1 THE MODIFIABLE AREAL UNIT PROBLEM

As Johnston et al. [20] argue, the modifiable areal unit problem (MAUP) is a special form of ecological fallacy, associated with the aggregation of data into areal units for geographical analysis. The term modifiable reflects the changeable nature of areal units and zones when used to capture and describe spatial phenomena. The MAUP is one of the most important problems that geographers and statisticians have to face. An appropriate method of analysis needs to address two questions: what is to be estimated, and how can the MAUP be tackled? The measure of area homogeneity provided by Holt et al. [21] is an example of manipulating areas while controlling for the MAUP. The effects of the MAUP can be divided into two components: the scale effect and the zoning effect [3]. The scale effect is the variation in results that can be obtained when data for a set of areal units are progressively aggregated into fewer and larger zones. For example, when small census units are aggregated into primary care catchment areas, health authority districts, or other administrative units, the results change with increasing scale. The zoning effect is the variation in numerical results arising from alternative groupings of the small areas into the same number of larger zones. For example, census units can be combined in millions of different ways to produce aggregations of health authority districts. Although all these zonings can have the same number of health authority districts, it is almost certain that any statistics calculated for different aggregations will differ considerably.

The original automated zoning procedure (AZP) of Openshaw [22] was developed to search for more advanced ways of capturing socioeconomic patterns and explore the MAUP effects. AZP is a computational procedure for aggregating a group of small areal units into a group of larger zones, while optimizing output objectives such as equal population, zonal compactness, or social homogeneity. In the mid-1990s, the original AZP was updated to a zone design system (ZDES) based on the work of Openshaw and Rao [23]. Rapid hardware improvements, coupled with increasing production of census output data, dramatically extended and improved the capabilities of the original ZDES system [24]. For example, a modified version of the original AZP was adopted for the creation of 2001 census output areas (OAs) in England and Wales [25], while zone-designed output geographies have been built for policymaking [2]. A similar system, the Spatial Analysis in a GIS Environment (SAGE) developed by Haining et al. [26] provides a collection of statistical tools (such as box plots, histograms, and scatter plots) for classification and regional building through a user-friendly environment built on the ARC/INFO® product (Environmental Systems Research Institute, Inc., Redlands, CA). SAGE provides some good statistical tools for socioeconomic analysis, but the initial allocation of areal units to zones is the main weakness of this system because it relies on a rather basic k-means classification procedure. With the exception of some exploratory spatial analysis of health data using SAGE [27], there has been no recent research in the use of zone design systems for public health policy.

15.3.2 The "Areal Units to Zones" System (A2Z)

Previous zone design systems tackled efficiently the twofold character of the MAUP, but they were heavily reliant on older-generation programming languages (such as Fortran 77) and the now dated environment provided by the ARC/INFO graphical user interface. It is argued here that, given the progress in computing environments and data availability, there is now need for a new zoning system that can stand independently and can cope with larger multivariate data sets and new objective functions. The new "areal units to zones" (A2Z) system has been developed in Visual Basic, using Map Objects 2.1 for the mapping needs of the system. The new A2Z system has been developed following the basic principles of the older systems, but it is based on an object-oriented architecture, providing extra functionalities for future updates. Generally, a zone design system consists of three operational parts. The contiguity-checking algorithm is the most important part, ensuring that all areas have been allocated to a zone and also that output zones are contiguous and leave no gaps. Equally crucial is the objective function that determines the rules of an optimum solution for a specific problem or task. Finally, the compactness rules provide a fundamental part of the zone design system, ensuring that the output zones have compact shapes. In this context, the new A2Z system applies a number of new algorithms for each of the three operational parts, as explained in the following subsections.

15.3.2.1 Checking Contiguity

A fundamental part of A2Z for ensuring contiguity stability of a zone is the algorithm that checks boundary changes and provides feedback on its current status. Such algorithms are generally borrowed from Graph Theory, and they are known as graph scanning (traversing) techniques, such as depth-first search (DFS) and breadth-first search (BFS) [28]. The A2Z system checks the contiguity of zones using these two approaches. Breadth-first search traverses a graph level by level and seems like a more organized order than the DFS. The traversal is implemented similarly to the nonrecursive implementation of DFS, but the stack (First In First Out) in DFS is replaced by a queue (First In Last Out) in BFS. In the zone design context, these techniques provide information for the contiguity stability of a zone, because any zoning system can be represented as a graph. The use of each approach is determined by two important factors. The first is the size of the data set (total number of areal units) and the second is the number of zones that the system is seeking to construct. Usually, the BFS is a memory-consuming algorithm because of its level-by-level nature, and it is best to be applied on a small number of areal units. On the other hand, the DFS is more appropriate for any data set size and it provides a fast tracing routine of output zones, including large numbers of areal units. In general, proper selection of the contiguity checking algorithm results in lower processing times, so that more resources can be allocated to other parts of the system, such as optimizing the objective functions and ensuring compactness of zone shapes.

15.3.2.2 Defining Objective Functions

Objective functions in zone design are usually specific to the aims of a particular case study. In the case study presented here, the target is simplified, because the main concern is the use of zone design as a health policy and planning tool. More specifically, the idea here is to rebuild the District Health Authorities (DHAs) according to two criteria: homogeneity of LLTI percentage in each ward and compactness of output zones. The implementation of such an objective function is loosely based on a k-means approach, supporting k number of variables. A standardized method for the indicators such as the signed χ^2 statistic or the percentage [29] is important in identifying concentrations of need. The k-means objective function can be expressed as follows:

$$d_{iz} = \sqrt{\left(\overline{x_{1z}} - x_{1iz}\right)^2 + \left(\overline{x_{2z}} - x_{2iz}\right)^2 + \ldots + \left(\overline{x_{kz}} - x_{kiz}\right)^2} \qquad (15.1)$$

where

d_{iz} = multivariate distance of 'i' areal unit

$\overline{x_{1z}}, \overline{x_{2z}}, \ldots \overline{x_{kz}}$ = mean of each indicator per zone (z)

$x_{1iz}, x_{2iz}, \ldots, x_{kiz}$ = current value of 'i' areal unit

The k squared differences $\left[\left(\overline{x_{1z}}-x_{1iz}\right)^2+\left(\overline{x_{2z}}-x_{2iz}\right)^2+\ldots\ldots+\left(\overline{x_{kz}}-x_{kiz}\right)^2\right]$ rep-resent the different variables and measure the multivariate distances (d_{iz}) between areal units (i) and the mean value of each indicator $\left(\overline{x_{1z}},\overline{x_{2z}},\ldots\ldots,\overline{x_{kz}}\right)$ per zone (z).

From the sum of distances per zone $\left(\sum_{i=1}^{n}d_{iz}\in z_j\right)$, Equation 15.2 calculates a local objective function score (dz_j) for each zone, and the results feed into Equation 15.3, for calculation of the global objective function score.

$$dz_j = \sum_{i=1}^{n}d_{iz}\in z_j \tag{15.2}$$

where

dz_j = local best score of each 'j' zone (z)
d_{iz} = distance of 'i' areal unit

$$BS = \frac{\sum_{j=1}^{m}dz_j}{m} \tag{15.3}$$

where

BS = Best score for an iteration
dz_j = local best score of each 'j' zone (z)
m = number of zones

An example of this approach is graphically represented in Figure 15.1(a) and Figure 15.1(b). These figures contain two zones (k_1 and k_2) and an areal unit (0) that needs to be allocated to one of them. Each point (areal unit 1, 2, 3, ...) is plotted based on the values of "Indicator A" and "Indicator B."

Areal unit "0" is about to be ascribed a label that describes which zone it belongs to. As Figure 15.1(a) shows, areal unit "0" is closer to the mean of zone one ($\overline{k_1}$), so it should receive label "1" because the algorithm is trying to minimize the distances between each areal unit and the zone's mean. However, in Figure 15.1(b), we have a different result, because the means of zone one and two have different values, and areal unit "0" is further away from the mean of zone one ($\overline{k_1}$), compared to the mean of zone two ($\overline{k_2}$), so the areal unit receives label "2."

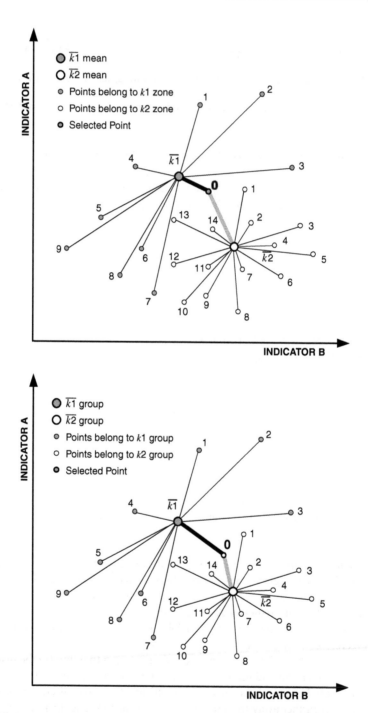

FIGURE 15.1 Top: Graphical representation of two indicators (A and B) — move of selected point "0" to k_1 group. Bottom: Graphical representation of two indicators (A and B) — move of selected point "0" to k_2 group.

	Circle		Square		Grid	
Shape	R		α α		α	
	$R = 1$		$\alpha = 1$		$\alpha = 1$	
Perimeter	$2\pi R$	6.28	4α	4	10α	10
Area	πR^2	3.14	α^2	1	$4\alpha^2$	4
Ratio	$\dfrac{(2\pi R)^2}{\pi R^2}$	12.56	$\dfrac{(4a)^2}{a^2}$	16	$\dfrac{(10a)^2}{4a^2}$	25

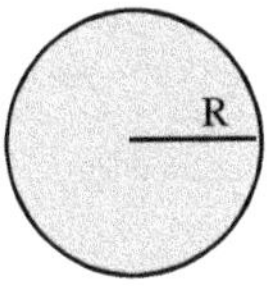

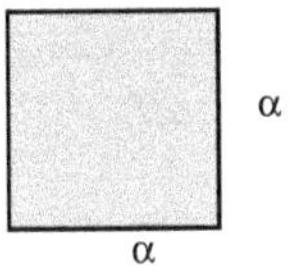

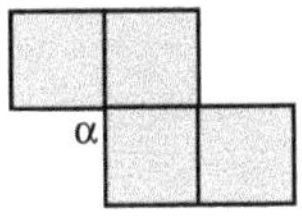

FIGURE 15.2 Ratios for three different shapes.

15.3.2.3 Shape Constraint Methods

Methods concerning the compactness of output zones have been attempted in the past by Alvanides and Openshaw [2] and Martin [25]. For example, the minimization of local spatial dispersion by Alvanides and Openshaw [1] is equivalent to a type of location–allocation problem. This method controls the shape calculating the distance between areal unit centroids and the output zone centroids. Unlike the objective function distances we examined earlier, shape constraints operate on geographical coordinates, dealing with the geometrical or population-weighted centroids of the areal units and zones. A geometrical alternative to the location–allocation constraint has been proposed by Professor D. Martin to the Office for National Statistics [30] for the design of the 2001 Census Output Areas. The minimization of the squared perimeter divided by the area of a zone effectively compares the ratio of each shape with the optimum ratio of a circle, as shown in Figure 15.2.

Both approaches have been used in various projects, producing compact zones within the zone design context. However, the calculation of compactness is a very time-consuming process, and when it is added to the contiguity-checking algorithm, the processing time increases dramatically. This is a result of current shape control methods calculating the compactness using geographical elements such as centroids, areas, and perimeters. On the other hand, the approach developed and described here takes into account the contiguity information of the areal units and builds a rule that applies to each areal unit the system examines. This contiguity constraint method improves the processing time of shape control calculations, because it focuses on the current areal unit and the two negotiating zones (local shape control), as shown in Figure 15.3. The algorithm first selects the areal unit about to be allocated and records all the neighboring areal units. If the number of the neighbors is above a user-defined threshold, then it is acceptable to move the areal unit to the target zone. Otherwise, the areal unit remains in the source zone.

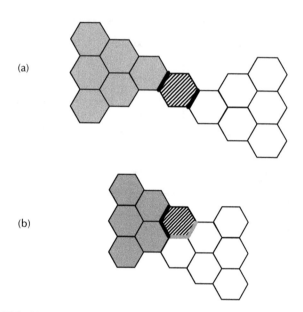

(a)

(b)

FIGURE 15.3 (a) The areal unit does not change zone (2 neighbors). (b) The areal unit changes zone (2 neighbors)

The user-defined threshold for the example presented in Figure 15.3 is 1, so the algorithm accepts changes of an areal unit only if it has more than one neighbor per zone. In Figure 15.2(a) there are two zones and an areal unit (hexagon) between them. The areal unit belongs to one of the zones, and because the algorithm records only one neighbor for each zone, the areal unit will remain in its source zone. On the other hand, in Figure 15.3(b) the areal unit has two neighbors per zone, and as a result it will move to the target zone. The measure of compactness by the contiguity constraint method is introduced here as an alternative approach for aggregating areal units. The advantage of this algorithm is the speed of compactness calculations, especially in large data sets. As mentioned above, the contiguity constraint algorithm does not use shape information and focuses on the contiguity information of areal units and zones.

15.4 DESIGNING HEALTH AUTHORITIES IN ENGLAND AND WALES

This section provides an example of original health analysis using zone design and demonstrates some of the potential capabilities of A2Z as a spatial analytical tool. The study area of England and Wales has been selected as the appropriate data set for designing health administrative regions based on the homogeneity of Limiting Long Term Illness percentages for each area. The data consists of the 9,930 census wards in England and Wales and each ward holds information on LLTI percentage and the District Health Authority (DHA) to which it has been allocated. The 195 DHAs were valid from 1974 to 1996 and were effectively created by aggregating

1991 Census Enumeration Districts for England and Wales. The majority of wards nest within DHAs, except for very few cases, such as the "Hogsthorpe" and "Crowmarsh" wards crossing the boundaries of two DHAs.

Figure 15.4(a) and Figure 15.4(b) illustrate the burden of LLTI for administrative areas in England and Wales using the standard deviation method. The gray-scaled areas are above LLTI average for England and Wales, while the dotted areas are below the average. The darker the gray shade or the thicker the dots, the more standard deviations above or below the national average an area is, respectively. At first glance, Figure 15.4(a) clearly highlights the two well-known Health Authorities that experience very high LLTI rates: Mid-Glamorgan in Wales and Durham in England. Figure 15.4(b) uses the same visualization method to highlight more detailed patterns given the smaller size of wards, with the DHA boundaries appearing as lines. By visually comparing the two maps, it quickly becomes obvious that DHAs are very heterogeneous in terms of LLTI rates. For example, a very clear pattern of high LLTI concentration for wards in West and Mid-Glamorgan, South Wales, has been divided amongst five DHAs.

The reconstruction of DHAs using homogeneity and compactness criteria is a spatial analytic approach, and for this reason the use of a zoning system is essential. In the LLTI example presented here, the general homogeneity Equation 15.1 discussed earlier takes a simple form, equivalent to the absolute difference between the mean LLTI indicator per zone (z) and the current value of areal unit 'i' as shown below:

$$d_{iz} = \left| \overline{x_{Az}} - x_{Aiz} \right| \qquad (15.4)$$

where

d_{iz} = distance of 'i' areal unit

$\overline{x_{Az}}$ = mean of 'A' indicator per zone (z)

x_{Aiz} = current value of 'i' areal unit

The A2Z system was used here to produce two sets of ward aggregations, starting from current DHAs. The first output is homogeneous zones without any shape constraint, as shown in Figure 15.5. LLTI is mapped for the unconstrained zones using the standard deviation method in Figure 15.5(a), showing below (dotted areas) and above (in gray shade) average regions. Comparing the new boundaries with the underlying LLTI patterns at the ward level in Figure 15.5(b), it is evident that the new regions generally follow LLTI patterns. However, it quickly becomes apparent that the shape of the unconstrained zones is highly irregular and therefore unsuitable for administrative purposes, such as the restructuring of DHAs in England and Wales.

A method for constraining the shape of output zones is needed, as discussed in the shape constraint section earlier, so that the homogeneous zones resulting from objective function 4 are also of compact shape. The contiguity constraint method

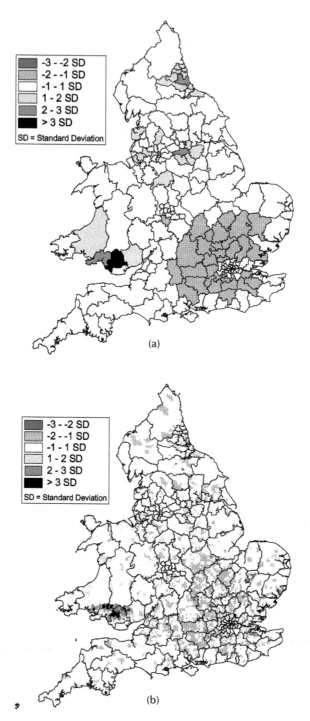

FIGURE 15.4 LLTI visualization using the Standard Deviation method, for (a) DHAs and (b) Wards.

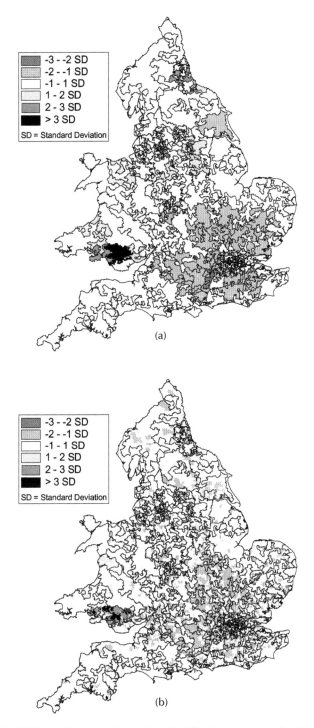

FIGURE 15.5 LLTI visualization using the Standard Deviation method, for (a) unconstrained zones and (b) Wards.

described earlier can be combined with the objective function in A2Z to create a second set of output zones, as shown in Figure 15.6(a). The mapping of the spatially constrained zones provides a better representation of designed health administrative areas, while satisfying the objective function for homogenous zones, at least compared to the official DHAs, as shown in Figure 15.6(b).

However, upon closer scrutiny the zones are actually more heterogeneous than the nonconstrained zones, as shown in the graph of Figure 15.7. The shape constraint method reduces the evolution of the homogeneity algorithm at the expense of the objective function score, compared to the unconstrained approach. The graph indicates that the homogeneity score before the aggregations is 101.36. For unconstrained zoning (dashed gray line) homogeneity improves rapidly as the number of the iterations increases, reaching 79.65 for 200 iterations. On the other hand, the contiguity constrain method (gray line) blocks any improvement of homogeneity after the hundredth iteration. In this example the compactness is a very important factor for the reconstruction of DHAs, and the homogeneity of a zone plays a secondary role. An in-depth view of boundary changes for each condition is illustrated in Figure 15.8. The Durham DHA is shown as it was until 1996 in Figure 15.8(a) and Figure 15.8(b). The homogeneous but highly irregular zone of Durham after 200 iterations is shown in Figure 15.8(c) and Figure 15.8(d). Finally, Figure 15.8(e) and Figure 15.8(f) illustrate the proposed homogeneous and compact zones. Comparing the unconstrained and constrained versions in Figure 15.8(d) and Figure 15.8(f), respectively, we can see that some of the high LLTI zones are lost in order to improve the shape of the zones. However, even at the expense of some degree of homogeneity the new constrained zones in Figure 15.8(e) seem to detect better the underlying LLTI rates, compared to the DHAs shown in Figure 15.8(a).

15.5 CONCLUSIONS

It is argued in this chapter that automated zone design could help health areas to be redesigned around patients' needs for better policy making. A new tool for designing zones from areal units was presented, taking into account the objective functions, as well as advanced methods for constraining the shape of the zones. This contiguity constraint method is very fast and seems to be performing well for designing health areas, but more advanced objective functions may be necessary to cover a variety of health needs at various scales. As the data from the 2001 U.K. Census is becoming available, it would be interesting to start rebuilding the new Strategic Health Authorities and compare them with the old DHAs.

Most medical facilities have zones from which the majority of their patients come, depending on such factors as distance or socioeconomic status. Questions derived from this kind of approach include, "What criteria might be employed in order to regionalize health care?" "How can one characterize health regions that already exist?" and "Should administrative units or patient flows be used to delineate health regions?" As Rushton [31] wrote, "The public and some public health professionals continue to believe that better cluster detection methods and better geographically referenced data will lead to knowledge that will benefit the public."

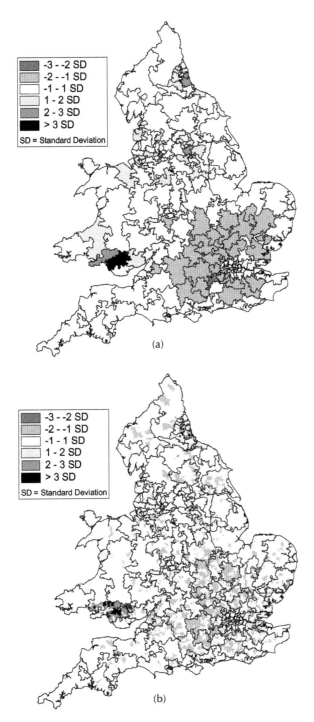

FIGURE 15.6 LLTI visualization using the Standard Deviation method, for (a) compact zones and (b) Wards.

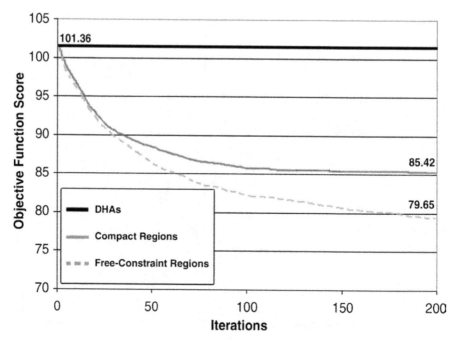

FIGURE 15.7 The improvement of homogeneity according the Objective Function Score.

ACKNOWLEDGMENTS

Census data and boundaries from CasWEB and U.K. Borders.

The first author acknowledges the support of the Hellenic State Scholarship Foundation (HSSF) during his postgraduate studies at Newcastle University.

FIGURE 15.8 The reconstruction of the DHA of Durham. (a) and (b): Abolished DHA of Durham, (c) and (d): Output Zone of Durham with free-constraint approach, (e) and (f): Output Zone of Durham with shape constraint control.

REFERENCES

1. Alvanides, S. and Openshaw, S., Zone design for planning and policy analysis, in *Geographical Information and Planning,* Stillwell, J., Geertman, S., and Openshaw, S., Eds., Springer-Verlag, Berlin, 1999, p. 299.
2. Alvanides, S., Zone design methods for application in human geography, Ph.D. Thesis, University of Leeds, Leeds, UK, 2000.
3. Openshaw, S., *The Modifiable Areal Unit Problem,* Geo Books, Norwich, 1984.
4. HMSO, Health Authorities Act 1995 (c. 17), Her Majesty's Stationery Office, Crown Copyright, London, 1995.
5. Department of Health, The NHS Plan: a plan for investment, a plan for reform, Stationary Office Cm 4818-I, 2000.
6. Macara, A., Managing for health: why health care? *Health Care Manage. Sci.,* 5, 239, 2002.
7. Department of Health, Shifting the Balance of Power: The Next Steps, Crown Copyright, 2002.
8. Higgs, G., Smith, D.P., and Gould, M.I., Realising 'joined-up' geography in the National Health Service: the role of geographical information systems?, *Environ. Plann. C Gov. Policy,* 21, 241, 2003.
9. Smith, D.P., Gould, M.I., and Higgs, G., (Re)surveying the uses of geographical information systems in health authorities 1991-2001, *Area,* 35, 74, 2003.
10. Department of Health, Consultation on the proposal to establish a new Health Authority for Avon, Gloucestershire and Wiltshire, Crown Copyright, 2001.
11. Department of Health, Consultation on the proposal to establish a new Health Authority for Somerset and Dorset, Crown Copyright, 2001.
12. More, A. and Martin, D., Quantitative health research in an emerging information economy, *Health Place,* 4, 213, 1998.
13. Moon, G., Gould, M., et al., *Epidemiology: An introduction,* Open University Press, Buckingham, UK, 2000.
14. Bryant, J., The importance of human and organisational factors in the implementation of computerised information systems: an emerging theme in European healthcare, *Br. J. Healthcare Comput. Inf. Manage.,* 15, 27, 1998.
15. Department of Health, Building the information core implementing the NHS Plan, Crown Copyright, 2001.
16. Bullen, N., Moon, G., and Jones, K., Defining localities for health planning: A GIS approach, *Soc. Sci. Med.,* 42, 801, 1996.
17. Higgs, G. and Gould, M., Is there a role for GIS in the 'New NHS,' *Health Place,* 72, 247, 2001.
18. Stiller, C.A. and Boyle, P.J., Effect of population mixing and socioeconomic status in England and Wales, 1979-85, on lymphoblastic leukaemia in children, *BMJ,* 313,1297, 1996.
19. Dickinson, H.O. and Parker, L., Quantifying the effect of population mixing on childhood leukaemia risk: the Seascale cluster, *Br. J. Cancer,* 81, 144, 1999.
20. Johnston, R.J., Gregory, D., and Smith, D.M., *The Dictionary of Human Geography,* Blackwell, Oxford, 1993
21. Holt, D., Steel, D.G., and Tranmer, M., Area homogeneity and the modifiable areal unit problem, *Geogr. Syst.,* 3, 181, 1996
22. Openshaw, S., A geographical solution to scale and aggregation problems in region-building, partitioning, and spatial modelling, *Trans. Inst. Br. Geogr.,* 2, 459, 1977.

23. Openshaw, S. and Rao, L., Algorithms for re-engineering 1991 Census geography, *Environ. Plann. A,* 27, 425, 1995.

24. Openshaw, S. and Alvanides, S., Applying geocomputation to the analysis of spatial distributions, in *Geographical Information Systems — Vol.1: Principles and Technical issues,* Longley, P.A., Goodchild, M.F., Maguire, D.J., and Rhind, D.W., Eds, Wiley, New York, 1999, p. 267

25. Martin, D., Optimising census geography: the separation of collection and output geographies, *Int. J. Geogr. Inf. Sci.,* 12, 673, 1998.

26. Haining, R.P., Wise, S.M. and Ma, J., The design of a software system for the interactive spatial statistical analysis linked to a GIS, *Comput. Stat.,* 11, 449, 1996.

27. Wise, S.M., Haining, R.P,. and Ma, J., Regionalisation tools for the exploratory spatial analysis of health data, in *Recent Developments in Spatial Analysis: Spatial Statistics, Behavioural Modelling and Neuro-Computing,* Fischer, M. and Getis, A., Eds., Springer-Verlag, Berlin, 1997, p. 83.

28. Mander, U., *Introduction to Algorithms: A Creative Approach*, Addison-Wesley, Reading, MA, 1989.

29. Simpson, S., Resource allocation by measures of relative social need in geographical areas: the relevance of the signed X2, the percentage, and the raw count, *Environ. Plann. A,* 28, 537, 1996.

30. Office for National Statistics, Geography for the 2001 Census in England and Wales, Martin, D., Ed., Census Customer Services, 2001.

31. Rushton, G., GIS to improve public health: guest editorial, *Trans. GIS,* 4, 1, 2000.

16 Tools in the Spatial Analysis of Offenses: Evidence from Scandinavian Cities

Vania A. Ceccato

CONTENTS

16.1 INTRODUCTION

Crime events are far from being random phenomena. They tend to occur in particular geographical areas in a city; they may occur at certain hours of the day and even in association with specific demographical, land use, and socioeconomic aspects of the population. As Hirschfield et al. [1] argue, the discovery of these patterns and regularities through crime analysis is the first step to more finely targeting resources to fight crime and formulate preventive strategies.

Recent literature suggests that geographical information systems (GIS) and spatial statistics can be used for urban planners in the toolbox designed to help in defining measures toward crime reduction in urban areas. Analysis of crime has been facilitated by the use of GIS in combination with spatial statistical techniques that are capable of handling spatially referenced offense data and integrating many types of data onto a common spatial framework, which together opens up new possibilities for better intervention practices in local planning [2–4]. Issues on crime data acquisition and data quality are examples of the remaining challenges for crime analysis and mapping. In a planning context, there is still a constant search of adequacy between choosing "appropriate tools" in accordance to the application goals, either for short or long-term decision-making.

After decades searching for answers in tables and pin maps, police officers and planners now aim at having more robust indicators of urban criminogenic conditions of the city. There is no single way to identify the dynamics of an offense, its spatial distribution over time and space, or the conditions that underlie its occurrence. Neither is the use of statistical packages and GIS the only way for providing a satisfactory basis for decision-making. What we argue here, however, is that this set of tools can, together with experts' knowledge and experience, contribute to a better understanding of the processes that are taking place in the city and provide support for short- and long-term strategies in local planning.

This chapter examines the potential of GIS in combination with spatial statistics in an exploratory analysis of urban geography of offenses in two Scandinavian cities. The term exploratory analysis implies here the use of techniques for detection of patterns in data (clusters) as well as statistical modeling. Techniques such as K-means portioning and Kulldorff's scan test are used to provide a simplified representation of where significant statistical concentrations of offenses occur across the city, while regression models are applied to explain such clusters. Three cluster techniques are applied to data on pickpocketing in Copenhagen, the capital of Denmark. This is followed by an attempt to explain patterns of vandalism using demographic, socioeconomic, and land use covariates in Malmö, the third largest Swedish city. The chapter concludes with a discussion of the strengths and limitations of these techniques for local planning.

These two Scandinavian cities were chosen for the following reasons. First, data availability was a decisive factor. In both Sweden and Denmark, the local police authorities systematically record offense data at a very detailed level of time and space. Second, Copenhagen and Malmö are part of the so-called Öresund region that includes southern parts of Sweden and the northern region of Denmark. The Öresund Bridge, which opened in July 2000, is the first fixed link between Sweden and Denmark, which replaces the boat traffic between Copenhagen and Malmö. Nordic and Baltic regions play an important role in international organized crime [5], and therefore it is reasonable to expect that a new transport link could potentially affect regional and local patterns of criminal activities in the region. For an extensive discussion of trends in offense patterns, see Ceccato and Haining [6]. Third, Malmö and Copenhagen have been targets of several governmental initiatives at all levels aimed at decreasing segregation and improving the quality of life of the citizens, including safety. In Malmö, for instance, since 1997, the State and the municipality have defined a strategy to meet the local needs of the so-called "problem areas." With regard to offense, most of these initiatives (e.g., *Storstadsatsningen*) are of a preventive character focusing on long-term structural changes. Fourteen development centers work to promote employment, which is believed to be crucial to decrease social exclusion, [7] and consequently discourage individuals from becoming offenders. Improvement of physical environment — which is an issue of controversy — is an example of a short-term intervention to improve urban quality and promote safety.

The structure of this chapter is as follows. Section 16.2 presents some guidelines on offense analysis using GIS and spatial statistics techniques. We discuss issues regarding data quality and the process of choosing the most suitable technique in relation to the application's goal. Three examples of techniques for detection of spatial concentrations of offenses are discussed in Section 16.3. A table summarizes the advantages and limitations of each technique from the user's point of view. In Section 16.4, the use of regression models to explain patterns of offenses using zone data is presented. Section 16.5 summarizes with a discussion about the potential of these techniques for planners to monitor, intervene, and strategically plan safety issues at neighborhood and municipal levels.

16.2 PREPARING A DATA SET FOR OFFENSE ANALYSIS

There are two important aspects to be aware of when working with crime analyses at the urban level. The first relates to the quality of the data set, while the second concerns issues of selecting the most suitable technique in relation to the application's goals.

16.2.1 QUALITY OF DATA

Data reliability is an important issue when mapping offenses. Underreporting is a known cause of lack of reliability in databases of offenses. According to Bowers and Hirschfield [8], levels of reporting tend to vary with the type of offense that has been committed and its seriousness. The British offense survey has shown, for instance, that burglary and theft of vehicle is far more likely to be reported than many other types of offenses, such as vandalism and domestic violence. There are also indications that the offense reporting level may be underestimated in areas where people think that it is not worthwhile reporting them, for example, in deprived areas with low social capital [9,10].

There are other problems of data quality that take place during the process of recording offenses. This can be caused by the lack of information about the event from the victim him/herself (not knowing exactly where the offense took place) or by the police officer failing to record the event properly (missing record on the exact location/time of the event). It may be the case that the police officer failed when entering the data on the system, at the first attempt, and the record becomes duplicated as soon as he or she makes the second attempt. This may create extra cases in those particular locations, and that, if not identified in advance, may contribute to "false hot spots."

Many of these problems of data quality relate to the lack of systematization of procedures when recording an offense. Despite the fact that there are conventions for recording offenses in many countries, including Scandinavia, differences still occur in practice. For instance, an assessment of the Swedish offense database over Skåne (Southern Sweden) has shown that large municipalities often have better offense records than the small ones. It is not rare that, in small towns, victims or the police officers only make an estimate of the offense's location (such as "in the surroundings of Jonsson's bakery," "Emma Larsson's garage," "in front of surgery medical center," which contributes to poor-quality records.

Similar reasoning explains the problems found when recording the exact time that an offense took place. This happens, for instance, when someone burglarizes a house while the owner is away. When there is no available information about the time that the event occurred, a range in hours is often the common practice (e.g., 12:00–16:00). However, for analysis purposes, this limits any space–time trend assessment. Aoristic models are used to interpolate time-related missing data (see, for instance, [11,12]).

Another issue related to offense data is related to the geocoding process. Geocoding is the process of matching records in two databases: the offense address database (without map position information) and the reference street map or any other "address dictionary" (with known map position information). The quality of the geocoding process depends very much on the quality of the offense records, the quality of the address dictionary, and the chosen method for geocoding. In cases where the matching of the exact offense location is not possible, a common practice is to choose a near location (such as midpoint of street) or the polygon centroid of a region (e.g., district polygon). This practice creates the so-called "dumping sites" for records [8], which is believed to generate false offense concentrations and consequently, a poor basis for any type of planning intervention. Ceccato and Snickars [13] estimate, for instance, that about 25% of all offenses committed in a Swedish neighborhood were attached to the polygons' centroids of the local commercial area instead of their "real locations."

Cases also exist where the offense site is unidentifiable, for example when it took place between A and B, on a bus, train, airplane, or through the Internet. Evidence from the Swedish database shows that it is unclear in some cases if the address, when reported, is actually the one where the individual was victimized. This may create "false hotspots," since a high number of records may be reported in airports, bus stations, and railway stations. One good way to identify these false hot spots is to try to check for long-term patterns. In case of false hot spots, they may disappear over time, since changes in the way the offense is reported may change, and this affects the choice for "dumping sites."

16.2.2 Type of Technique and Application's Goal

Another key issue when dealing with crime mapping refers to the process of identifying the most appropriate technique in accordance with what the user wants to achieve. According to Craglia et al. [2], crime mapping has essentially three main areas of application: dispatching, community policing, and offense analysis and resource planning (Table 16.1). Each tends to operate at different geographical scales, involving different actors (e.g., police officers, planners, community experts), and has different requirements in terms of data quality and currency and analytical

TABLE 16.1
Typology of Offense Mapping Applications

Application	Data	Geography	Function
Dispatching	Seconds/minutes	x,y coordinate	Visualization
Community policing	Hours/days	Neighborhood/district	Mapping/some analysis
Offense analysis and resource planning	Weeks/months/years	City	Analysis/modeling

Source: From Craglia, M., Haining, R., and Wiles, P., *Urban Stud.,* 37, 712, 2000. With permission.

capabilities. Its application also differs, as suggested by Haining [14, pp. 37–38], in terms of time horizons, from the short-term tactical to long-term strategic deployment of resources. Tactical deployment of resources is often focused on a very narrow and specific set of objectives (e.g., sudden upsurge in street robberies), requiring rapid data collection followed by relevant data processing, perhaps a hot spot analysis to identify the areas in need. Strategic deployment of resources is based on long-term data series and on analyses of the underlying factors (e.g., demographic, socioeconomic, land use) that might help to characterize and explain crime patterns.

On a daily basis, police officers may make use of precise data to dispatch patrol cars to the scene of crime. Community policing requires user-friendly systems for officers to enter the location and characteristics of a reported offense. According to Rich [15], officers in the United States have become producers of maps rather than simply users. Using information collection and automated mapping, officers are able to "walk and use" the system, which requires a limited expertise. This may require a need for rapid data processing, perhaps by using pin maps or cluster detection techniques, for hot spot detection. This is fundamental to identifying areas that need attention and prioritizing resources. There are examples where this may involve other actors such as planners, experts, neighborhood volunteers, and residents (see, for instance, [13]).

An assessment of urban criminogenic conditions could combine different tools and types of data, from simple monitoring of frequency diagrams to more sophisticated techniques. Ratcliffe and McCullagh [11] exemplify, for instance, how the combination of hotspot analysis within a GIS environment, a hot spot perception survey of police officers, and small focus groups can be used to assess the dissemination information on high-risk offense areas. Another way is by using techniques that allow different types of data to calculate risks of being a victim of crime across the city. Tracking changes over time is also a very important issue, not only in monitoring how the offense risks vary over time but also because these changes should potentially affect the police actions and security measures on short- and long-term interventions. A map showing that residential burglary is concentrated in neighborhoods A and B during the day and in C and D during the night may be crucial for avoiding intervention in the "wrong neighborhood" and, consequently, waste of resources. Maps showing changes over time are important to highlight not only how an offense prevention program has reduced offenses in the target area but also whether or not the program has been successful in avoiding offense displacement to the surrounding areas.

Crime is often regarded as the tip of the iceberg of other long-term social problems. The role of long-term strategic planning often involves the coordinated work of the police with public sector agencies and other local actors. In this context, offense mapping goes beyond the detection of patterns and tries to explain why certain areas have a high risk for offense through modeling or combination of several data sets and techniques. Bowers and Hirschfield [8, p.5] state "mapping can be used to make useful inferences about the underlying processes that are causing particular types of offense cluster to form ... it can be used as evidence of the likely presence of a particular process." In the next section, three examples of cluster detection techniques in offense mapping are compared, and this is followed by a discussion of their potential and limitations in the context of local planning.

16.3 TECHNIQUES FOR DETECTION OF SPATIAL CONCENTRATIONS OF OFFENSES

Areas with high spatial concentrations of offenses are often generated by dominance of certain types of land uses in the city (such as a concentration of pubs, restaurants, tourism-related places) but also by the relationship between activities and places (e.g., pickpocketing in central urban areas or drug selling points near schools and clubs for young people). There are many different statistical techniques designed to identify spatial concentrations of an event [16–18]. Cluster statistical techniques "aim at grouping cases together into relatively coherent clusters" [1]. This section presents three different methods for cluster detection, using point and area-based data, and reviews issues and challenges associated with such techniques.

16.3.1 NEAREST NEIGHBOR HIERARCHICAL CLUSTERING TECHNIQUE

The nearest neighbor hierarchical (NNH) clustering technique identifies groups of incidents that are spatially close. It clusters points together on the basis of a criterion. The clustering is repeated until either all points are grouped into a single cluster or else the cluster criterion fails [17]. For this example, CrimeStat® was utilized. This is a spatial statistics program for the analysis of offense incident locations, developed by Ned Levine & Associates under grants from the National Institute of Justice [1] (available at http://www.icpsr.umich.edu/NACJD/offencestat.html#SOFTWARE). Because this package is used in many police departments in the United States as well as by criminal justice and other research institutes, it has been decided to assess its applicability in the Scandinavian context. For the purpose of this case study, we focus on pickpocketing over Copenhagen, as it is among the twenty most common offenses in the period 2000 and 2001. The data set used in this analysis was extracted from the Copenhagen Policy Authority's database on offenses from 2000 to 2001. As Figure 16.1 shows, pickpocketing is very concentrated in inner city areas of Copenhagen, excluding the municipality of Frederiksberg.

The CrimeStat NNH technique uses a nearest neighbor method that defines a threshold distance and compares the threshold to the distances for all pairs of points.

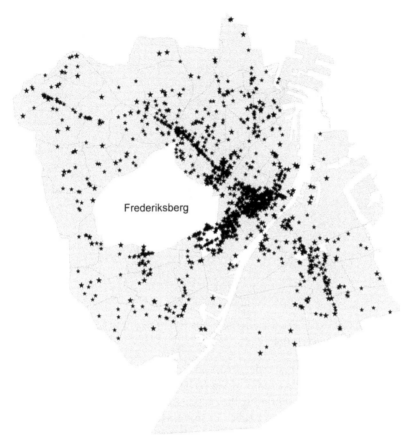

FIGURE 16.1 Pin map of pickpocketing over Copenhagen city, 2000 to 2001. Frederiksberg does not belong to Copenhagen municipality.

Only points that are closer to one or more other points than the threshold distance are selected for clustering. This threshold distance is a probability level for selecting any two points (a pair) on the basis of a chance distribution. In this first criterion, we have chosen the default value for the threshold distance (with probability of 0.5), which means that if the data were spatially random, approximately 50% of the pairs will be closer than this distance. However the number of clusters is dependent on the threshold distance and the minimum number of points in each cluster. In this case, changing the threshold distance from the default 0.5 to the minimum value (the likelihood of obtaining a pair by chance would be 0.001%), does not affect the number of clusters. The second criterion is the minimum number of points that should be included in any cluster. Since we wanted to detect clusters that would reveal vulnerable microenvironments in the city, we used the default of 10 as the minimum cluster size.

Figure 16.2 illustrates the first- and second-order clusters of pickpocketing over Copenhagen using two minimum cluster sizes. Decreasing the number of points per cluster from 10 to 5 increases the number of clusters found from 14 (13 first order,

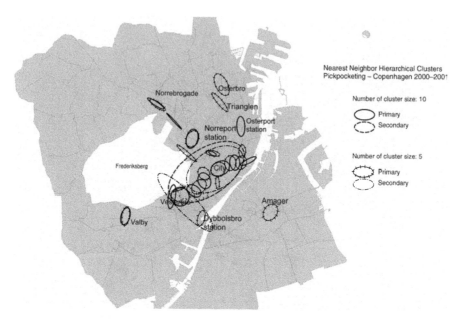

FIGURE 16.2 Nearest neighbor hierarchical (NNH) clusters of pickpocketing over Copenhagen, varying the cluster's size.

1 second order) to 22 (20 first order, 2 second order). Therefore, the user's knowledge of the study area in making sense of the NNH outputs is fundamental in choosing these criteria. The third criterion is the output size of the clusters (in the form of a deviational ellipse) that the user specifies in terms of standard deviations. We used the default value of one standard deviational ellipse "since other values might create an exaggerated view of the clusters" [18].

As these findings illustrate, the NNH technique can identify in detail geographical environments where pickpocketing is concentrated — an important piece of information for short- and long-term intervention. What is evident in this pattern is how clusters of pickpocketing follow main streets (e.g., Norrebrogade), stations (e.g., Osterport station, Norreport station), and local centers (e.g., Trianglen, Österbro), most of which are concentrated in the inner city areas of Copenhagen. This is confirmed by the form and location of the second-order cluster. These places are mostly constituted by either transport links (such as main streets) or transport nodes (such as train stations) — public places that lack "capable guardians" despite being crowded places. Travelers who could in theory be considered as informal guardians may in practice be ineffective. Most people have no sense of ownership in places like train stations and often do not want to get involved, either to intervene during the act or later as witnesses. The same reasoning could be applied to main streets, with a large flow of people passing through during the day. Therefore, transport links and transport nodes of any kind are typical examples of poorly guarded places and highly attractive to motivated offenders for committing pickpocketing.

Another advantage of the NNH technique is that it can be applied to an entire data set (e.g., from the neighborhood level to the county level), which facilitates comparisons between different areas and different cluster levels over time. This means that there is no need to break down the database into different levels. This technique allows the user to detect clusters at several levels: from microurban environments, through first-order clusters (e.g., street corners), to a more comprehensive view of the whole city or county by checking second-, third-, and higher-order clusters.

However, the NNH fails to detect clusters based on attributes other than the observations' location (e.g., offense). NNH technique detects areas where a lot of pickpocketing is committed, regardless of underlying distribution of population or characteristics of land use, for instance. Moreover, as it is suggested, the size of the grouping area is dependent on the sample size, which does not provide a consistent definition of a hot spot area, since a cluster should be dependent on environment and not on the sample size. Finally, the total number of clusters is dependent on the minimum size cluster, which in its turn varies across users, their experience, and knowledge to make sense of the results from each output.

16.3.2 K-MEANS PORTIONING CLUSTERING TECHNIQUE

The K-means clustering technique is a portioning procedure where the data are grouped into K groups defined by the user. The routine searches to find the best positioning of the K centers and then assigns each point to the center that is nearest. Unlike NNH technique, all observations are assigned to clusters, and therefore there is no hierarchy in the procedure, which creates clusters at one level only [17,18]. Since we already knew from the results of NNH technique that there were 13 first-order clusters when using the default values for all criteria, we decided to set as 13 the total K groups in CrimeStat, so we could check whether the clusters had the same geography as the ones produced by the NNH technique.

When many clusters look concentrated in a geographical area such as in Copenhagen city center, a smaller separation is suggested. This will tend to subdivide more concentrated clusters, reducing the distance of each point from the cluster center. In the case of pickpocketing in Copenhagen, the difference in the geography is visually detected in the results when the default value 4 is decreased to 2 (Figure 16.3). As with the NNH method, the K-means clusters also vary, depending on the user's knowledge and experience in identifying the "right number of clusters" or the "right separation" for a specific area. There is always a risk of setting either too many clusters, which will result in clusters that don't really exist, or too few, which will lead to poor differentiation among environments that are different in nature. For an extensive discussion of this issue, see Murray and Grubesic [19].

Compared with the NNH clusters, the K-means clusters are generally larger in size, especially those in the peripheral parts of the city (see, for instance, the clusters in Valby, Amagerbrogade, and Emdrup). At least three new locations appear as clusters in the K-means map that were not clustered in the NNH method, regardless of the variation in the criteria. Both NNH and K-means techniques are better seen as exploratory tools for refining clusters of high values. This implies that if police officers or planners have previous knowledge of where there should be "cluster of high values" then these techniques could be used to verify if these hot spots actually

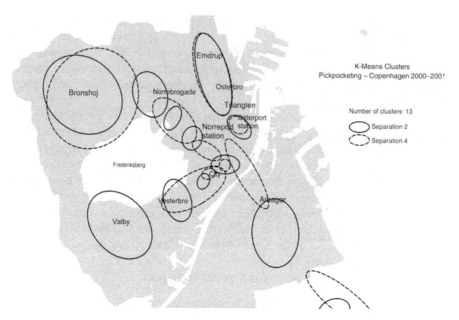

FIGURE 16.3 K-Means clusters of pickpocketing over Copenhagen, varying the cluster's center.

correspond to their perception. In many cases, these outputs generate new questions for the user when "unexpected" hot spots are identified. Therefore, the user's experience is crucial to differentiate "possible real clusters" from the ones that are a product of the statistical procedure or of the data quality. As Craglia et al. [2, p. 716] suggest, these techniques are useful for short-term intervention, since they provide a good basis as to where offenses are concentrated. However, "from a strategic point of view, there is often a need to go beyond the pure count of events," which these techniques are unable to provide. In the next section, Kulldorff's scan text is discussed, using the pickpocketing data set in polygon-based format.

16.3.3 Hot Spots of Offense: Exploring Time Scale with Area-Based Data

In this section, we assess Kulldorff's scan test to detect clusters of pickpocketing over Copenhagen in relation to its nighttime and daytime population. First, we split the pickpocketing data set in 24 slices, corresponding to 24 hours of the day to check the variations in the geography of this offense over time. Then, we discuss the reasons why choosing the "right" denominator when detecting patterns of clusters over time is an important issue for planning purposes.

In order to detect geographical clusters of pickpocketing for the 24 hours of the day, Kulldorff's scan test was used [20] in the data set of Copenhagen. This software has a number of techniques routinely used in spatial epidemiology but could be used for virtually any application searching for measures of relative risk. Kulldorff's scan test has a rigorous inference theory for identifying statistically significant clusters

[21]. The tests use the Poisson version of the scan test, where the number of events in any area is assumed to be Poisson distributed. This adjusts for heterogeneity in the background population. The spatial scan statistic imposes a circular window on the map, which is in turn centered on each of several possible centroids positioned throughout the study area. For each centroid, the radius of the window varies continuously in size up to a maximum window size that includes 50% of the spatial units. The circular window is flexible both in location and size and is moved across the maps to search out the most significant clusters irrespective of size. The spatial scan statistic uses a large number of distinct geographical circles, with different sets of neighboring polygons within them, each being a possible candidate for a cluster. Each spatial unit is represented by a centroid, which determines, for any given window, whether the spatial unit is inside or outside the window. The likelihood function is maximized over all windows, identifying the window that constitutes the most likely cluster (that is to say, the cluster that is least likely to have occurred by chance). Its distribution under the null hypothesis and its corresponding p-value is obtained by repeating the same analytic exercise on a large number of replications of the data set generated under the null hypothesis, in a Monte Carlo simulation (for more detailed information on the spatial scan test, see [20,22]).

Clusters of pickpocketing vary in size depending on the time of day and the denominator with which they are compared. Figure 16.4 shows the frequency of pickpocketing by (a) hours of day in Copenhagen, as well as two examples of clusters resulting from Kulldorff's scan test, standardized by (b) daytime and nighttime population and (c) total population. The first slice was chosen because it refers to the lowest frequency counts of pocket picking, between 7:00 and 8:00 in the morning. The second slice refers to one of the peak hours for this offense, between 2:00 and 3:00 in the afternoon. This example shows that clusters of pickpocketing are unchanged in size when allowing for variation in total population in each area.

However, if we assume that pickpocketing is a function of the daily variation of total people who work and live in a certain area, then the denominator with which the offenses are compared should also reflect this change. Nighttime population was used as a denominator for the sample between 7:00 and 8:00 in the morning, while daytime population was used for the afternoon sample. As Figure 16.4 illustrates, the afternoon cluster was susceptible to change in the denominator and, as a consequence, the cluster became larger than the one standardized by the total population. Despite the fact there has been no large geographical change in the cluster location in the case of pickpocketing (no other clusters appeared, for instance), this may have many practical implications since decisions can be taken based on inaccurate information. For police tactical work, this means that patrols may be sent to a high-risk area that is smaller than it should be. In cities where there is a great variability in space–time clusters, targeting resources to the "right area" may be a difficult task for planners working with strategic distribution of resources.

Lack of demographic and land use data used to create the ratios often limits crime analysis over time. When they are available, they may not be appropriate to all types of offenses. For pickpocketing, robbery, and other violent crimes, the standardization is commonly performed using population totals, and less commonly, day- and nighttime population. However, for offenses such as vandalism, car-related

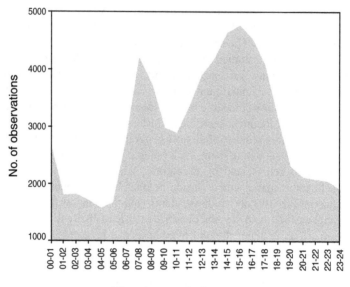

a

FIGURE 16.4 Cluster of pickpocketing in Copenhagen: (a) Frequency of pickpocketing by hours of day. (b) Standardized by night- and day-time population. (c) Standardized by total population.

thefts, and residential burglary, there is no agreement on which denominator should be used for the standardization when the time dimension is taken into account. For car theft, for instance, the total number of cars in each zone might be a good indicator but such data is rarely available by day- and night-time periods. Table 16.2 summarizes the main advantages and disadvantages of each cluster technique.

16.4 TOWARD EXPLANATIONS OF OFFENSE PATTERNS: MODELING VANDALISM IN MALMÖ, SWEDEN

A common question among experts and those involved in security issues in strategic local planning is to what extent the relative risk of being a victim of crime varies across the city and why the risk is greater in certain areas than in others. In this section, we suggest a set of procedures that first provide a notion of risk variability for vandalism across the city, followed by an attempt to explain such distribution, linking vandalism to socioeconomic and land use covariates. The Malmö data set used in this analysis was extracted from Skåne Policy Authority's database on offenses in 2001. In the Malmö case, vandalism involves "offenses on physical targets (e.g., causing damage to cars, walls, buildings, including graffiti) and disturbance (e.g., by starting a fire)."

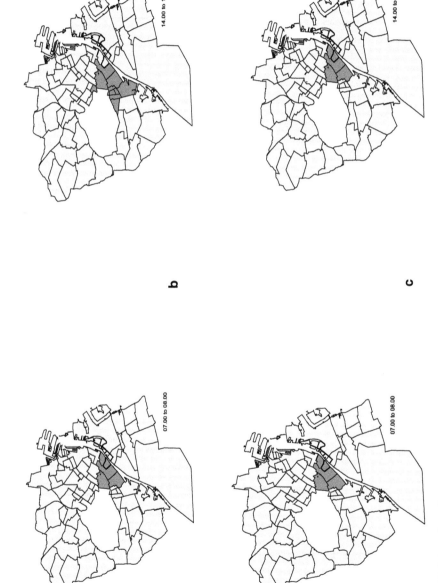

FIGURE 16.4 (continued)

TABLE 16.2
Cluster Detection Techniques: Advantages and Disadvantages

Technique Type	Definition	Advantages	Disadvantages
Hierarchical technique: Nearest neighbor hierarchical clustering	NNH clusters observations together on the basis of a threshold distance between their locations. Number of clusters is highly dependent on the threshold distance and the minimum number of points in each cluster chosen by the user	Identification of small scale clusters (street, local commercial centers, etc.) Multilevel clustering detection (e.g., street, district, city, county levels) that allows different policing strategies	Detects clusters based only on observations (e.g., offenses, x-y co-ordinates); attributes associated with the observation are not incorporated in the routine Hot spots are dependent on number of observations Minimum size cluster may vary across users
Partioning technique: K-Means portioning clustering	K-means divides the observations into a predetermined number of clusters chosen by the user Separation value may also have influence on the pattern	User defines the number of clusters Output is suitable for larger geographical areas	User defines the number of clusters *a priori;* therefore it is highly dependent on user's knowledge and experience There is no hierarchy in the routine
Risk-based technique Kulldorff's scan test	Kulldorff's scan test detects clusters based on an underlying denominator variable, such as population, number of households.	Has a rigorous inference theory for identifying statistically significant clusters (21) Standardization provides a robust measure of how offenses vary over an area Does not depend on polygon size Provides a general view of concentration patterns	Difficulty in including detailed environmental data (x,y data when working with polygon data, for instance) Using polygon data, the results may be too crude given the high degree of heterogeneity within zones and across them

Figure 16.5 shows the map of relative risk for vandalism in Malmö, using a polygon framework. This defines the expected number of acts of vandalism for each polygon, under the assumption that vandalism occurs randomly across the city. The standardized vandalism rate (SVR) for polygon i is given by:

$$SVR(i) = [O(i) / E(i)] \times 100 \qquad (16.1)$$

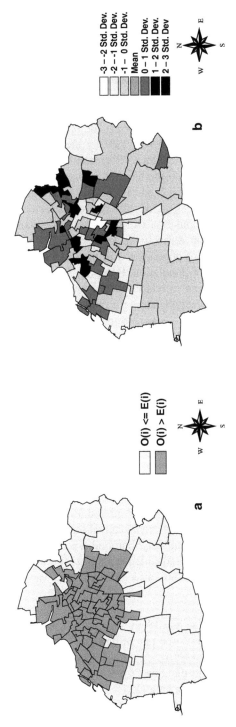

FIGURE 16.5 (a) Standardized vandalism rates. (b) Spatial pattern of residuals — ordinary least square regression model.

where O(i) is the observed number of cases of vandalism, and E(i) is the expected number of cases of vandalism. An average vandalism rate for Malmö was obtained by dividing the total number of offenses by the total size of the chosen denominator. For vandalism, the best denominator suggested is the area of the polygon (see [23]). Since the polygon area varies greatly from the Malmö city center to the periphery, the total population in each polygon was used in this study (minimum population size was 998, the maximum was 7836, the mean was 3748, and the standard deviation was 1617). For each polygon i, this average rate is multiplied by the size of the chosen denominator in polygon i to yield E(i). The observed number of incidences of vandalism in each polygon is later divided by the expected number and then multiplied by 100. Any polygon with an SVR greater than 100 has a vandalism rate greater than would be predicted on the basis of its area (Figure 16.5(a)).

As suggested by Craglia et al. [2] and Haining [14] this statistic may not be the best estimate of the relative risk across the region, since counts for areas with relatively small numbers of population will be sensitive to small errors in reporting offenses and sensitive to random errors in the occurrence of offenses. Rates computed on areas with small numbers of population will therefore be less robust than those computed on large numbers of population. However, the SVR still provides a local measure of crime concentration in relation to the population in the area. Not surprisingly, most of the more vulnerable areas for vandalism in Malmö are concentrated in the central areas of the city — where a mixture of land use determines daily activities and the vulnerability of the area for vandalism. For long-term intervention, more knowledge would be needed on the underlying processes that are causing this particular type of offense to cluster in the central areas and vary greatly in other parts of the city.

In order to try to explain the relationship between vandalism rates and differences in demography, socioeconomic, and land use composition of the city, the ordinary least square linear regression model was fitted. Earlier research has emphasized the relationship between vandalism, social disorganization risk factors [24,25], and low guardianship. Individuals living in areas with high rates of disorder or crime tend to lose their sense of commitment to the neighborhood. Individuals living in problem areas may refrain from local social life, and this breaks down formal and informal social control and involvement at the neighborhood level [26–28]. This in turn leads to more crime and disorder. Although social capital seems to affect crime in general, its effect may depend on crime and neighborhood type. Levels of crime are significantly higher than expected in disadvantaged areas with low levels of social capital [29]. Based on this existent literature, a set of explanatory variables were drawn, as the proportion of:

- Population younger than age 18 (X_1)
- Population with (at least) one parent born abroad (X_2)
- Population born abroad (X_3)
- Population moving into the area (X_4)
- Population moving out the area (X_5)
- Privately owned single family houses (X_6)
- Average income per household (X_7)

- Local leisure associations by population (X_8)
- "Neighborhood Watch Schemes" by population (X_9)
- Bus stops by population (X_{10})
- Central area — dummy (X_{11})
- Commercial areas — dummy (X_{12})

Since the set of SVR values shows a highly skewed distribution, the raw SVR was transformed using its square root transformation to produce a data set that is more nearly normal. The regression analysis and the creation of the lagged variables was implemented in SpaceStat 1.91 [30], because the software has regression modeling capabilities that are appropriate for spatial analysis. SpaceStat provides several statistics measuring the fit of the model, including diagnostic tests, such as tests for multicollinearity among independent variables and tests on model residuals (normality, heteroscedasticity, and spatial autocorrelation). In order to perform tests for spatial autocorrelation in the residuals, a binary weight matrix was used for accounting for the spatial arrangement of the data.

Model results show that the land use variables were statistically significant in explaining the pattern of relative risk in Malmö. The presence of bus stops seems to deter vandalism, while the dynamic of inner city areas (especially the commercial area) is responsible for high vandalism rates. However, we should be careful in drawing conclusions from these results, since the model shows spatial dependence on the residuals (Moran's I is significant), which is a violation of classic regression assumptions (Table 16.3). Martin [31] suggests that, if significant spatial dependence

---

TABLE 16.3
Results of Classic and Spatial Lag Models — SVR

Classic Model — Ordinary Least Square Estimation	Spatial Lag Model — Maximum Likelihood Estimation
Y = Square Root of the Standardized Vandalism Rate	Y = Square Root of the Standardized Vandalism Rate
$Y = 15.72 + (-5.18)X_{10}** + (6.87)X_{11}** + (4.78)X_{12}*$ $\quad\quad\ (-4.91)\quad\quad\ (4.47)\quad\quad (1.92)$	$Y = 6.48 + 0.65W\_y + (-3.64)X10** + (3.35)X11**$ $\quad\quad\ (7.01)\quad\quad (-4.43)\quad\quad\ (2.67)$
(t-values in brackets) * significant at the 5% level **significant at the 1% level	(z-values in brackets) * significant at the 5% level **significant at the 1% level
$R^2 \times 100 = 46.5\%$ R^2 (adjusted) $\times 100 = 44.07\%$ Log Likelihood –194.87 Akaike Information Criterion 397.75	$R^2 \times 100 = 59.1\%$ R^2 (adjusted) $\times 100 = 67.71\%$ Log Likelihood –181.36 Akaike Information Criterion 370.73
Normality of errors — Jarque-Bera 1.49 Prob 0.51 Multicollinearity condition 2.74 Heteroskedasticity — Breusch Pagan 7.76 Prob 0.06 Moran's I (error) 0.29 Prob 0.00	Heteroskedasticity — Breusch Pagan 2.32 Prob 0.31 Lagrange Multiplier test (error) 0.20 Prob 0.65 Lagrange Multiplier (lag) 30.84 Prob 0.00

is found on the residuals, this indicates that some source of variation has been omitted from the model or that the functional form of the model is not correct. As Figure 16.5(b) shows, areas with higher vandalism rates than predicted by the model tend to occur in groups. One way to account for this spatial dependence on residuals is by applying a spatial lag model. A spatial lag model treats spatial dependence as a spatial diffusion process. Thus, in this particular case, the spatial lag model tests the explanation that vandalism rates are spatially autocorrelated, because offenses from areas with high vandalism rates spill over into adjacent areas through people's spatial interaction. Patterns of offending do not recognize district boundaries, and motivated offenders do not only operate in their own districts. Therefore, the spatial lag model was used in recognition that the effects on vandalism ratios, of high or low levels might extend beyond the boundaries of the particular spatial unit. The spatial lag variable (Y_w) is automatically computed in SpaceStat as the average of vandalism rates in adjacent polygons (Table 16.3).

By comparing the fit of the spatial lag model with the classic model, we realize that the overall fit of the model was substantially improved by the inclusion of the spatial lag variable (Table 16.3). Now approximately 60% of the variation in the vandalism rate is accounted by two variables in the model, mostly by the presence of bus stops and "being in the inner city areas." Martin [31] suggests that the inclusion of the lag variable improves the model fit at the same time that it reduces the magnitude of the effects of each independent variable. The most extreme case is the variable X_{12} (dummy for commercial areas) that becomes no longer significant in the spatial lag model. The model results also show that the problem with heteroscedasticity has been reduced, while the spatial dependence on the residuals is no longer significant.

High rates of vandalism are found in central areas, especially in less guarded places, possibly with fewer bus stops. These findings are quite general but are indicative of the processes that generate vandalism in central areas. In the case of Malmö, the presence of people in public places, around bus stops in central areas, seems to deter vandalism. These results may produce insights that may be helpful in long-term strategies, in pointing the way in terms of how resources should be targeted, particularly to vandalism in central areas.

16.5   FINAL CONSIDERATIONS

This chapter has illustrated how GIS and a set of spatial statistics techniques can be used to detect and aid in explaining the geography of selected offenses in two Scandinavian cities. A discussion of the results of three cluster techniques applied to pickpocketing data is presented. This is followed by the attempt to explain patterns of vandalism using demographic, socioeconomic, and land use covariates. These case studies also show how results are highly dependent on the employed criteria and/or data chosen for each technique.

The cases of pickpocketing and vandalism provide evidence of the complexity of factors underlying the geography of crime in urban areas. Although in Copenhagen transport nodes, such as train stations, seem to create just the "right" criminogenic conditions for pickpocketing, in Malmö, bus stops and the dynamics of their surroundings seem to deter vandalism, particularly in central areas. Despite the limitations in comparing offense data between countries [32], future research should focus on identifying trends of offenses in urban areas across countries and search for local factors that may be responsible for particular patterns of crime.

Much work still remains to be done in solving problems of data quality. There is a general consensus that underreporting and lack of systematization of procedures when recording and geocoding offenses are still the common problems when mapping and assessing the geography of crime. There is also a need to make users (e.g., police officers, experts, planners) aware of what can (or cannot) be done using spatial statistics and GIS. The combination of techniques and different data sources in GIS provides an approach for providing a better knowledge base for decision-making. This chapter illustrates the importance of employing a variety of techniques and experimenting different criteria in each method to crosscheck the spatial patterns that seem to exist at the first attempt. The user's knowledge and experience is therefore important in the process of making sense of different outputs.

ACKNOWLEDGMENTS

This research was undertaken while Vania Ceccato was a visiting fellow in the Department of Geography at the University of Cambridge, England. The support of the Marie Curie Fellowship Scheme (Grant reference HPMF-CT-2001-01307) and STINT — The Swedish Foundation for International Cooperation in Research and Higher Education (Dnr PD2001-1045) are gratefully acknowledged by the author. The author would also like to express her thanks to the municipality of Malmö, Länsförsäkringar Skåne, the Skåne Police Authority, and Copenhagen Police Authority for providing the data set used in this analysis.

REFERENCES

1. Hirschfield, A., Brown, P., and Todd, P., GIS and the analysis of spatially-referenced crime data: experiences in Merseyside, UK. *Int. J. Geogr. Inf. Syst.,* 9, 191–210, 1995.
2. Craglia, M., Haining, R., and Wiles, P., A comparative evaluation of approaches to urban crime patterns, *Urban Stud.,* 37, 711–729, 2000.
3. Messner, S.F., Anselin, L., Baller, R., Hawkins, D.F., and Tolnay, S.E., The spatial patterns of homicide rates: an application of exploratory data analysis. *J. Quant. Criminol.,* 15, 423–450, 1999.
4. Ceccato, V., Haining, R., and Signoretta, P., Exploring offence statistics in Stockholm city using spatial analysis tools, *Ann. Assoc. Am. Geogr.,* 92, 29–51, 2002.
5. Galeotti, M., Cross-border crime in the former Soviet Union. International Boundaries Research Unit (IBRU) Boundary and territory briefing, Durham, UK, 1995.
6. Ceccato, V. and Haining R., Crime in border regions: The Scandinavian case of Öresund, 1998–2001. *Ann. Assoc. Am. Geogr.,* 94, 807–826, 2004.
7. Berg, M., En tyckande bomb: invandring och enklaver. *Veckans affärer (special),* 28–30, 2003.
8. Bowers, K.J. and Hirschfield, A., Introduction, in *Mapping and Analysing Crime Data: Lessons from Research and Practice,* Hirschfield., A., and Bowers, K.J., Eds., Taylor & Francis, London, 2001.

9. Farr, M., How does crime reporting, crime levels and crime mix vary by type of neighbourhood? A national perspective, presented at Understanding crime and the neighbourhood: concepts, evidence and police design, Bristol, 2003.

10. Ceccato, V. and Haining, R., Collective resources and vandalism: evidence from a Swedish city, *Urban Stud.*, in press.

11. Rattcliffe, J.H. and McCullagh, M.J., Aoristic crime analysis. *Int. J. Geogr. Inf. Sci.*, 12, 751–764, 1998.

12. Rattcliffe, J.H. and McCullagh, M.J., Crime, repeat victimisation and GIS, in *Mapping and Analysing Crime Data: Lessons from Research and Practice*, Hirschfield., A., and Bowers, K.J., Eds., Taylor and Francis, London, 2001.

13. Ceccato, V. and Snickars, F., Adapting GIS technology to needs of local planning. *Environ. Plann. B Plann. Design*, 27, 923–937, 2000.

14. Haining, R., *Spatial Data Analysis: Theory and Practice*, Cambridge University Press, Cambridge, UK, 2003.

15. Rich, T., *The use of computerised mapping in crime control and prevention programs*. NIJ Research in Action, 1995, (available from http://www.ncjrs.org/txtfiles/riamap.txt).

16. Everitt, B., *Cluster Analysis*, John Wiley & Sons, New York, 1974.

17. Levine, N., "Hot Spot" Analysis I. in *Crime Stat: Software Guide*, Levine, N., Ed., 2002.

18. Levine, N., "Hot spot" Analysis II. in *Crime Stat: Software Guide*, Levine, N., Ed., 2002.

19. Murray, A.T. and Grubesic, T.H., Identifying non-hierarchical spatial clusters. *Int. J. Ind. Eng.*, 9, 86–95, 2002.

20. Kulldorff. M.A., A spatial scan statistic. *Commun. Stat. Theory Methods*, 26, 1481–1496, 1997.

21. Haining, R. and Cliff, A., Using a Scan statistic to map the incidence of an infectious disease: measles in the USA 1962–1995. *Proceeding of GEOMED Conference, 2001, Inserm, Paris.*

22. Hjalmars, U., Kulldorff, M., Wahlquist, Y. and Lannering, B., Increased incidence rates but no space-time clustering of childhood malignant brain tumors in Sweden. *Cancer,* 85, 2077–2090, 1999.

23. Wikström, P.O., *Urban Crime, Criminals, and Victims: the Swedish Experience in an Anglo-American Comparative Perspective.* Springer-Verlag, Stockholm, 1991.

24. Park, R.E. and Burgess, E.W., *The City,* University of Chicago Press, Chicago, 1925.

25. Shaw, C.R. and McKay, H.D., *Juvenile Delinquency and Urban Areas.* University of Chicago, Chicago, Press, 1942.

26. Skogan, W.G., *Disorder and Decline: Crime and the Spiral of Decay in American Neighbourhoods,* Free Press, New York, 1990.

27. Perkins, D.D., Meeks, J.W., and Taylor, R.B., The physical environment of street blocks and resident perceptions of crime and disorder: implications for theory and measurement, *J. Envir. Psychol.*, 12, 21–34, 1992.

28. Kelling, G.L. and Coles, C., *Fixing Broken Windows: Restoring Order and Reducing Crime in Our Communities*, Free Press, New York 1996.

29. Sampson, R.J. and Groves, W.B., Community structures and crime: testing social disorganization theory, *Am. J. Sociol.*, 97, 774–802, 1989.

30. Anselin, L., SpaceStat tutorial a book for using SpaceStat in the analysis of spatial data, West Virginia University, Morgantown, 1992.

31. Martin, D., Spatial patterns in residential burglary: assessing the effect of neighbourhood social capital, *J. Contemp. Criminal Justice*, 18, 32–146, 2002.

32. Aebi, M., Barclay, G., Jehle, J., and Killias, M., *European Sourcebook of Crime and Criminal Justice Statistics: Key Findings,* Council of Europe. 1999.

17 Sustainable Hazards Mitigation

Tarek Rashed

CONTENTS

17.1 INTRODUCTION

Sustainability, as already introduced in several chapters of this book, is a concept that refers to the progression toward improved quality of life in a way that maintains the environmental, social, and economic processes on which life depends, both now and in the future [1]. The World Summit on Sustainable Development convened in Johannesburg (August 2002) marked more than two decades of increasing attention, which this concept has been gaining on the agendas of international, regional, and local communities. A major concern for these communities has been the questions of human/environment relations and the future of people and ecosystems in the face of global environmental change [2–4]. A great deal of effort has been directed to the sustainability of urban ecosystems, in part because cities represent the locus of a diversity of environmental problems with negative consequences that potentially affect millions of people, and in part because of their importance for economic growth and the well-being of present and future generations [5–7].

An important dimension of urban sustainable development relates to how well a city is successful in reducing hazards impacts. Hazards mitigation is a collective term that refers to the different actions taken to reduce risks to life, property, and social and economic activities from hazardous events [8]. Contemporary theoretical, empirical, and policy work on hazards and disasters gives its primary attention to the relation between the risks to which urban societies are exposed and the human practices that may increase, decrease, or reallocate these risks [7,9–20]. The notion of "interaction" is firmly entrenched at the heart of this work, representing hazards as dynamic phenomena that involve people not only as victims but also as contributors and modifiers. Consequently, many researchers and practitioners in the hazards community are changing their emphasis regarding the best way to mitigate against and cope with the complex impacts of hazardous events. This change in emphasis entails a shift from studying the geophysical or technological nature of the events *per se,* to an approach that seeks explanations of disasters primarily in the characteristics of the social, ecological, and built-up environments that experience these events [13–15,18,21].

This "renewed" approach to urban hazards mitigation rests itself intuitively within a broader framework of sustainability, because reducing losses from hazardous events is not a problem that can be solved in isolation through a traditional urban planning model. Rather, it requires an understanding of the magnitude of shock that a given urban system is prepared to absorb and remain capable of operating, and of the means for building a more resilient system through management models that take into account the long-term impacts of mitigation efforts on current and future generations. Sustainable hazards mitigation addresses these concerns by linking wise management of land use and natural resources with local economic and social resiliency, so that actions taken to reduce losses from extreme events are consistent with the following principles of sustainability [7]:

1. Environmental quality, which implies that human activities in a particular locale should not reduce the carrying capacity of the ecosystem for any of its inhabitants

2. Quality of life, which entails many issues such as income, education, health care, employment, legal rights, and other standards that local communities should define

3. Plan for disaster resiliency, which means that a locale is able to withstand any extreme event without suffering devastating losses

4. Economic vitality, which recognizes that vital local economies are essential to tolerating damage and disaster losses

5. Inter- and intra-generational quality, which means not precluding a future generation's opportunity for satisfying lives by exhausting resources in the present generation, destroying necessary natural systems, or passing along unnecessary hazards

6. Participatory process, which adopts a consensus-building approach among all the people who have a stake in the outcome of the decision being pondered

Realizing these goals of sustainable hazards mitigation requires clear ways for specifying as precisely as possible the characteristics of social and physical settings that increase or decrease the vulnerabilities of local communities to different types of hazards. Therefore, adopting a sustainable approach to hazards mitigation directs attention to the role of science and technology, not only in assessing the broad impacts of hazards, but also in helping integrate the three pillars of environmental, social, and economic sustainability in cities [22,23]. Because hazards almost always deal with communities within a defined spatial context (i.e., urban or regional space), geospatial technologies top the list of technology options available for sustainable hazards mitigation. The value of geospatial technologies, especially GIS and remote sensing (RS), in supporting sustainable hazards mitigation in cities arises directly from the benefit of integrating technologies designed to support spatial decision-making into a field with a strong need to address numerous critical spatial decisions [24].

Hazards researchers and practitioners have realized the potential value of GIS and RS for a long time. There is no doubt that these technologies have dramatically improved our efforts to understand the dynamics of hazardous events. This is true when we examine how these tools have helped increase our understanding of the physical character of several hazards and how they have become now essential ingredients of the forecasting and warning efforts of almost all hazards. However, GIS and RS have not had a higher level of positive impacts in support of other mitigation efforts, particularly in assessing the vulnerability and resilience of community [25]. In this latter context, the use of GIS and RS has been criticized for numerous theoretical and technological shortcomings that prohibit optimal vulnerability assessments, thus affecting any subsequent short- and long-term mitigation strategies [24,26–29]. The need to improve upon the current limited role of GIS and RS is more warranted in the context of sustainable hazards mitigation, in which the objectives expand beyond mitigating against hazards to achieving community or regional challenging goals of sustainability.

The purpose of this chapter is twofold: first to present a vision of vulnerability in the context of sustainable hazards mitigation in cities, and second, to provide a critical assessment of the current state of the art in using GIS and RS technologies as they apply to the questions of community resilience in the face of urban hazards. To attain its purpose, the chapter first introduces some important definitions necessary for subsequent discussion, followed by a synoptic discussion of urban vulnerability as it relates to sustainable hazards mitigation. The chapter then highlights the role of GIS and RS in hazards mitigation. It illustrates some examples of how these technologies are currently used to support risk reduction efforts and sheds light on some of the gaps and obstacles that slow down their full utilization in the strategic and operational planning and management of sustainable hazards mitigation. Next, the chapter presents examples of the efforts being made toward addressing current limitations. In its conclusions, the chapter proposes a suggested geospatial approach to vulnerability analysis and attempts to articulate a preliminary agenda of the key considerations that need to be taken into account for a better application of GIS and RS to sustainable hazards mitigation.

17.2 DEFINITIONAL CLARIFICATION

Hazards are often described by their origins, be they natural (e.g., earthquake, tornado), technological (e.g., chemical spill, dam failure), or human made (e.g., a terror attack, war). However, as Cutter [30] notices, this classification is losing favor among the research community because many hazards have more complex origins. The discussion presented in this chapter focuses mainly on hazards originated by extreme natural events. Nevertheless, many of the concepts being discussed apply well to other kinds of hazards because, as argued below, the effects of hazards relate more to the urban context within which they occur than to the source of the hazards themselves.

A clarification of hazards-related terminology is warranted for any attempt to assess the role of geospatial technologies to sustainable hazards mitigation. This is because academic and professional work on hazards is multidisciplinary and multidimensional, and this has led to differences in priorities and divergent views within the hazards community [7,14,30,31]. As a result, key concepts such as hazard, disaster, risk, and vulnerability are used interchangeably, or imply different things to different researchers. Providing a precise definition for each of these key concepts is essential not only for clarification purposes, but also for the understanding of the range of disaster management activities.

An urban "hazard" can be defined as a potential interaction between urban settlements and extreme events that constitutes a threat to society. A "disaster" is the actual occurrence of an event that has a large impact on society in terms of loss of life or injuries, property damages, or environmental losses. Urban "vulnerability" refers to the inherent weakness in certain aspects of the urban environment that are susceptible to harm due to certain biological, physical, or design characteristics. It is generally defined as a measure of coping abilities of human and physical systems in the urban environment. Urban "risk" indicates the degree of potential losses that could result from an interaction between hazards and community. Risk can be thought of as a product of the probability of hazards occurrence and the degree of vulnerability (i.e., risk = hazard × vulnerability).

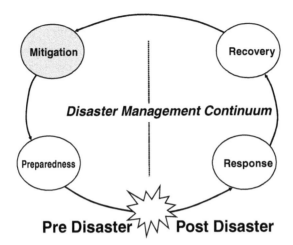

FIGURE 17.1 The disaster management continuum.

What are the implications of these definitions? In fact, there are several, but I shall focus here on the most important two for the sake of brevity. First, they draw a clear distinction between hazards as threatening events and disasters as after-the-fact situations. This difference is crucial to the understanding of the spectrum of activities conducted along the disaster management continuum. Figure 17.1 represents the four main phases of the disaster management continuum: response, recovery, preparedness, and mitigation (see Oakley [32], for a detailed description of the activities included in each phase). Response refers to the range of actions (e.g., rescue, evacuation, firefighting, sheltering) taken immediately before, during, and just after a disaster to save lives and minimize property damage. Recovery refers to actions that help return infrastructural systems to minimum operating standards (e.g., damage assessment, debris removal, crisis counseling). Preparedness includes a variety of measures aimed at assuring that the community is prepared to react during an emergency (e.g., identify responder organizations, design and implement plans for emergency situation, conduct evacuation exercises). Finally, mitigation refers to the ongoing efforts that communities must make on a continuous basis to reduce loss of life, livelihood, and property caused by threatening events.

Of the four phases described above, mitigation is the only phase that deals with hazards, while the remaining ones entail activities that occur just before or after disasters. In the United States, the Federal Emergency Management Agency (FEMA) defines mitigation as "any sustained action taken to reduce long-term risk to human life and property from hazards." This definition highlights the long-term impact that effective mitigation can produce, and endorses an active and adaptive view of the responsibility of human society in the face of dangers [33]. It recognizes both the long-term certainty of hazardous events and the high degree of uncertainty in the short and medium terms.

The challenge, then, is to identify the paths to which hazards mitigation should be directed to realize risk reduction goals. The answer to this question is the second important implication that can be inferred from the definitions mentioned above and concerns the formula, risk = hazards × vulnerability. This formula, originally introduced by the United Nations Disaster Relief Office (UNDRO), implies that risks of a particular community are determined by two crucial components of the disaster complex: (1) the physical characteristics of the hazardous events, and (2) other social, economical, and political factors that determine the vulnerability of that particular community. Hence, there are two possible paths to mitigating hazards: either by reducing vulnerability or by modifying, where possible, the hazard. The latter path, modifying hazards, has been the only track for hazards mitigation for years. An example of this is the attempts to control hazardous events through engineering solutions such as levees, dams, and other protection structures [17,18,34,35]. This track, however, was subject to considerable criticism, which viewed the modification of hazards as a short-term relief that would ultimately exacerbate the hazards in the future [34]. Tobin and Montz [36] mention the levee effect as a classical example of the consequence of using technology to modify hazards, showing how the construction of levees and other flood control structures has given a false sense of safety, leading to an increased development in hazardous sites.

Consequently, reducing vulnerability becomes the only alternative path for hazards mitigation (Figure 17.2). The vulnerability approach to hazards mitigation acknowledges that the patterns of damage provoked by hazardous events depend upon a set of factors that may or may not be related to the hazardous events themselves [16,37]. Examples of these factors include the conditions of human settlements and infrastructure, public policy and administration, the wealth of a given society and its organized abilities, as well as gender relations, economic patterns, and ethnic or racial stratification. The collective effect of these factors shapes the adaptive and coping capacities of the urban system and determines the extent to which a society can tolerate damage from extreme events. Hence, the vulnerability path to hazards mitigation, by placing these factors in the central reality of the disaster, suggests that disasters do not originate from "outside" the urban place, nor do they erupt accidentally within it. Rather, and as discussed in the following section, they represent a product of everyday social life and ongoing urban dynamics that acts upon the society and controls its mutual relationship with the environment [7,11,12,14–16,18,19,31,38–41].

17.3 VULNERABILITY ANALYSIS AND SUSTAINABLE MITIGATION IN CITIES

17.3.1 VULNERABILITY AND THE URBAN COMPLEX

Remarkably, none of the principles stated earlier in this chapter for sustainable hazards mitigation maintains an explicit focus on the geophysical process of a particular natural or technological hazardous event. Rather, all of the six principles direct attention to societal conditions that promote or reduce safety. From this perspective, the potential damage from disasters in urban areas is mostly dependent upon differential vulnerability within and between societies. Accordingly, vulnerability analysis becomes essential: (1) to explain the differential losses between people, urban ecosystems, and physical features due to disasters at the local level;

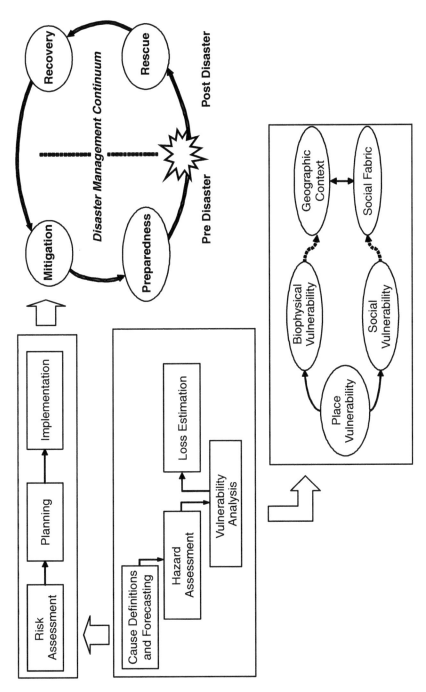

FIGURE 17.2 The scope of vulnerability analysis in hazards mitigation.

(2) to evaluate the ability of urban systems to absorb the impact of disasters (i.e., system resistance) while continuing to function and recover from losses (i.e., system resilience); and ultimately (3) to determine the best options available to mitigate against hazards.

Borrowing from Hewitt's ecological analysis of risk [37,42,43], Mitchell et al.'s contextual framework of hazards [44], Cutter's hazards-of-place model [13,45], and Mileti's systems approach to disasters [7], it can be asserted that patterns of vulnerability to hazards in the urban environment are contingent upon the physical, technological, social, economic, and political realities of the urban place. In order to better assess the vulnerability of a place, one must build a local-based knowledge of the relative contribution of each of these realms to the overall resilience profile of the urban community in that place. As shown in Figure 17.2, an analysis of place vulnerability that incorporates such local knowledge provides a basis for a better assessment of the impacts of hazards in that place and will ultimately help devise prosperous and sustainable hazards mitigation policies.

In both the ecology [46–48] and sustainability science [1,3,22,23,49] literatures, general systems theory [50] is typically used to bridge gaps between different domains toward understanding the dynamics of complex systems and their implications on sustainability. Such complex systems' thinking is likewise useful in the context of vulnerability analysis and sustainable hazards mitigation in cities. It brings together a number of rigorous concepts for exploring the resilience of urban communities to hazardous events across a variety of scales and time dimensions. Perhaps of utmost importance is the holistic systems thinking of the urban place as a *holon*. This term is borrowed from Arthur Koestler, who borrowed it from the Greek language to express the idea that any system, whether large or small, complex or simple, is made up of parts to which it is the whole (a *suprasystem*) and at the same time is part of some larger whole of which it is a *component or subsystem* [51,52]. The notion of holon requires the observer to attend to both the subsystems of that focal system (i.e., the system chosen to receive primary attention) and the suprasystem of which the focal system is a part (or the significant environment to which the focal system is related), in order to fully understand it [53].

One of the main implications of conceptualizing urban places as holons is that there exist mutual interactions between the various components of an urban place governed by a "systemic" relationship of cause and effect. [54–57]. For example, the survival of individuals in the urban place would be impossible without the necessary policy to organize their lives, and the necessary engineering to provide structures to house people, transportation to move them, and production and recreational facilities. Similarly, differences in social characteristics and the relative densities of urban population not only affect the political trends but also impact the configuration of urban places, and the availability of transportation, shelter, and natural resources. Another key implication is to recognize that a change in any component of the urban ecosystem, for example due to a hazardous event (e.g., ground shaking, flood) or due to mitigation measures (e.g., damming a river or draining a bog), may alter the entire system in an unpredictable manner [58]. Therefore, it may not be useful to seek an understanding of the mutual relationships between vulnerability, sustainability, and hazards mitigation by searching for linear,

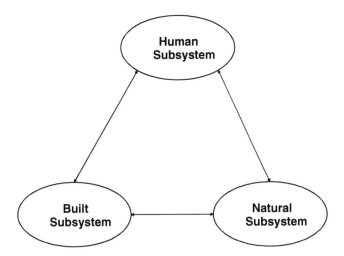

FIGURE 17.3 The dynamic nature of the urban system.

one-directional cause–effect relationships within the urban context. Rather, vulnerability and sustainable mitigation may be better understood in terms of the systemic relationships between the urban components.

17.3.2 FORMS OF VULNERABILITY IN CITIES

In order to demonstrate the systemic notion of causality and how it affects the resilience of urban communities, the urban place can be partitioned among conceptual foci of three subsystems: *the human subsystem, the built subsystem,* and *the natural subsystem.* Figure 17.3 reconfigures these foci as a three-component "human-built-natural" conceptual framework of the urban place. As shown in the figure, these three subsystems of the urban place are intricately interrelated and depend greatly upon each other. They thus provide the contextual "filters" to the understanding of the factors and constraints that either exacerbate or attenuate various patterns of vulnerability in the urban environment, and which should be targeted by sustainable hazards mitigation policies. Examples of the forms of vulnerability (and their underlying factors) that may emerge from the systemic interrelations within the urban place are given below. These examples have been partitioned among the conceptual foci of the urban place shown in Figure 17.3.

17.3.2.1 Human-Built

A number of factors and phenomena amplifying the vulnerability and undermining the sustainability of urban communities can be considered through the human-built filter. Perhaps the most obvious is related to urban crowdedness, which has direct negative consequences on vulnerability such as the segregation of people into different neighborhoods on the basis of social characteristics (e.g., ethnicity, occupation, income) [59,60]. Directly aligned with urban crowdedness is urban sprawl, where urban activities intrude into the previously nonurbanized landscape [61].

Urban sprawl typically results in dramatic transformations in land cover at the fringe zone that are typically economically inefficient and entail potential problems in terms of disaster impacts [59]. Another example is the accessibility within urban areas, which represents a spatial construct influencing how people can escape the danger, and how emergency services can reach affected areas [62,63]. Accessibility depends on the status of transportation networks, and on information about who needs to be evacuated and where each needs to be routed to reach safety [24]. A third example is the response capacity of the urban system, as indicated by the availability of emergency services (e.g., hospitals, fire stations), and the quality of utilities (e.g., gas, water, electricity). Response capacity determines how well a community is prepared to cope with any major accident affecting its local settings [37,64]. Recent studies have shown that characteristics of the urban morphology also influence the function of the city in the immediate aftermath and during the recovery from disaster impacts [24,55]. Included among these characteristics are such factors as the age of buildings, building materials, and construction types as overall proxies for the quality of construction of structures in a local place. Also included are insurance legislations, building codes, and preparedness plans as overall proxies for the appropriateness of mitigation measures undertaken by a given urban community.

17.3.2.2 Human-Nature

The theory of "social marginalization" states that global processes such as population growth and near landlessness conditions in many cities pressure people with certain social and demographic characteristics to cluster for survival in marginal urban places [65,66]. These places are around features that entail actual or potential ongoing problems for local residents such as threats of disasters, for example the tectonic zones in Los Angles and the reclaimed land in Tokyo [60,67]. The social fabric of these marginal places is typically shaped by particular socioeconomic status, age structure, and ethnic characteristics [11,37,66]. Mileti [7, pp. 22] argues that "differences in these characteristics result in a complicated system of stratification of wealth, power, and status ... which in turn results in an uneven distribution of exposure and disaster losses." Understanding and assessing these factors require a more integrative approach that combines views from engineering and the social and political sciences to help identify new and interesting relations that may not be revealed with either one of these views taken separately (e.g., relationship between socioeconomic status, building codes, and the structural weakness of buildings).

17.3.2.3 Built-Nature

Vulnerability relates to a number of concerns that speak to the core of sustainable policies regarding the alteration of ecosystems and biophysical conditions of the natural subsystem within urban areas. This alteration results in negative changes in the characteristics of the natural system and can produce extreme events [14,68]. Examples include the degradation of soils, forests, water, air, and climate [69]. As a result of such changes, specific hazards have particular manifestations in urban areas [8,41]. For example, flooding in the urban areas is exceptional because the urban fabric consists of extensive impervious surface. Similarly, seismic hazards and landslides are highly significant in the urban context because they induce other problems that are urban specific such as fires, and toxic and chemical releases. It is therefore crucial for any assessment of urban vulnerability to incorporate metrics that describe the ecological patterns (e.g., cohesion, fragmentation) of different materials comprising urban landscape. Generating these metrics can help in examining the extent to which environmental issues are considered in sustainable public policies that deal with urban hazards.

17.4 GIS AND RS FOR VULNERABILITY ANALYSIS

17.4.1 LIMITATION IN CURRENT GIS/RS-BASED VULNERABILITY ASSESSMENT MODELS

The discussion in this chapter has thus far focused on making a case that proper vulnerability assessments are the means to building a more resilient urban system and a guiding planning principle in devising sustainable hazards mitigation policies. This line of reasoning draws to attention that successful GIS and RS applications to sustainable hazards mitigation largely depend on the extent to which these technologies are effectively used in the context of vulnerability analysis and the development of mitigation measures. "Effective use" means utilizing GIS and RS as objective ways of specifying the characteristics of social and physical settings that increase or decrease the resilience of a particular local community in the face of threatening hazards. It also implies using these technologies as quantitative means to assessing the effects of hazards policies and urban planning models on the vulnerability profile of the community both on the short- and long-terms. Current vulnerability assessment models fall short in this regard [13,23,25,26,70], though there is no doubt that a commendable progress has been made by researchers and practitioners in the field toward a better understanding of societal/hazards relations.

Drawing from a review of the literature, the limitations in current vulnerability models stem from the following key factors [10,11,20,39,40,45,55,60,63,69,71–76]:

- There are divergent views of the meaning of vulnerability and approaches to its assessment.
- The majority of vulnerability models are hazard-specific and rarely incorporate a spatial dimension.
- Many vulnerability models are essentially descriptive and too broad to be of operational value to planners and disaster managers.
- There is a lack of conceptual development on the most appropriate metrics to measure and compare the relative vulnerability among places and groups of people.
- Few models, if any, suggest ways to assess the sensitivity of a community's vulnerability to both the short- and long-term effects of the mitigation measures these models recommend.

The rest of this section is devoted to elaborating more on these limiting factors.

17.4.1.1   Divergent Views of Vulnerability

Although vulnerability was long recognized as an essential concept in hazards research [17,37,77,78], there is little consensus among researchers, planners, and disaster managers regarding the best way to undertake vulnerability analysis. Therefore, assessing urban vulnerability can generally be regarded as an ill-structured problem (i.e., a problem for which there is no unique, identifiable, objective optimal solution) [25,79]. The literature shows a number of contrasting definitions of what vulnerability means, as well as numerous conflicting perspectives on what should or should not be included within the broad assessment of vulnerability in cities. For example, Cutter [13, pp. 531–532] lists more than a dozen different definitions of vulnerability. Likewise, many discrepancies are found when we examine the available models of vulnerability. On the one hand, there are sociopolitical models that direct remarkably little treatment to the geographic space within which patterns of vulnerability are shaped [10,11,19,80]. On the other hand, there are more technically oriented models in which vulnerability is assumed to be closely tied to the physical properties of the geophysical hazards shown on the maps [75,81,82]. These models limit the analysis of urban vulnerability to the technically narrow concept of risk that describes how an urban place is exposed to hazards, rather than how it could cope with their impacts. What is missing in these models is a broad-based assessment of vulnerability in terms of people's resilience and coping capacity, partly because disaster managers do not fully appreciate the value of this information and partly because methodologies for including it in vulnerability analyses have not been worked out [60].

17.4.1.2   Hazards-Specific Models

There is a need to move beyond the hazard-specific focus in existing vulnerability assessment models toward a broader approach that acknowledges the cumulative impact of hazards within contemporary world cities. As cities in the world become geographically more dispersed and increasingly complex and diverse with respect to population, infrastructure, and the built environment, more and new kinds of urban hazards are brought about by the increasing dependence of communities on technology and more complex interactions within the urban systems. Emerging views in the hazards community realize that a better comprehension of the collective impact of multiple hazards on urban areas is warranted [4,7,12,70,77,83–85]. A multiple hazards approach to vulnerability is necessary within this context to assess the long-term impact of mitigation strategies on community vulnerability and to ensure that strategies implemented to mitigate one hazard does not amplify the vulnerability to another hazard.

17.4.1.3   The Descriptive Nature of Current Models and Lack
of Comparative Metrics

Although one cannot say that there exists a nonvulnerable urban community, there are clearly places that are more vulnerable than others at any given regional or national context. For a disaster manager, the goals are to identify those population groups and urban places that are the most vulnerable, and then, to develop immediate programs to reduce their vulnerabilities. For an urban planner, the goals expand well beyond short-term solutions toward setting-up management models that take into account the long-term impacts of mitigation efforts on current and future generations. In either case, this requires "quantifiable" indicators to concentrate on those people and places most in need. One of the problems associated with the ill-structured nature of vulnerability analysis is that many existing models are essentially too descriptive and too broad to help in this regard. Further, in their attempts to identify the "root causes" of disasters, these models come close to analyzing everything in the urban setting on the perfectly reasonable grounds that everything is relevant. However, the pragmatics of policy and management operations, set within constrained budgets and competing interests, require a focus on a smaller number of factors that allow planners and decision-makers to focus on the more vulnerable communities in their midst, and thus to help develop measures that could prevent hazards from becoming major human disasters.

17.4.1.4 Lack of Policy-Sensitive, Long-Term Models

Radke et al. [29] divide the contribution of geospatial technologies to disaster management into three axes. One axis represents the use of these technologies in the assessment of hazards, for example, mapping flood zones, predicting the path of a tornado, monitoring fires, or forecasting the spatial and temporal extent of a terrorist biological warfare attack. A second axis represents the "short-term" actions taken during the rescue, recovery, and preparedness phases of disaster management (shown earlier in Figure 17.1). A third axis reflects the "long-term" actions, such as discovery, planning, mitigation, management, and policy. Much of the work utilizing GIS and RS techniques in hazards-related contexts has been focused so far on the first two axes (assessing hazards and short-term disaster management). In many cases, the focus on mapping is still a major part of the effort [9,24,27,86–88]. Little has however been done along the third axis, where vulnerability assessment and sustainable hazards mitigation fall, especially in assessing the long-term impact of mitigation policies and their effects on the resilience of urban systems. For example, what is the impact of "smart-growth" policies that call for urban compactness and vertical growth on vulnerability? How long does it take for a change in building codes to reflect on the overall safety of a city?

17.4.2 Examples of Recent Efforts Being Taken to Address These Limitations

Fortunately, several efforts within the hazards and GIScience communities are on the move to address many of the key issues highlighted above, fueled to some degree by recent advances in technologies for geospatial data acquisition, communications, and data access. While much of these efforts do not deal explicitly with the question of vulnerability analysis, nor are they directly oriented toward sustainable hazards mitigation, they nevertheless represent cornerstones in what can be thought of as a foundation for a geospatially oriented solution to the dilemma of vulnerability analysis.

One example of these efforts is the advances that took place in GIS-based disaster simulation models, a remarkable example of which is a GIS-based multihazards and loss simulation model (HAZUS-MH, http://www.fema.gov/hazus/hz_meth.shtm) that has been released very recently (2004) by the Federal Emergency Management Agency in the United States. HAZUS-MH can generate an estimate of the consequences to a city or region of either a deterministic or probabilistic scenario of an earthquake, a flood, and/or a wind hazard [89]. The resulting loss estimate generally describes the scale and extent of damage and disruption that may result from potential hazards in a given area. This includes estimates for physical damage (e.g., building collapse, content damage), social losses (e.g., number of deaths and injuries), and direct and indirect economic losses (e.g., loss in productivity and functionality). While HAZUS-MH is essentially a "risk" (but not vulnerability) assessment tool, its multihazards focus and ability to generate quantitative estimates of different disaster scenarios can provide a starting point to reveal and compare the vulnerabilities of places [25]. Similar tools are available worldwide such as RADIUS (Risk Assessment tools for DIagnosis of Urban Areas against Seismic disasters, http://www.geohaz.org/radius/), EPEDAT (Early Post-Earthquake Damage Assessment Tool, http://www.eqe.com/), and RiskLink-DLM (Detailed Loss Module, http://www.rms.com/). Unlike HAZUS, these are hazards-specific models but could certainly be modified to address a range of other hazards.

A second example relates to several efforts being undertaken to define sets of comparative indicators for vulnerability. Rashed and Weeks [90] proposed an index of social vulnerability based on the wealth of urban communities and weighted by such factors as race, age, and gender, which are used as proximate determinants of access to resources among urban communities. Cutter et al. [91] proposed a more sophisticated social vulnerability index that incorporates over 40 demographic and socioeconomic variables derived from the U.S. population census using a factor analytic approach. Cova [63,92] developed an index of evacuation vulnerability. Lindsay [93] compiled a composite index of health vulnerability for Canadian provinces during disasters, which combines over 12 different health-related variables.

A third example concerns the use of remote sensing satellites to derive physical indicators of vulnerability. A considerable number of recent studies have been conducted to utilize satellite sensor data in the analysis of urban change [94,95–103]. While they do not explicitly speak of vulnerability, there is no doubt that the fruitful findings of these studies can help enrich our understanding of the physical and socioeconomic derivers of changes in urban land cover. They also add to our understanding of the implications of these changes on urban vulnerability as it relates to land use practices, resource management, and hazards mitigation policies in cities. Some of these studies went further beyond the characterization of change and its causes and attempted to integrate remotely sensed data with models of socioeconomic and demographic processes, including social vulnerability [90,104–108].

Finally, few efforts have been undertaken toward building reliable models for evaluating mitigation strategies. For example, Gupta and Shah [109] introduced a decision support tool, the strategy effectiveness chart (SEC), to evaluate different mitigation strategies based on their cost-effectiveness. The SEC first prioritizes risks and mitigation goals according to four sectors: residential, commercial/industrial, lifelines, and government. It then assesses the effects of each mitigation measure on these sectors to create a composite indicator of the effects of each mitigation strategy on all the sectors at a given study area. The resultant score could be translated into dollar values, which can be easily interpreted by decision-makers. While this attempt is to be applauded, a more comprehensive evaluation of sustainable mitigation strategies is yet to be developed. Such evaluation needs to reflect a complete understanding of the interaction between the human and the physical systems and how all the pieces can be put together to support long-term sustainability goals.

17.5 A PROPOSED GIS/RS-BASED APPROACH TO VULNERABILITY ANALYSIS

A decade ago, Coppock [27] wrote about GIS applications to disaster management, documenting several issues that were conceived of as major obstacles in this domain during the 1990s. Coppock broke down these issues into four major areas: (1) deficiencies of widely available commercial GIS software during this era with respect to risk assessments and socioeconomic data modeling; (2) lack of large volumes of appropriate data typically required in emergency applications; (3) inability to meet the needs of intended users adequately; and (4) lack of appropriate methods that are based on a sound understanding of the phenomena under consideration.

Now, many of these issues do not seem to be problematic any more. Many of the socioeconomic and housing censuses all over the world are complied into GIS databases. Large volumes of datasets are collected at fine spatial and temporal scales, many of which are on web-based servers and virtually accessible to millions of potential users. Advances in commercial GIS software, combined with the power of the Internet for rapid sharing and analysis of data, have created an exciting range of possibilities for those working in collaborative settings and concerned with crises management and response to disasters. Several tools have been developed to simulate disaster impacts and to virtually mimic the activities undertaken during the event of a disaster. Moreover, considerable effort has been made in developing data models specifically designed as frameworks for collecting and sharing geospatial information. For example, a series of data models developed by ESRI® (Environmental Systems Research Institute, Inc., Redlands, CA, http://support.esri.com/dataModels) for defense and intelligence, local governments, transportation, land parcels, census of population, environmental regulations, and hydrology collectively provide a rich start for conceptualizing and inventorying the range of data layers needed for disaster management. Likewise, recent advances in remote sensing systems, both commercial and state-owned, provide immeasurable benefits for monitoring disaster situations in real time, as done during the events of September 11 [110,111]. Thus, there is no doubt that the current state of the art in GIS and RS for emergency response and disaster management has changed considerably since the comments made by Coppock back in 1995.

The question is, then, what needs to be done so that we are able to say the same about the limitations that are currently impeding proper GIS and RS applications to vulnerability analysis and sustainable hazards mitigations? It can be argued that the answer to this question is not related to data availability, nor is it due to problems in current GIS software. Rather, the answer stems from an intellectual issue which lies in bridging the gap between the two fields of hazards management and GIScience. This is not to say that data or software are not important to vulnerability assessment, but it is to stress the need for new methodologies that are more suitable for dealing with the problem of urban vulnerability and specifically oriented to sustainable mitigation.

There is a need for a comprehensive approach to vulnerability analysis that is able to balance two competing demands. The first demand is offering a replicable way for researchers as well as planners and decision-makers undertaking local mitigation efforts to generate concrete profiles of vulnerable communities and to monitor changes in these profiles over time. The second is being able to bring together divergent perspectives on urban vulnerability in order to test related theories and hypotheses, thus improving our understanding of the linkage among various contextual factors that produce vulnerability patterns.

A number of frameworks and research agendas for vulnerability analysis have been recently proposed by Cutter [70,112], Polsky et al. [113], and Turner et al. [23] They all share a common theme in emphasizing the need for comparative indicators that reflect the forces that either amplify or attenuate vulnerability, and the role of geospatial technologies in aiding the decision-making process. I build below upon these ideas to propose a new design structure for a vulnerability assessment procedure that has been identified from among the competing options.

The proposed procedure is centered on a dynamic causal model (Figure 17.4) that adopts a systems-thinking approach and attempts to explain how vulnerability patterns arise from adverse interactions between and among the components of the urban system. Causal models can be oriented in one of two ways: starting with a set of causes and examining their consequences, or starting with a set of consequences and moving down to their causes. I propose to use the latter path through a distinctly spatial approach to vulnerability analysis. The underlying rationale for selecting this path stems from the fact that major losses from disasters in urban places do not necessarily result from the immediate impact of the triggering hazard. Instead, they are more likely to arise due to a "chain" of other hazards that may be induced by the triggering event (e.g., fire, chemical release, landslides). Rather than being uncertain or accidental, this interpretation of risks suggests that high impacts from urban hazards are to be expected in those urban areas where the society and nature interrelations are adverse (e.g., unsustainable and not balanced). From a spatial perspective, this means that the problem of vulnerability can be conceptualized as a problem of searching a particular geographic region for evidence of such unsustainable relationships.

Figure 17.4 takes this core idea and attempts to translate it into an empirical spatial approach to vulnerability assessment that utilizes current advances in geospatial techniques. In the proposed spatial approach, the effects of past and hypothetical disaster experiences from multiple hazards are simulated for a particular region. Each simulation will show how potential damages or losses (risks) from a simulated hazard are distributed across the region. When several simulations are run using a single set of date, then variation in simulation results becomes a function of the

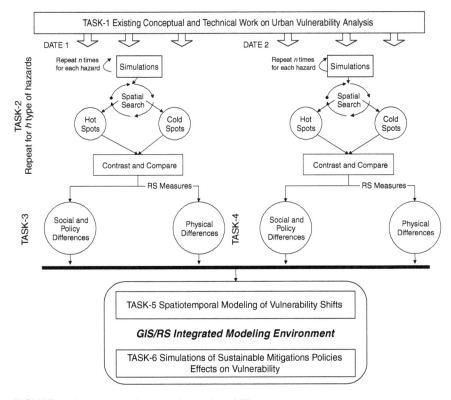

FIGURE 17.4 A proposed approach to vulnerability assessment.

type, location, and magnitude of the hazard being simulated (recall, risk = hazard × vulnerability). In such case, finding the most vulnerable places within an urban region becomes a matter of: (1) ranking urban places based on the severity of losses calculated from each simulation, and then (2) searching the region for those places that maintain relatively higher ranks across all simulation scenarios. These places are deemed the most vulnerable, because maintaining a higher rank across different scenarios implies that a place is always experiencing severe losses regardless of the hazard type, originating source, or magnitude. Hence, the losses in that place can directly be attributed to its vulnerability. Once places with higher vulnerability are located (the hot spots), spatiotemporal comparisons to places with lower levels of vulnerability (the cold spots) can be conducted to identify the differences and commonalities in the social, physical, and political characteristics of their communities. These differences are then utilized to improve our understanding of the relative importance of the various factors influencing vulnerability.

The design structure of the proposed approach as shown in Figure 17.4 could be implemented to: (1) address the limitations of current vulnerability assessment models, (2) recognize the divergent perspectives on urban vulnerability, (3) be multihazards based, (4) incorporate policy and more explicit planning components, (5) generate quantitative parameters that allow for the comparison of vulnerability across temporal and spatial scales, and (6) involve a spatiotemporal modeling engine for urban dynamics that would allow us to collect evidence to support or reject the alternative hypotheses concerning the causal linkages between vulnerability, and the social and physical characteristics of urban places, as well as, the effects of planning policies.

17.6 SUMMARY AND CONCLUDING REMARKS

Striving for sustainable hazards mitigation requires results from cutting-edge research that can help link researchers with practitioners and decision-makers. The primary goal of this chapter was not to propose a grand theory for urban vulnerability, nor was it to conduct a comprehensive assessment of GIS and RS contributions to the full spectrum of activities carried out in sustainable hazards mitigation. Rather, the goal was to point out the promise that GIS and remote sensing technologies hold for incorporating the assessment of vulnerability into the strategic and operational planning of hazards mitigation. Taken from this perspective, the principal contribution of this chapter has been to lay essential groundwork to propose an empirical GIS/RS-based methodology directed toward understating spatial and temporal patterns of vulnerability in urban areas.

In the assessment of natural hazards, Mileti [7] ranks the need for translating the technology of assessing vulnerability into procedures for easy use by local governments as one of the high-priority needs for hazards research. In order to reduce long-term risks from urban hazards, state and local policy makers and ordinary citizens need to understand not only that a particular area is subject to particular hazards, but also that, with existing or proposed land use and its implied land cover patterns in an area, certain levels of loss can be expected. This kind of information is hard for disaster analysts in state and local governments to convey, because it is not available [7] and is not likely to become available without investment in research to develop and disseminate vulnerability analysis tools to local governments. Susan Tubbesing [114], a senior analyst in the Earthquake Engineering Research Institute (EERI), articulated this issue in a very recent comment:

> Unfortunately, insufficient knowledge in several key research areas hampers the development of tools needed to secure society against catastrophic earthquake losses ... unless effective technologies are developed and implemented to reduce existing risks ... the costs of future earthquakes and other natural and technological disasters will continue to escalate.

One of the urgent needs for the hazards community is the development of a methodology that is capable of providing consistent assessment of vulnerability (as opposed to risk) that can be integrated with similar assessments of environmental quality to foster holistic sustainable hazards mitigation. There is a need for new approaches such as the one proposed herein to build on current GIS loss estimation models and expand their capabilities beyond the assessment of risks to the assessment of vulnerability. There is also an urgent need to test the utility of RS data as a quantitative means of measuring aspects related to the spatial structure of urban places that influence vulnerability.

Finally, any attempt to implement a GIS/RS-based approach to assess vulnerability must be based in a careful evaluation of the phenomena in hand and careful consideration of information needs and information flows, because the outcomes may directly affect people's lives. Rejeski [28] identifies four important factors that one should take into account in this task: *believability, honesty, decision utility,* and *clarity.*

The first factor, believability, examines whether the models used in the analysis and supporting data are properly chosen. The second factor, honesty, implies the accuracy of the analysis outcomes and examines whether uncertainties in the analysis are being assessed and conveyed to the end user. Remote sensing and GIS may add additional problems unique to the technology, data integration, and analysis. Since expensive and politically sensitive decisions may be based on their outputs, some means is required to model uncertainty in the final products. The third factor is decision utility, which speaks to the question of whether the analysis provides a clear base for action. Decision utility ensures that the methodology of assessing vulnerability is sufficiently robust and extends beyond risk assessment to risk management (i.e., preparedness for relief). Finally, clarity addresses the role of remote sensing, GIS, and cartographic design criteria for communicating the outcome of developed models. It is one challenge to model vulnerability and the impacts of mitigation policies, but it is another to convey the results to the intended audience in a meaningful way.

REFERENCES

1. Beer, T. and Ismail-Zadeh, A., Eds., Risk science and sustainability: science for reduction of risk and sustainable development in society, in *Series II: Mathematics, Physical and Chemistry,* Dordrecht, 2003.
2. ICSU, Science and Technology for Sustainable Development, International Council for Science, Mexico City, 2002.
3. UNEP, Poverty and the Environment: Reconciling Short-Term Needs with Long-Term Sustainability, United Nation Environment Program: UNEP Publications, Goals. Nairobi, Kenya, 1995.
4. Turner, B.L. II, et al., Science and technology for sustainable development: illustrating the coupled human-environment system for vulnerability analysis: three case studies, *Proc. Natl. Acad. Sci. U.S.A.,* 100, 8080, 2003.
5. Pincetl, S. et al., Toward a Sustainable Los Angeles: "Nature's Services" Approach, A Second Year Report to the John Randolph Haynes and Dora Haynes Foundation, Center for Sustainable Cities, University of Southern California, Los Angeles, 2003.
6. Dow, K., Social dimensions of gradients in urban ecosystems, *Urban Ecosystems,* 4, 255, 2000.
7. Mileti, D.S., *Disasters by Design: A Reassessment of Natural Hazards in the United States,* Joseph Henry Press, Washington, DC, 1999.
8. Burby, R.J., Natural hazards and land use: an introduction, in *Cooperating with Nature: Confronting Natural Hazards with Land Use Planning for Sustainable Communities,* Burby, R.J., Ed., Joseph Henry Press, Washington, DC, 1998, pp. 1–26.
9. Alexander, D., A survey of the field of natural hazards and disasters studies, in *Geographical Information Systems in Assessing Natural Hazards,* 1st ed. Carrara, A. and Guzzetti, F., Eds., Kluwer Academic Publishers, Dordrecht, The Netherlands, 1995, pp. 1–19.

10. Blaikie, P. et al., *At Risk, Natural Hazards, People's Vulnerability, and Disasters*, Routledge, London, 1994.
11. Bolin, R., Jackson, M., and Crist, A., Gender inequality, vulnerability and disaster: issues in theory and research, in *The Gendered Terrain of Disaster: Through Women's Eyes*, Enarson, E.P. and Morrow, B.H., Eds., Praeger, Westport, CT, 1997, pp. 27–51.
12. Burby, R.J., Cooperating with nature: confronting natural hazards with land use planning for sustainable communities, in *Natural Hazards and Disasters*, Washington, DC, Joseph Henry Press, 1998.
13. Cutter, S.L., Vulnerability to environmental hazards, *Prog. Hum. Geogr.*, 20, 529, 1996.
14. Dow, K., Exploring differences in our common feature(s): the meaning of vulnerability to global environmental change, *Geoforum*, 23, 417, 1992.
15. Godschalk, D.R., *Natural Hazards Mitigation: Recasting Disaster Policy and Planning*, Island Press, Washington, DC, 1999.
16. Hewitt, K., Interpretations of calamity from the viewpoint of human ecology, in *Risks & Hazards Series,* 1, Allen & Unwin, Boston, 1983.
17. Jones, D., Environmental hazards in the 1990s: problems, paradigms and prospects, *Geography*, 78, 161, 1993.
18. Mitchell, J.K., Hazards research, in *Geography in America*, 1st ed. Gaile, G.L. and Willmott, C.J., Eds., Columbus, Merrill Publishing Company, OH, 1989, pp. 410–424.
19. Quarantelli, L.E., Disaster studies: an analysis of the social historical factors affecting the development of research in the area, *Int. J. Mass Emerg. Disasters*, 5, 285, 1988.
20. White, G.F., Paths to risk analysis, in *Environmental Risks and Hazards*, Cutter, S.L., Ed., Englewood Cliffs, NJ, 1994.
21. Meyer, W.B. and Turner, B.L., II, Human transformation of the earth, in *Ten Geographic Ideas that Changed the World*, 1st ed., Hanson, S., Ed., Rutgers University Press, New Brunswick, NJ, 1996.
22. Cash, D.W. et al., Science and technology for sustainable development: knowledge systems for sustainable development, *Proc. Natl. Acad. Sci. U.S.A.*, 100, 8086, 2003.
23. Turner, B.L., II et al., Science and technology for sustainable development: a framework for vulnerability analysis in sustainability science, *Proc. Natl. Acad. Sci. U.S.A.*, 100, 8074, 2003.
24. Cova, T.J., GIS in emergency management, in *Geographical Information Systems*, Longley, P.A., Goodchild, M.F., Maguire, D.J. and Rhind, D.W., Eds., John Wiley & Sons, New York, 1999, pp. 845–858.
25. Rashed, T. and Weeks, J., Assessing vulnerability to earthquake hazards through spatial multicriteria analysis of urban areas, *Int. J. Geogr. Inf. Sci.*, 17, 547, 2003.
26. NRC, Reducing Disaster Losses through Better Information, Report, National Research Council, National Academy Press, Washington, DC, 1998.
27. Coppock, J.T., GIS and natural hazards: an overview from a GIS perspective, in *Geographical Information Systems in Assessing Natural Hazards*, 1st ed. Carrara, A. and Guzzetti, F., Eds., Kluwer Academic Publishers, Dordrecht, The Netherlands, 1995, pp. 21–34.
28. Rejeski, D., GIS and risk: a three-culture problem, in *Environmental Modeling with GIS*, Goodchild, M.F., Parks, B.O. and Steyaert, L.T., Eds., Oxford University Press, Oxford, UK, 1993, pp. 318–331.
29. Radke, J. et al., Application challenges for GIScience: implications for research, education, and policy for risk assessment, emergency preparedness and response, *URISA J.*, 12, 15, 2000.
30. Cutter, S.L., The changing nature of risks and hazards, in *American Hazardscapes: The Regionalization of Hazards and Disasters*, Cutter, S.L., Ed., Joseph Henry Press, Washington DC, 2001, pp. 157–165.

31. Cutter, S.L., Ed., *Environmental Risks and Hazards*, Prentice Hall, Englewood Cliffs, NJ, 1994.
32. Oakley, D.J., A National Disaster Preparedness Service, in *Natural Disasters: Protecting Vulnerable Communities*, Merriman and Browith, Eds., Thomas Telford, London, UK, 1993, pp. 31–46.
33. Fitzpatrick, K. and LaGory, M., *Unhealthy Places: The Ecology of Risk in the Urban Landscape*, Routledge, London, 2000.
34. Cutter, S.L., Societal responses to environmental hazards, *Int. Soc. Sci. J.*, 48, 525, 1996.
35. Tobin, G.A. and Montz, B.E., *Natural Hazards: Explanation and Integration*, Guilford Press, New York, 1997.
36. Tobin, G.A. and Montz, B.E., Natural hazards and technology: vulnerability, risk, and community response in hazardous environments, in *Geography and Technology*, Brunn, S.D., Cutter, S.L. and Harrington, J.W.J., Kluwer Academic Publishers, Eds., Dordrecht, 2004, pp. 547–570.
37. Hewitt, K., *Regions of Risk: a Geographical Introduction to Disasters*, Longman, Harlow, 1997.
38. Enarson, E.P. and Morrow, B.H., *The Gendered Terrain of Disaster: Through Women's Eyes*, Praeger, Westport, CT, 1998.
39. Kates, R.W., Human adjustment, in *Ten Geographic Ideas that Changed the World*, 1st ed. Hanson, S., Ed., Rutgers University Press, New Brunswick, NJ, 1996, pp. 87–107.
40. White, G.F., *Natural Hazards, Local, National, Global*, New York, 1974.
41. Alexander, D., *Natural Disasters,* Chapman & Hall, New York, 1993.
42. Hewitt, K., The Idea of calamity in a technocratic age, in *Interpretations of Calamity from the Viewpoint of Human Ecology*, Hewitt, K., Ed., Allen & Unwin, Boston, 1983, pp. 3–32.
43. Hewitt, K. and Burton, I., The Hazardousness of a Place: a Regional Ecology of Damaging Events, Published for the University of Toronto Dept. of Geography by University of Toronto Press, 1971.
44. Mitchell, J.K., Devine, N., and Jagger, K., A contextual model of natural hazards, *Geogr. Rev.*, 79, 391, 1989.
45. Cutter, S.L., Mitchell, J.T., and Scott, M.S., Revealing the vulnerability of places: a case study of Georgetown County, South Carolina, *Ann. Assoc. Am. Geogr.*, 90, 713, 2000.
46. Capra, F., *The Web of Life: A New Scientific Understanding of Living Systems,* Anchor Books, Doubleday, New York, 1996.
47. Naveh, Z. and Lieberman, A., *Landscape Ecology: Theory and Application*, Springer-Verlag, Heidelberg, 1994.
48. Forman, R.T.T. and Gordon, M., *Landscape Ecology*, John Wiley & Sons, New York, 1986.
49. Meo, M., Ziebro, B., and Patton, A., Tulsa turnaround: from disaster to sustainability, *Nat. Hazard. Rev.*, 5, 1, 2004.
50. Bertalanffy, L.V., *General System Theory: Foundations, Development, Applications*, Penguin University Books, London, 1968.
51. Koestler, A., *The Act of Creation*, Dell, New York, 1967.
52. Koestler, A., Janus: A summing up. In *Beyond Reductionism: New Perspectives in the Life Sciences*, Koestler, A. and Smythies, J.R., Eds., Beacon, Boston, 1979.
53. Anderson, R.E., Carter, I., and Lowe, G.R., *Human Behavior in the Social Environment: A Social Systems Approach*, Aldine De Gruyter, New York, 1999.
54. Menoni, S., Chains of damages and failures in a metropolitan environment: some observations on the Kobe earthquake in 1995, *J. Hazard. Mater.*, 83, 101, 2001.

55. Menoni, S., et al., Measuring the seismic vulnerability of strategic public facilities: response of the health care system, *Disaster Prev. Manage.*, 9, 29, 2000.

56. Menoni, S., Petrini, V., and Zonno, G., Seismic Risk Evaluation through Integrated Use of Geographical Information Systems and Artificial Intelligence Techniques, European Commission, Environment and Climate Programme- Report No. ENV4-CT96-0279, Brussels, 1999.

57. Menoni, S. and Pergalani, F., An attempt to link risk assessment with land use planning: a recent experience in Italy, *Disaster Prev. Manage.*, 5, 6, 1996.

58. Campbell, B., *Human Ecology: The Story of Our Place in Nature from Prehistory to the Present*, Aldine de Gruyter, New York, 1995.

59. Girard, C. and Peacock, W.G., Ethnicity and segregation: post-hurricane relocation, in Hurricane Andrew, in *Ethnicity, Gender, and the Sociology of Disasters*, Peacock, W.G., Morrow, B.H. and Gladwin, H., Eds., Routledge, London, 1998, pp. 191–265.

60. Wisner, B., Marginality and vulnerability: why the homeless of Tokyo don't 'count' in disaster preparation, *Appl. Geogr.*, 18, 25, 1998.

61. Cadwallader, M., *Urban Geography: an Analytical Approach*, Prentice Hall, New York, 1996.

62. Degg, M., Perspectives on urban vulnerability to earthquake hazards in the Third World., in *Environment and Housing in Third World Cities*, Main, H. and Williams, S.W., Eds., John Wiley & Sons, Chichester, UK, 1994, pp. 24–48.

63. Cova, T.J. and Church, R., L., Modelling community evacuation vulnerability using GIS, *Int. J. Geogr. Inf. Sci.*, 11, 763, 1997.

64. Walker, G., Industrial hazards, vulnerability and planning in Third World Cities, with reference to Bhopal and Mexico City, in *Environment and Housing in Third World Cities*, Main, H. and Williams, S.W.C., Eds, Wiley, Chichester, UK, 1994, pp. 49–64.

65. Susman, P., O'Keefe, P., and Wisner, B., Global disasters, a radical interpretation, in *Interpretations of Calamity*, Hewitt, K., Ed., Allen & Unwin, Winchester, MA, 1983, pp. 22–40.

66. Wisner, B., Disaster vulnerability: scale, power, and daily life, *GeoJournal*, 30, 127, 1993.

67. Main, H. and Williams, W., Marginal urban environments as havens for low-income housing, in *Environment and Housing in Third World Cities*, Main, H. and Williams, S.W.C., Eds, Wiley, Chichester, UK, 1994, pp. 151–170.

68. Clarke, K.C. and Couclelis, H.M., Developing an Integrated Modeling Environment for Urban Change Research: Methodological and Theoretical Challenges, A proposal submitted to 1998 US-NSF urban initiative solicitation, UCSB/NCGIA, Santa Barbara, CA, 1999.

69. Kreimer, A., Munasinghe, M., and Preece, M., Reducing environmental vulnerability and managing disasters in urban areas, in *Environmental Management and Urban Vulnerability*, Kreimer, A. and Munasinghe, M., Eds., World Bank Discussions Papers, Washington, DC, 1992.

70. Cutter, S.L., A Research agenda for vulnerability science and environmental hazards, *IHDP Update: Newsletter for the International Human Dimensions Programme on Global Environmental Change*, 2, 8, 2001.

71. Bolin, R. and Stanford, L., Constructing vulnerability in the First World: The Northridge earthquake in Southern California, 1994, in *The Angry Earth: Disaster in Anthropological Perspective*, Oliver-Smith, A. and Hoffman, S.M., Eds., New York, 1999.

72. Liverman, D.M., Vulnerability to global environmental change, in *Global Environmental Change*, Kasperson, J.X. and Kasperson, R.E., Eds., United Nations University Press, London, 2001, pp. 201–216.

73. Mitchell, J.K., Urban vulnerability to terrorism as hazard, in *The Geographical Dimensions of Terrorism*, Cutter, S.L., Richardson, D.B., and Wilbanks, T.J., Eds., Kluwer Academic Publishers, New York, 2003, pp. 17–25.

74. Uitto, J., The geography of disaster vulnerability in megacities: a theoretical framework, *Appl. Geogr.*, 18, 7, 1998.

75. NOAA, Community vulnerability assessment tool: New Hanover County, North Carolina case study, 1999.

76. Milton, E.J., Image endmembers and the scene model, *Can. J. Remote Sensing*, 25, 112, 1999.

77. Alexander, D., *Confronting Catastrophe: New Perspectives on Natural Disasters*, Oxford University Press, Oxford, UK, 2000.

78. White, G.F. and Haas, J.E., *Assessment of Research on Natural Hazards*, MIT Press, Cambridge, MA, 1975.

79. Sinnott, J.D., A model for solution of ill-structured problems: implications for everyday and abstract problem solving, in *Everyday Problem Solving: Theory and Applications*, Sinnott, J.D., Ed., Praeger, New York, 1989, pp. 72–99.

80. Dynes, R.R. and Drabek, T.E., The structure of disaster research: its policy and disciplinary implication, *Int. J. Mass Emerg. Disasters*, 12, 5, 1994.

81. FEMA-NIBS, HAZUS: *Users's Manual and Technical Manuals*, vols 1–3 (4 vols). Washington DC, 1999.

82. UN, *Mitigating Natural Disasters: Phenomena, Effects, and Options: a Manual for Policy Makers and Planners*, UNDRO (United Nations Disaster Relief Organization), New York, 1991.

83. Hill, A.A. and Cutter, S.L., Methods for determining disaster proneness, in *American Hazardscapes*, Cutter, S.L., Ed., Joseph Henry Press, Washington, DC, 2001, pp. 13–36.

84. Turner, B.A. and Pidfeon, N.F., *Man-Made Disasters*, Butterworth-Heinemann, Oxford, UK, 1997.

85. Mitchell, J.K., Megacities and natural disasters: a comparative analysis, *GeoJournal*, 49, 137, 1999.

86. Carrara, A. and Guzzetti, F., Eds., Geographical information systems in assessing natural hazards, in *Advances in Natural and Technological Hazards Research*, Kluwer Academic Publishers, Dordrecht, The Netherlands, 1995.

87. Smith, K., *Environmental Hazards: Assessing Risk and Reducing Disaster*, Routledge, London, 1992.

88. Choudhry, S. and Morad, M., GIS errors and surface hydrologic modeling: an examination of effects and solutions, *J. Surveying Eng.-Asce*, 124, 134, 1998.

89. Whitman, R.V. and Lagorio, H.J., The FEMA-NIBS Methodology for Earthquake Loss Estimation, 1998.

90. Rashed, T. and Weeks, J., Exploring the spatial association between measures from satellite imagery and patterns of urban vulnerability to earthquake hazards, *Int. Arch. Photogrammetry, Remote Sensing Spatial Inf. Sci.*, Vol XXXIV-7/W9:1, 44, 2003.

91. Cutter, S.L., Boruff, B.J., and Shirley, W.L., Social vulnerability to environmental hazards, *Soc. Sci. Q.*, 84, 242, 2003.

92. Cova, T.J. and Johnson, J.P., A network flow model for lane-based evacuation routing, *Transportation Research Part A: Policy and Practice*, in press.

93. Lindsay, J.R., The determinants of disaster vulnerability: achieving sustainable mitigation through public health, *Nat. Hazard.*, 28, 291, 2003.

94. Costa, S.M.F.D. and Cintra, J.P., Environmental analysis of metropolitan areas in Brazil, *ISPRS J. Photogrammetry Remote Sensing*, 54, 41, 1999.

95. Chen, S., Zheng, S. and Xie, C., Remote sensing and GIS for urban growth in China, *Photogrammetric Eng. Remote Sensing*, 66, 593, 2000.

96. Kwarteng, A.Y. and Chavez, P.S., Change detection study of Kuwait City and environs using multi-temporal Landsat Thematic Mapper data, *Int. J. Remote Sensing*, 19, 1651, 1998.

97. Batty, M. and Howes, D., Predicting temporal patterns in urban development from remote imagery, in *Remote Sensing and Urban Analysis*, Donnay, J.-P., Barnsley, M.J., and Longley, P.A., Eds., Taylor & Francis, London, 2001, pp. 186–204.

98. Ward, D., Phinn, S.R., and Murray, A.T., Monitoring growth in rapidly urbanization areas using remotely sensed data, *Prof. Geogr.*, 52, 371, 2000.

99. Yang, X. and Lo, C.P., Using a time series of satellite imagery to detect land use and land cover changes in the Atlanta, Georgia metropolitan area, *Int. J. Remote Sensing*, 23, 1775, 2002.

100. Madhavan, B.B., et al., Appraising the anatomy and spatial growth of the Bangkok metropolitan area using a vegetation-impervious-soil model through remote sensing, *Int. J. Remote Sensing*, 22, 789, 2001.

101. Weber, C. and Puissant, A., Urbanization pressure and modeling of urban growth: example of the Tunis metropolitan area, *Remote Sensing Environ.*, 86, 341–352, 2003.

102. Rashed, T., et al., Measuring temporal compositions of urban morphology through spectral mixture analysis: toward a soft approach to change analysis in crowded cities, *Int. J. Remote Sensing*, 26(4), 688, 2005.

103. Herold, M., Goldstein, N., and Clarke, K., The spatiotemporal form of urban growth: measurement, analysis and modeling, *Remote Sensing Environ.*, 86, 286, 2003.

104. Weeks, J. et al., The fertility transition in Egypt: intra-urban patterns in Cairo, *Ann. Assoc. Am. Geogr.*, 94, 74, 2004.

105. Weeks, J.R. et al., Measuring the environmental context of urban fertility in Arab countries: a "top-down" approach to fertility in Cairo, Egypt, *Int. J. Popul. Geogr.*, 2005, submitted.

106. Weeks, J., Remote sensing in demography, in *Encyclopedia of Population*, Demeny, P. and McNicoll, G., Eds., New York, 2005 (in press).

107. Pesaresi, M. and Bianchin, A., Recognizing settlement structure using mathematical morphology and image texture, in *Remote Sensing and Urban Analysis*, Donnay, J.-P., Barnsley, M.J., and Longley, P.A., Eds., Taylor & Francis, London, 2001, pp. 55–67.

108. Phinn, S., et al., Monitoring the composition of urban environments based on the vegetation-impervious surface-soil (VIS) model by subpixel techniques, *Int. J. Remote Sensing*, 23, 4131, 2002.

109. Gupta, A. and Shah, H., The strategy effectiveness chart: a tool for evaluating earthquake disaster mitigation strategies, *Appl. Geogr.*, 18, 55, 1998.

110. Cutter, S.L., GI science, disasters, and emergency management, *Trans. GIS*, 7, 439, 2003.

111. Cahan, B. and Ball, M., GIS at ground zero: spatial technology bolsters World Trade Center response and recovery, *GEO World*, 15, 26, 2002.

112. Cutter, S.L., The vulnerability of science and the science of vulnerability, *Ann. Assoc. Am. Geogr.*, 93, 1, 2003.

113. Polsky, C. et al., Assessing Vulnerabilities to the Effects of Global Change: An Eight-Step Approach. Research and Assessment Systems for Sustainability Program Discussion Paper 2003–05., Environment and Natural Resources Program, Belfer Center for Science and International Affairs, Kennedy School of Government, Harvard University, Cambridge, MA, 2003.

114. Tubbesing, S., Securing society against catastrophic earthquake losses, *Nat. Hazard. Observer*, XXVII, 1, 2002.

Part III-A

Learning from Practice:
GIS as a Tool in Planning
Sustainable Development

Urban Dynamics

18 Urban Multilevel Geographical Information Satellite Generation

Sébastien Gadal

CONTENTS

18.1 INTRODUCTION

The exploitation of satellite data in urban land planning is sometimes unpredictable because of the specific needs of these practices compared to others. In common practice, morphological, environmental, and social aspects are needed to describe the characteristics of urban zones. However, remote sensing image processing methodologies, generating social and environmental information from satellite data as multispectral classification and textural filters, give a zone description of the urban environment mostly limited to the land use [1], the land cover, or the build densities zone's description [2]. Furthermore, the socio-environmental level generated is dependent on the spatial resolution of satellite imagery. This inadequacy limits the use of remote sensing data for multilevel urban and socioeconomic database information systems development. New methodologies based on geometrical, logical set

313

filters, and thermal imagery characteristics have been developed and utilized to generate and integrate socio-environmental and economical information databases at three different spatial levels [3]. The settlement level (the interface between imagery, society, and social characteristic of territories) deals with information on urban form, economical and social functions, levels of life, and equipment [4]. The meso-metropolitan level deals with demographic information (urban and human densities, human development indicators, the types of social or economical zone activities, and water pollution). The global level concerns the environmental information, global densities and localizations of populations [5], land cover and land use [6], spatial structures, and urban form dynamics [7,8].

The first method requires morphological operators and symbolic recognition (description of the geometrical properties such as the surface, the compactness, etc.). The objective of this method is to automatically generate a vectorial map of every build elements-settlements and a descriptive geometrical database for the geographical objects. From this descriptive geometrical database, a classification processing is then used to extract the different urban forms and built elements settlements typology.

The second methodology uses logical set theory and textural filters to separate and identify functions of build elements at settlement level; it allows recognition of spatial structures and urban forms at a global level. Thirdly, a set of methodologies based on interpretation of urban thermal gradient permits characterization of social domains and production of demographical indicators. All these methodologies generate environmental and social urban databases which may be useful in supporting territorial control and urban planning, as it will be shown with reference to a case study developed for the Morocco Atlantic Metropolitan (MAM) Area (Kenitra-Rabat-Casablanca).

Production, update, and availability of multilevel urban geographic databases are some of the main problems many land-planning agencies and local governments from the Maghreb have to face in daily practice. While socio-demographic census data exist, like in Morocco, and can be used and implemented at two different spatial levels (prefectures and districts), the accuracy, pertinence, and efficiency are not properly exploited yet for the MAM's territorial control and urban planning. For these reasons, the use of satellite data has been chosen and tested as the basis for the socio-environmental multilevel information system implementation. The use of satellite imagery for the socio-environmental and multilevel GIS implementation constitutes an advantage, because it makes it possible to produce, to associate, to merge, and to generate several types of socio-spatial information, as shown in the remainder of this chapter.

18.2 CONTRIBUTION OF DATA SATELLITES FOR URBAN GEOGRAPHIC INFORMATION SYSTEM (UGIS): ACCURACIES OF SOCIOECONOMIC AND DEMOGRAPHIC STATISTICAL INFORMATION

The accuracy of socioeconomic and demographic statistical databases is dependent on the special context, which varies for different countries. In general, three sets can be distinguished.

- The information collections based on exhaustive censuses are generally produced or updated every 8 to 10 years. They are usually based on administrative spatial units [9]. The spatial units generally serve as a base for annual samplings updating. This mode of information collection represents the majority of the cases in the world. Generally, the studies on the urban processes based on the demographic and socioeconomic statistics show social practices of the populations, their dynamics, and their distributions on the territory. They also allow approaching the economic dimension of the urban territory. They give a social, human, and economic representation of urban territories, describing them geographically by highlighting the territorial organizations. The plurality of the data, indicators, and statistical variables allows encircling the variety of the human facts and describing the urbanization dynamics. The statistical relationships of economical and human variables and the preparatory statistical methods verify the aptness of UGIS information. They examine the significance of the practices and the socioeconomic dynamics while characterizing them. They have descriptive and heuristic characters. As economic and socio-anthropological measure, the statistical data get at the same moment the driving elements and the actors of these geographic dynamics. Therefore, they offer the geographer a means to encircle the explanatory factors. However, they face difficulties so that it is often necessary to look for socio-anthropological and cultural factors. Hence, questions arise about the geographic scale aptness to report urban processes such as their choice in the studies concerning this geographic process. The interest in this descriptive statistical analytical method and this structuring information mode should take into account a very high number of variables. It allows refining the measure of the urban state process of the geographic space portion under study. The fitted multilevels method links the urban levels in various geographic scales and allows understanding of whether the urban process is at a stage of development that is only local or embraces the whole region or country. Thus, it can be defined as an indicator of geographic and urban spatial distribution. The normalization of the statistical measure extrapolated at the national or regional level reports, certainly, the urban level tendency and the average level of territory at that scale. It does not represent, however, the differential character of the urban process in its full character. Thus, issues arise about the administrative spatial unity choice and the most relevant geographic scale.
- Other "conventional" data are produced from the administrative registers such as the registry office or statutory: building permission, cadastre, etc. While they often refer to the same concepts, these data often have different semantic meanings and refer to different spatial units [10].
- The data produced from the spatial remote sensing imageries. Remote sensing data give a physical description of the urban territory, from which it is possible to extract environmental and social indicators.

18.3 INTERESTS OF REMOTE SENSING DATA FOR GEOGRAPHICAL AND STATISTICAL DATABASES GENERATION

Satellite imagery does not succeed in reporting all the above aspects per se. It describes, it measures the visible physical aspects — by remote sensing — of the consequences that result from geographic processes. Unlike studies made with socioeconomic data, urban studies using satellite data nevertheless show the influence of the geographical context which relates the urban processes through the situation, the localization, the neighborhood, the spatial differentiation, etc. Indeed, the use of demographic and socioeconomic statistical data in the UGIS supplies an aspatial representation of the urban territory. They can be geo-referenced, but they are not necessarily spatial in essence. The cartographic transcription of urban processes, *a posteriori*, remains constrained by their membership in an area with boundaries that may be the result of a statistical sampling technique or an administrative one: the municipality, the region, etc. Satellite data are not affected by this limit. In some other rare cases, on the other hand, the studies based on ground observations present the inconvenience to be too punctual to give a systematic description. A monograph is indeed often too local to allow more that a confirmation or a local analysis of an aspect of the urban processes.

Nevertheless, if conceived as a separate element becoming integrated into the processing line, the field studies based and planned as sampling technique for statistical methods can be an indispensable aid to the expert to validate or invalidate his results or identify unrecognized geographic objects. These methods can help in understanding elements of social and cultural dynamics difficult to recognize by only image processing or data analysis [11,12]. Urban econometric and demographic analyses supply evident advantages, but they also present a certain number of deficiencies: failing to properly deal with the spatial dimension, and in the integration of physical, social, and human environment with cultural urban characteristics. Available statistical data are often inadequate when adapted to the problem at hand. Often, geographers or urban planners have to adapt the method and the geographic level of interest within the research focus or the urban planning project to the available statistical data, and not the opposite. The question is then how to construct geographies with information sources which are not made for the geographer or the urban planner? In the past, geographers and urban planners coped, like it or not, with this established limit, restricting the access to a part of the geographical reality.

Paradoxically, the problem is less important for geographers or urban planners in developing countries, where information is poorest, ill assorted, or still unavailable. Hence, the geographer or the urban planner has to create his own information for its work. This fact gives place to the experimentation with a number of spatiotemporal information production methodologies in developing countries. The result of these research efforts is that these "African methods" are nowadays beginning to replace classical urban spatial sampling techniques even in Western countries [6,13]. "African" sampling technique methods have the advantage of being less expensive, easier to implement from an administrative point of view, and often more effective. Often in developing countries, data are rarely updated and many geographers or planners hardly deal with the information production techniques with evident limit in the knowledge building process. Moreover, the statistical information production is often not tested, putting the expert in a face-to-face dependence on the statistical agencies for the exploitation of socioeconomic information. It is then difficult to estimate validity of the data used and, eventually, the suitability for their work, with evident limitation in interpreting the results.

18.4 SPACE IMAGERY, URBAN DYNAMICS, AND UGIS

Satellite remote sensing data allow approaching the various aspects of the territorial dynamics along with the social and cultural, environmental, historic and physical aspects. Remote sensing techniques are able to supply a measure of the human and society impacts onto the environment. Indeed, it is this interaction between the human and physical geographies that made the territory. The remote sensing images and the aerial photographs describe the territory, its structures, it organization, its dynamics, its landscapes, its morphologies, the marks and signs of its history. Remote sensing techniques allow geographers and urban planners to study the territory in a "trans-disciplines" way or, if required by the study at hand, in a specific manner.

Whatever logic of analysis geographers and planners are using, they need to analyze the themes by their spatial aspects. This is indeed peculiar in geographic and urban planning analyses made by means of remote sensing data. The multiplicity and the variety of the information offered by the satellite images, like the complexity of the physical and spatial mark structures of the urban dynamics, require the development of several information extraction and analysis methods by means of the image-processing techniques. Each method supplies a series of information, describing a state or a geographic process. These techniques can be effectively used when the objective is to detect, to recognize, to identify, to extract, to quantify, and to qualify urban structures and dynamics. They allow, for every geographic object describing urban growth and structures, implementation of one or several methodologies to extract simple or complex information characterizing the urban territory, informing about the dynamics and the actual phenomena.

However, the analysis of a part of the geographic space by remote sensing techniques, which search urban forms and spatial organizations, makes it necessary to know beforehand the phenomena, the objects, and the geographic places which structure it, in order to understand what satellite or airborne image may be useful to describe them. This approach is valid only if we limit ourselves to the detection and analysis of a given geographic phenomenon, which in this particular case is the urban dynamic. The analysis of the territory according to the geographic process theory using radiometric measures of the geographic space reality has the advantage of giving to the analyst reading, analysis, and interpretation grids which are made from quickly available sources at several levels of resolution and which supply multiple types of information. The deductive approach, which defines element recognition by theorized heuristics, has, as the other advantage, to focus the image processing work on detection of the geographic objects and the spatial entities.

18.5 AN APPROACH OF URBAN DYNAMICS
BY UGIS SATELLITE DATA GENERATION

18.5.1 THE MOROCCO ATLANTIC METROPOLITAN AREA EXAMPLE:
THE AVAILABLE AND INTEROPERABLE UGIS QUESTION

The levels of social indicators and geographic information and the pattern of the administrative boundaries are not suitable to properly represent urban dynamics and socioeconomic processes; information is aggregated too much, and it does not integrate the geographic and the environmental dimensions. This type of geographic multilevel information implemented in GIS is not efficient for urban planning purposes. In addition, discontinuities introduced by administrative boundaries do not give a suitable view of social and environmental processes. Administrative divisions made with population and territory controls in mind are not suitable for an integrated land management and a security control of the sprawling metropolitan area. Local urban databases in Rabat and Casablanca are available at planning agencies, but cannot be integrated to the same geographic information database system. However, the most urgent problem for GIS implementation, apart from the inadequacy of the socio-spatial information level of representation, is the obsolescence of databases supplied by census and land planning agencies, with their updating problems. The strong rates of urban growth and the fast transformation of society make data obsolete every three months [14]. The costs of updates by traditional survey techniques and of data production at a suitable spatial level make it impossible to implement the relevant geographic and social information levels in GIS.

18.5.2 A MULTIDIMENSIONAL APPROACH

The study of the urban processes by remote sensing for urban and land planning purposes has been developed by integrating several methods, each of them relating to a set of semiological and semantic information testing one of its aspects or objects, such as the industrial parks, communication infrastructure, built elements, etc. In other words, the ways spatial information was produced depended on several characteristic objects: detection of the built elements and their economical and social functions, communication infrastructures, and social segregations; urban concentrations recognition at the infraterritorial and regional levels; and recognition of the morpho-landscape and geographic structures, hierarchies and spatial interrelations. The different information and analysis methods were based on several available spectral sources, such as panchromatic and thermal infrared imageries. Each of these two spectral data types supplies at different spectral and spatial resolutions an electromagnetic brilliance measure that is a radiometric biophysical radiation of the surface and of the urban fabric. Panchromatic and infrared thermal data imageries bring a series of information, often additional, sometimes redundant, of the same spatial reality, which is subdivided as imprints of a geographic, geologic, social, anthropological, historic, ecological, or political reality, etc. The deductive approach used in this method, as to say the recognition of geographical elements defined by a theorized heuristics, has the other advantage to focus the image processing work on the detection of the geographic objects and the spatial entities that show and explain urban dynamics. Descriptive elements can be classified in three categories:

- The descriptive elements of selected objects (as built, urban concentrations, roads, etc.)
- Regionalized descriptive elements (objects such as landscaped units, geographic zones, etc.)
- Spatial descriptive elements (reporting forms of organizations, structures, hierarchies of the physical environment)

Besides putting in evidence a larger number of geographic descriptive elements, objects of urban dynamics, and structures (refining, in this way, the analysis and the heuristics), the multimethods approach has the potential for the methods to validate each other, in a complementary logic. For example, the combinatorial extraction and classification model of built objects recognizes townships, while the morpho-landscaped recognition model recognizes all the types of built objects with the exception of townships. This last informative model, thanks to the types and the forms of landscapes recognition, allows replacing in the physical context built patterns recognized with the extraction and classification of the built objects combinatorial model. The method has a double role: complementarity and validation of the other models used.

The overall image processing method can be resumed in three phases as follows:

- A phase of data optimization, which is a preprocessing step, intended "to improve" the satellite data for the needs of the following analysis processing
- An image and analysis-processing phase based on NOAA [15,16] and LANDSAT [17] thermal infrared satellite images
- An image and analysis processing phase using SPOT panchromatic satellite images relying on three heuristic detection models, two morpho-landscape models, and a geometrical morphology model of the building and road objects

The available data and the methods developed to analyze the urban dynamic allow extracting a certain amount of information, suitable to describe the phenomenon with enough accuracy. This was experimented in Morocco along the Kenitra-Rabat-Casablanca's urban axis, the MAM.

Figure 18.1 and Figure 18.2 illustrate the methodologies of Urban Multilevel Geographical Information Satellite Generation, while Figure 18.2–18.5 illustrate a few examples of Multilevel Geographical Urban Information produced by RS.

18.6 TECHNICAL AND OPERATIONAL PROBLEMS: THE INFORMATION PARADIGM QUESTION

The singular and normative aspect of the urban form in emergence study for urban planning by remote sensing raises the technical operational problem of the "informative"

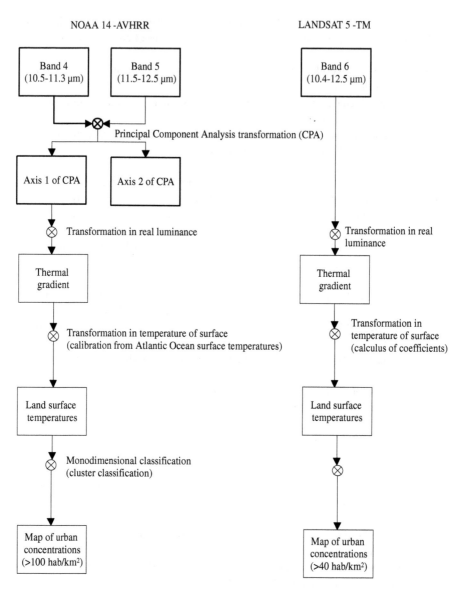

FIGURE 18.1 AVHRR and LANDSAT 5 TM thermal data processing lines. (From Gadal, S., *Recognition of Metropolization Spatial Forms by Remote Sensing*, Eratosthenes, Lausanne, Switzerland, 2003. With permission.)

paradigm. The airborne and satellite data used as almost unique information bring geographers and planners to a situation of a new environment of expertise and a new way of working because of the few general experiences acquired in the developing countries in the field of urban study, and in the absence of established and well-tested methodologies.

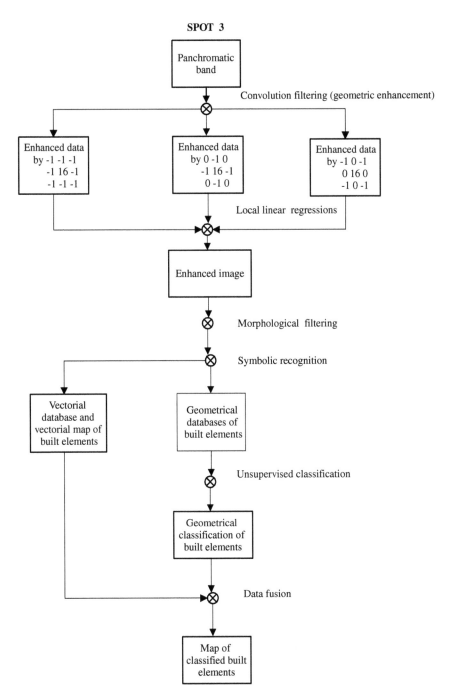

FIGURE 18.2 SPOT panchromatic data processing lines. (From Gadal, S., *Recognition of Metropolization Spatial Forms by Remote Sensing*, Eratosthenes, Lausanne, Switzerland, 2003. With permission.)

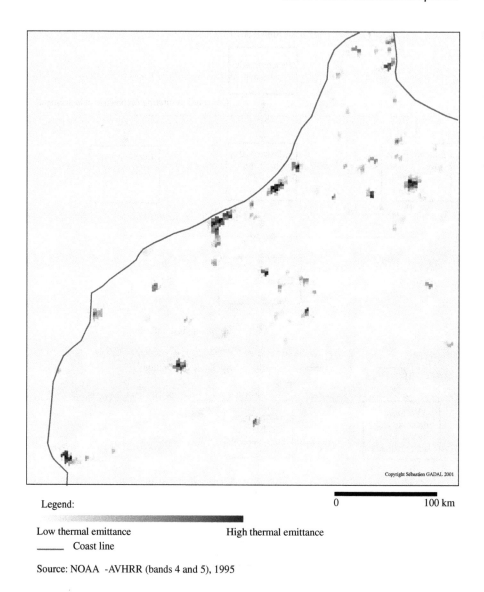

Legend: 0   100 km

Low thermal emittance   High thermal emittance
_____ Coast line

Source: NOAA -AVHRR (bands 4 and 5), 1995

FIGURE 18.3 Urban and socio-demographic concentration recognition. (From Gadal, S., *Recognition of Metropolization Spatial Forms by Remote Sensing*, Eratosthenes, Lausanne, Switzerland, 2003. With permission.)

However, image-processing methods present several advantages. On the one hand, they allow an increased independence from the statistical agencies, diffusing and interpreting geographical information totally tested and capable of tackling knowledge and data scarcity or lack. They also offer the possibility of a sort of multiple-logic reasoning and interpretation, because the results stemming from spatial methods integration give a much wider and more reliable description of the

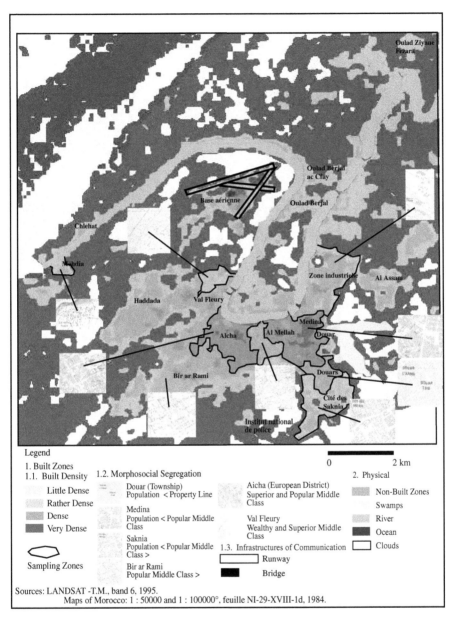

FIGURE 18.4 Built density and morphological segregation. (From Gadal, S., *Recognition of Metropolization Spatial Forms by Remote Sensing*, Eratosthenes, Lausanne, Switzerland, 2003. With permission.)

actual geographic situation. They build a capacity for dynamic urban analysis on almost real-time inquiry anywhere on the Earth, thanks to the use of multiple data sources that can be produced without regard to the political, geographic, environmental, and climatic conditions. These characteristics are fundamental for an operational urban

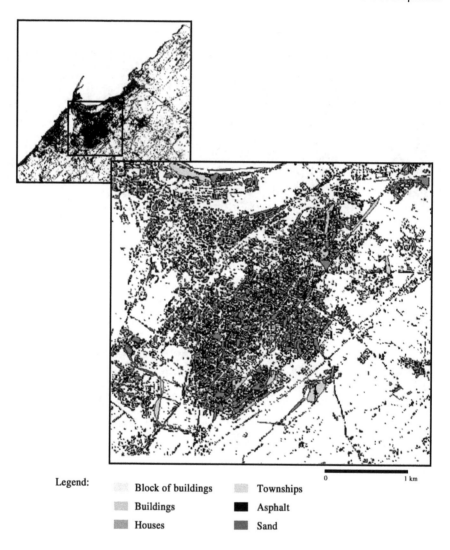

Legend:

	Block of buildings		Townships
	Buildings		Asphalt
	Houses		Sand

0   1 km

Data: SPOT 3 panchromatic image, 1994   Copyright Sébastien Gadal 2001-2004

FIGURE 18.5 Example of vector map generation and built elements classification on geo-metrical criteria. (From Gadal, S., *Recognition of Metropolization Spatial Forms by Remote Sensing*, Eratosthenes, Lausanne, Switzerland, 2003. With permission.)

territory analysis. They support a better understanding and a deeper territorial knowl-edge on the strategic stakes in the sociocultural, economic, territorial, technological, and political settings that urban dynamics infer on the individual and collective plans. These societal and territorial strategic stakes are of high importance with regards to urban dynamics, particularly in developing countries [3,18].

The reproducibility of image-processing methods in urban analysis allows com-paring in a systematic way the various forms of worldwide urban dynamics [19–21].

The urban dynamic processes study by remote sensing methods can also be approached within the framework of the "information / territory / knowledge" paradigm [3]. It poses two problems: on the one hand, that of the data characteristic and, on the other hand, that of the implementation of reproducible image processing in urban analysis chains. Both are going to determine the reliability and the aptness of the necessary knowledge for interpretation and decision-making. They answer the question of what we measure, how this measure is computed, and in what it is relevant, with regard to the topic of research and to the knowledge we extract. In this way, information contained in the data, quite as the existence of a preliminary formulation of a theory, is fundamental to achieving the project of geographic knowledge necessary to support effective sustainable planning.

The monospectral panchromatic data have several advantages, if compared to the multispectral data used. In fact, it is difficult, from the latter, to extract roads, buildings, urban zones by their radiometries. Built objects such as roads, network infrastructures, and urban concentrations have multiple and different radiometries that do not necessarily characterize the urban domain or the road networks. The spectral reflectance multiplicity for the same object makes the recognition and the extraction of these objects difficult [22]. It makes the reproducibility of the methods unpredictable on other urban settlements. The information that seems exploitable on the panchromatic images is, rather, the texture. It easily combines the morphological and geometrical information at a 100-m^2 spatial resolution. The three types of spatial information (texture, morphology and geometry) are characterized, as main assets, by being relatively stable for the same object in time and on different geographic spaces. The problem is the variability of the spectral reflectance for the same class of objects and the difficulty of separating it for different classes of objects. The thermal infrared information offers about the same qualities of temporal and geographic temporality as textures [23] and morphological and geometrical information [3,13]. This characteristic is particularly true in the urban spaces. This informative stability constitutes an asset in the implementation of reproducible methodologies for urban territory analysis, although the social-indicators integration is appropriate for each of them. Nevertheless, the thermal infrared satellite data are only weakly operational during summer climatic periods in the subtropical and desert regions; the urban zone's radiometries become confused with rocky zones or characterized by the absence of vegetation. The same consideration stands for the panchromatic images in an equatorial zone during rainy season. The strong cloudiness, the atmosphere moisture content, and the ground degrade the electromagnetic signal.

18.7 CONCLUSION

The problem of the control of space organization shape, the urban territories and the metropolization phenomena conjugates with a doctrinal revolution of the planning and research behaviors in which the information has a central role. Satellite data that are currently available turn out to be insufficient to state all the dimensions of the urbanity for the surveillance of the urban territories or for the control of the settlement future form. Ikonos 2, Orbview 2-3, Eros 1A, Spot 5 or future Pleiades satellite images have a greater geometrical and spectral resolution [24,25] that shows certain anthropo-socio-cultural aspects of interest for the geographers or urban planners. The difficulty recognizing and identifying urban dimensions will be still reduced near 2010–2015 with the 0.5 × 0.5 m spatial resolution images' availability in the civil market and new sensors' generation from military origin conceived exclusively for detection, recognition, and identification of urban territories, geographic objects, and individuals composing. These airborne and satellite sensors, previously dedicated to target identification and civilian-military discrimination, in the urban environment and real-time observation, are going to increase in notable proportions the perception of the urban geographic and anthropological environment and make visible phenomena and dynamics of metropolization that appear durably as not perceptible on the scale of the individual. Together with the enhancements in spatial resolution, the duplicable cost reduction and the increase in temporal resolution are going, in the near future, to turn remote sensing data into useful socio-environmental, urban, and morphological information in urban GIS technologies.

REFERENCES

1. Shaban, M. and Diskshit, O., Land use classification for urban areas using spatial properties, paper for the International Geoscience and Remote Sensing Symposium 99, Germany, 1999.
2. Fashi, A., Assaf, M., and Azerzaq, M., Cartography of the built densities from the satellite images: application in Casablanca, *Bulletin de la Société française de photogrammétrie et de télédétection*, 144, 11, 1996.
3. Gadal, S., *Recognition of Metropolization Spatial Forms by Remote Sensing*, Eratosthenes, Lausanne, Switzerland, 2003, 421.
4. Jensen, J. and Cowen, D., Remote sensing urban/suburban infrastructure and socioeconomic attributes, *Photogrammetric Eng. Remote Sensing*, 65 (5), 611, 1999.
5. Mesev, V., The use of census data in urban image classification, *Photogrammetric Eng. Remote Sensing*, 64 (5), 431, 1998.
6. Baudouin, Y., Cavayas, F., and Marois, C., Towards a new method of inventory and update of the activity/use of the ground in urban zones, *Can. J. Remote Sensing*, 1, 28, 1995.
7. Bourcier, A., Vaguet, O., and Vaguet, A., Urban morphology and dynamics to Coimbatore, South of India, *Photo-interprétation*, 1, 23, 1996.
8. Chatelain, A., Analysis of the urban morphologies and their evolution from remote sensing data, *Informatique et sciences humaines*, 44, 83, 1978.
9. Moriconi-Ebrard, F., *Geopolis: Comparing Cities of the World*, Anthropos, Paris, 1994, 246.
10. Blaser, T., Contribution of the remote sensing in the conceptualization and in the update for land planning, Ph.D. thesis, Ecole Polytechnique Fédérale de Lausanne, Lausanne, Switzerland 1992.
11. Chen, S. and Xie, C., Remote sensing and GIS for urban growth analysis in China, *Photogrammetric Eng. Remote Sensing*, 66 (5), 593, 2000.
12. Gar-On Yeh, A. and Li, X., Measurement and monitoring of urban sprawl in a rapidly growing region using entropy, *Photogrammetric Eng. Remote Sensing*, 67 (1), 83, 2001.
13. Barr, S. and Barsley, M., Application of structural pattern-recognition techniques to infer urban land use from ordnance survey digital map data, paper for 3rd International Conference on GeoComputation, Bristol, 1998.

14. Webster, C., Urban morphological finger sprints, *Environ. Plann. B*, 23 (3), 279, 1996.
15. Franca, G. and Xue, Y., Dynamic aspects study of surface temperature from remotely-sensed data using advanced thermal inertial model, *Int. J. Remote Sensing*, 15, 2517, 1996.
16. Li, Z. and Becker, F., Feasibility of land surface emissivity determination from AVHRR data, *Remote Sensing Environ.*, 43, 67, 1993.
17. Sthor, C. et al., Classification of depression in landfill covers using uncalibred thermal-infrared imagery, *Photogr. Eng. Remote Sensing*, 60 (8), 1019, 1994.
18. Rakodi, C., Ed., *The Urban Challenge in Africa. Growth and Management of Its Large Cities*, United Nations University Press, Tokyo.
19. Dureau, F. and Duchemin, J.P., The use of SPOT for the demographic observation in urban zones, Research Report, CNES-ORSTOM, 1988.
20. Dobson, J. E. et al., Landscan: a global population database for estimating population at risk, *Photographic Eng. Remote Sensing*, 66 (7), 849, 2000.
21. Lavalle, C. et al., Towards an urban atlas. Assessment of spatial data on 25 European cities and urban areas, EEA, Copenhagen, 2002, 131, http://reports.eea.eu.int/environmental_issue_report_2002_30/en.
22. Couloigner, I. and Ranchin, T., Mapping of urban areas: a multiresolution modeling approach for semi-automatic extraction of streets, *Photogrammetric Eng. Remote Sensing*, 66 (7), 867, 2000.
23. Bellagente, M., Gamba, P., and Savazzi, P., Fuzzy texture characterization of urban environments by SAR data, paper for the International Geoscience and Remote Sensing Symposium 99, Germany, 1999.
24. Park, J.H. et al., The potential of high resolution remotely sensed data for urban infrastructure monitoring, paper for the International Geoscience and Remote Sensing Symposium 99, Germany, 1999.
25. Hepner, G. et al., Investigation of the integration of AVIRIS and ISFAR for urban analysis, *Photogrammetric Eng. Remote Sensing*, 64 (8), 813, 1998.

19 Urban Scenario Modeling and Forecast for Sustainable Urban and Regional Planning

José I. Barredo, Carlo Lavalle, and Marjo Kasanko

CONTENTS

19.1 INTRODUCTION

Throughout the world, and in particular in Europe, processes related to urbanization, development of transport infrastructures, industrial constructions, and other built-up areas, are severely influencing the environment, and are often modifying the landscape in an unsustainable way [1]. The main aim of the monitoring land use cover dynamics (MOLAND) project, which is coordinated by the Institute for Environment and Sustainability of the European Commission's Joint Research Centre, is to provide up-to-date, standardized, and comparable information on the past, current, and likely future land-use development in Europe.

As part of MOLAND, an urban growth model has been developed. This model is used to assess the likely impact of current spatial planning and policies on future land-use development. To date, the MOLAND database has covered more than 40 urban areas, transport corridors, and extended regions. The aim of MOLAND is to

329

assess, monitor, and model past, present, and future urban and regional development from the viewpoint of sustainable development and natural hazards, by setting up GIS databases for cities and regions. MOLAND has defined and validated a methodology in support of assessing the impacts of European sectoral policies with territorial and environmental implications.

The aim of the chapter is to disseminate some results of a modeling framework that has been developed to help spatial planners and policy makers to analyze a wide range of spatial policies and their associated consequences in spatial patterns of land use. An application case study for land-use simulation in Udine (Italy) is illustrated as an example of the approach. The core of this methodology consists of dynamic spatial models that operate at both the micro and macro geographical levels. At the macro level, the modeling framework integrates several component submodels, representing the natural, social, and economic subsystems typifying the area studied. These are all linked to each other in a network of mutual reciprocal influence. At the micro level, cellular automata (CA)-based models determine the fate of individual parcels of land, based on their individual institutional and environmental characteristics, as well as on the types of activities in their neighborhoods. The approach permits the straightforward integration of detailed physical, environmental, and institutional variables, as well as the particulars of the transportation infrastructure.

19.1.1 THE MOLAND PROJECT

The implementation of MOLAND is divided into three phases — corresponding to the three specific aims — called "Change," "Understand," and "Forecast."

In the "Change" phase of MOLAND, detailed GIS databases of land-use and transport networks are produced for each study area. The databases are typically for four dates (early 1950s, late 1960s, 1980s, late 1990s) for urban areas, or (in the case of larger areas) for two dates (mid-1980s, early 2000s), at a mapping scale of 1:25,000. The MOLAND land-use legend, which is an extended and more detailed version of the coordination of information on the environment (CORINE) land cover legend, includes approximately 100 land-use classes. The database is complemented with socioeconomic data sets and statistics.

In the "Understand" phase of MOLAND, the emphasis is on spatial and statistical analysis of urban and regional development. Central to the analysis is the computation of different types of indicators of urban and regional development. These indicators are used to assess and compare the study areas in terms of their progress toward sustainable development. The databases have also been used to support strategic environmental assessments (SEA) of the impact of transport links on the landscape.

Under the "Forecast" phase of MOLAND — the one with which this chapter is concerned — a generic model for simulating urban growth, based on dynamic spatial systems, has been developed. The aim here is to predict future land-use development under existing spatial plans and policies and to compare alternative possible spatial planning and policy scenarios (including the scenario of no-planning). In this chapter the main emphasis is on the description of the dynamic spatial model and its results for urban and regional planning policies assessment.

19.1.2   TOWARD A SUSTAINABLE PHYSICAL PLANNING

Sustainability is a broad and multidimensional concept that comprises several elements. It involves the maintenance of natural resources, including land, and spatial patterns of land use that must be ecologically, socially, and economically beneficial [2]. The spatial dimension of sustainability can be focused on the dynamics of land uses and its derived consequences, such as fragmentation of natural and agricultural areas and urban decentralization.

Once physical space has been used for built-up areas or infrastructures, it may be impossible to reclaim. Despite that, there is clear evidence regarding the unsustainable development of urban areas in Europe. Built-up areas have expanded by 20% during the last two decades, which is much faster than the 6% of population growth [3]. Although population growth in some urban areas has now stabilized, as in the case of Udine presented here, urban development around the periphery of principal urban centers continues, demonstrating a decentralization of urban land uses. Rising standards of living and increased distances between residential areas and places of employment have contributed to an increase in traffic and the infrastructure needed to accommodate it. These trends are causing increasing losses of agricultural land and the fragmentation of natural areas in most of Europe [3].

A predominant feature of planning policies for major conurbations in many countries is the concentric evolution of development. Examples of this phenomenon can be seen in London, Madrid, and Paris, where many key services and employment opportunities attract millions of long-distance commuters every day. A proposal under the European Spatial Development Perspective (ESDP) is to encourage a polycentric evolution of development, whereby dispersed urban areas in a country would be connected to each other to help dissipate pressures across a wider area and revive neglected regions, in particular rural areas [3].

Taking into account the aforementioned precedents, physical planning becomes a strategic aspect for more sustainable policies at urban and regional levels. The focus of physical planning is the "optimization of the distribution of land uses in an often limited space, focusing on land use allocation" [2, p. 66; 4, p. 84]. However, how to measure the impact of physical planning actors is not an easy task without tools which embrace the complexity of urban land-use dynamics. To this end, dynamic spatial models can serve as a tool for the realistic simulation of urban processes under several planning hypotheses. Furthermore, the results of such models can be used to produce indicators about fragmentation, access to green areas, time used by commuters, etc. Those indicators can provide us with information for the assessment of more sustainable planning strategies in an integrated approach.

19.1.3   SPATIAL DYNAMIC SYSTEMS FOR URBAN SCENARIO SIMULATION

The estimation of future impacts on land-use development of existing spatial plans and policies and the consideration of alternative planning and policy scenarios for impact minimization are of particular interest for urban and regional planners. In the last decade CA have gained popularity as a modeling tool for the simulation of spatially distributed processes. Since the pioneer work of Tobler [5], several approaches have been proposed for modifying standard CA in order to make them suitable for urban simulation [6–17]. The results of the previous applications are promising and have shown realistic results in cities of different continents.

CA are a joint product of the science of complexity and the computational revolution [18]. Despite their simplicity, CA are models which deal with processes that show complexity or, in other words, with complex systems. CA have been defined as very simple dynamic spatial systems, in which the state of each cell in an array depends on the previous state of the cells within a neighborhood and is produced according to a set of transition rules [14]. What is surprising in CA is their potential for modeling complex spatiotemporal processes despite their very simple structure. Very simple CA can produce surprisingly complex forms through a set of simple deterministic rules. Cities studied as dynamic systems show some complexity characteristics that can be modeled using CA-based applications in an integrated approach.

Effective urban and regional planning and management requires both spatial data on current conditions and ability to foresee the likely consequences of projects and policies. The approach presented in this chapter is aimed specifically at enhancing the use of GIS and spatial dynamic models for urban planning assessment. We propose a generic model of urban dynamics that will support the realistic exploration of urban futures under a variety of planning and policy scenarios.

19.2 METHODS: THE MODEL FOR URBAN DYNAMICS

The CA-based model proposed in this chapter comprises several factors that drive land-use dynamics in a probabilistic approach. Previous studies in the urban CA arena have shown that the transportation network and land-use suitabilities are the determinant factors of the "visual urban form" [11, p. 338]. These factors drive, to a great degree, the growth of the city: vacant areas in a city with high accessibility and the right suitability conditions are highly prone to urbanization. In addition, the land-use zoning regulation is also a factor that influences the land-use allocation in a city, since it establishes the legal framework for future land uses. The process of urban land-use dynamics can be defined as a probabilistic system in which the probability that a place in a city is occupied by a land use is a function of accessibility, suitabilities, zoning status, and the neighborhood effect measured for that land use. All these factors, in addition to a stochastic parameter, have been included in the urban CA-based model. The stochastic parameter has the function of simulating the degree of stochasticity that is characteristic in most social and economic processes.

The model used in this application is an improved version inspired by the model developed by White et al. [11]. In this new CA-based model an extensive number of states are considered, including several types of residential land use. Other improvements are discussed later in this section. In this model the probability that an area changes its land use is a function of the aforementioned factors acting together at a defined time. However, the factor that makes the system work like a nonlinear system is the iterative neighborhood effect, whose dynamism and iterativity can be understood as the core of the land-use dynamics. The iterative neighborhood effect is founded in the "philosophy" of standard CA, where the current state of the cells and the transition rules define the configuration of the cells in the next time step. From the described approach, a constrained urban CA model has been designed and developed for the simulation of urban land-use dynamics [14,19,20]. It has the following specificities.

The digital space in the CA-based model used for this study consists of a rectangular grid of square cells, each representing an area of 100 m × 100 m. This is the same size as the minimum area mapped in urban areas in the MOLAND's land-use data sets. In the model, each cell can assume a state. The model uses a number of cell states representing land-use classes in which the studied city is subdivided. Some classes represent fixed features in the model, that is, states which are assumed not to change and which therefore do not participate in the dynamics. They do, however, affect the dynamics of the active land-use classes, since they may have an attractive or repulsive effect in the cell neighborhood. Examples of fixed features are: abandoned areas, road and rail networks, airports, mineral extraction sites, dump sites, artificial nonagricultural vegetated areas, and water bodies. Another group is passive functions, that is, functions that participate in the land-use dynamics, but whose dynamics are not driven by an exogenous demand for land; they appear or disappear in response to land being taken or abandoned by the active functions. Examples of passive functions are: arable land, permanent crops, heterogeneous agricultural areas, forest, pastures, and shrublands. The active functions are the urban land-use classes. These functions are forced by demands for land generated exogenously to the model in response to the growth of the urban area. They usually are: residential continuous dense urban fabric, residential continuous medium dense urban fabric, residential discontinuous urban fabric, residential discontinuous sparse urban fabric, industrial, commercial, and public and private services. Construction sites represent a transitional state between one function and another. It is remarkable that the model is able to simulate an extensive number of urban land uses, including four types of residential land use. This aspect is one of the differences of this urban model with respect to other CA-based models previously developed (e.g., [9,12,13,17]).

In standard CA, the fundamental idea is that the state of a cell at any time depends on the state of the cells within its neighborhood in the previous time step, based on the predefined transition rules. In this CA-based model this aspect is modified as follows. A vector of transition potentials (one potential for each function) is calculated for each cell from the suitabilities, accessibilities, zoning status and neighborhood space effect, and the deterministic value is then given a stochastic perturbation using a modified extreme value distribution, so that most values are very slightly modified, while a few others are changed significantly. The probabilistic function is thus obtained by the equation:

$$ {}^{t}P_{K,x,y} = \left(1 + {}^{t}A_{r,K,x,y}\right) \cdot \left(1 + S_{K,x,y}\right) \cdot \left(1 + {}^{t}Z_{K,x,y}\right) \cdot \left({}^{t}N_{K,x,y}\right) \cdot {}^{t}v \qquad (19.1) $$

where:

$^tP_{K,x,y}$ CA transition potential of the cell (x, y) for land use at time t
$^tA_{r,K,x,y}$ Accessibility of the cell (x, y) to infrastructure element r for land use K
at time t
$S_{K,x,y}$ Intrinsic suitability of the cell (x, y) for land use K
$^tZ_{K,x,y}$ Zoning status of the cell (x, y) for land use K at time t
$^tN_{K,x,y}$ Neighborhood space effect on the cell (x, y) for land use K at time t
v Scalable random perturbation term at time t; it is defined as:
$v = 1 + [-\ln(rand)]$, where $(0 < rand < 1)$ is a uniform random variable,
and a is a parameter that allows the size of the perturbation to be calibrated.

The transition rule works by changing each cell to the state for which it has the
highest potential. However, it is subject to the constraint that the number of cells in
each state must be equal to the number demanded in that iteration. Cell demands
are generated outside the model. During each iteration all cells are ranked by their
highest potential, and cell transitions begin with the highest-ranked cell and proceed
downwards until a sufficient number of cells of a particular land use has been
achieved. Each cell is subject to this transition algorithm at each iteration, although
logically most of the resulting transitions are from a state to itself, that is, the cell
remains in its current state.

In the urban CA-based model described herein, the neighborhood space is
defined as a circular region around the cell with a radius of eight cells. The neighborhood thus contains 197 cells that are arranged in 30 discrete distance zones. The
neighborhood radius represents 0.8 km. This distance delimits an area that can be
defined as the influence area for urban land-use classes. Thus, this distance should
be sufficient to allow local-scale spatial processes to be captured in the model
transition rules. In the urban CA-based model, the neighborhood effect is calculated
for each of the 13 function states (passive and active) to which the cell could be
converted. It represents the attraction (positive) and repulsion (negative) effects of
the various states within the neighborhood. In general, cells that are more distant in
the neighborhood will have a smaller effect, a positive weight of a cell on itself
(zero-distance weight) represents an inertia effect due to the implicit and monetary
costs of changing from one land use to another. Thus each cell in a neighborhood
will receive a weight according to its state and its distance from the central cell.
The neighborhood effect is calculated as:

$$^tN_{K,x,y} = \sum_c \sum_l w_{K,L,c} \cdot {}^tI_{c,l} \tag{19.2}$$

In Equation 19.2, $^tN_{K,x,y}$ is the contribution of neighborhood effect in the calculation of the transition potential of cell (x, y) for land use K at time t. $w_{k,L,c}$ is the
weighting parameter expressing the strength of the interaction between a cell with
land use K and a cell with land use L at a distance c in the neighborhood. And $^tI_{c,l}$
is a binary function returning a value of 1 if a cell l is in the state L, or returning a
value of 0 if it is not in the state L.

The accessibility factor represents the importance of access to transportation networks for various land uses for each cell, again one for each land use type. Some activities, such as commerce, require better accessibility than others, for instance residential discontinuous sparse urban fabric. Accessibilities are calculated as a function of distance from the cell to the nearest point in the transport network as follows:

$$^{t}A_{r,K,x,y} = \frac{1}{1 + \dfrac{D_r}{a_{r,K}}} \tag{19.3}$$

In Equation 19.3, $^{t}A_{r,K,x,y}$ is the accessibility of cell (x, y) to infrastructure element r for land use K at time t; D_r is the distance between cell (x, y) and the nearest cell $(x,'y')$ on infrastructure element r; and $a_{r,K}$ is a calibrated distance decay accessibility coefficient expressing the importance of good access to infrastructure element r for land use K.

Finally, each cell is associated with a set of codes representing its land-use zoning status for various land-use classes, and for various periods. Then each cell is also associated with its suitability. These suitabilities are defined as a weighted sum or product of a series of physical, environmental, infrastructural, historical, and institutional factors. For computational purposes, they are standardized to values in the range 0–1 and represent the inherent capacity of a cell to support a particular activity or land use.

Due to the combined effect of suitabilities, accessibilities, zoning status, and the neighborhood effect, every cell is essentially unique in its qualities with respect to possible land-use classes. It is in this highly differentiated digital space that the dynamics of the model take place. In constrained CA, the land-use demands are generated exogenously to the cellular model [11] such as in this case. Demands reflect the growth of a city rather than the local configurational dynamics captured by the urban CA. Thus in the present model, cell demands for each land use type are generated exogenously to the CA. In integrated models for regional and urban dynamics, the demands for land uses are calculated from demographic, economic, and transport growth trends and feedbacks to the CA spatial model.

19.2.1 An Application Case Study for Udine, Italy

The objective of this study is to develop a future urban simulation for the city of Udine (Italy) based on the current land-use planning policies (master plans) implemented by local authorities in the Friuli-Venezia Giulia (FVG) Region. The Udine case study is a pilot study in the framework of a project that is currently being carried out including the whole FVG Region. The total surface of this region is 7850 km². Land use is mapped over a period of 40 years: 1950, 1970, 1980, and 2000. Land-use data sets and additional socioeconomical and environmental data will be used to assess urban development trends and environmental pressure, to characterize areas, highlighting their strengths and weaknesses, and to assess the impacts of policies. Developing scenarios of growth will serve as major input to formulate and evaluate mid/long-term strategy for sustainable development.

This section describes the methodology and data used in the Udine pilot study case. This study follows a similar methodology as the one used for scenario simulation in other cities such as Dublin [20] and Lagos in Nigeria [19]. The methodological approach followed is based in several phases. Initially, the CA model was calibrated by using historical (1980) and reference (2000) data sets that were compiled for Udine (Figure 19.1).

19.2.2 CALIBRATION OF THE MODEL

The simulation for the period 1980–2000 initiates using the historical data sets for the year 1980, in order to test the simulation results using the reference data sets for 2000. With this approach, we tested the simulation by comparing the results with actual land-use data sets. Often the testing of the simulation results has been considered as a weakness in urban CA modeling [12,21]. An empirical and practical way for testing the model is to use historical data sets. Once the results of the calibration are satisfactory, the future simulation of land use can be done using the parameters of the calibrated model, assuming that the interactions between land-use classes (parameters of Equation 19.2) will remain stable during the studied period.

The increase or decrease in the number of cells for each land-use class in the modeled period (1980–2000) has been calculated from the historical and reference data sets. On the other hand, in the case of future simulations, the addition of planned transport links could represent an interesting approach for predicting the impact of new infrastructures on the territory, or for evaluating different development alternatives. In addition to this, it is also possible to simulate the urban land-use evolution using different zoning regulation data sets in order to study the spatial impact of planners' actions (Figure 19.2).

Figure 19.1 shows the land-use in Udine for 2000. It is noticeable that the more prominent land-use classes for Udine in the studied period are residential discontinuous urban fabric and industrial. Accurate simulations in the present urban CA-based model depend on several factors, among which one of the most important is the calibration of the weighting parameters of Equation 19.2, which define the neighborhood effect. The weighting parameters are calibrated in order to minimize the differences between the simulated and actual land-use maps for 2000. The schema of weighting parameters for any pair of land use classes is based on a rational evaluation of the land-use patterns in the city and their evolution. The calibration of the model is based on an interactive procedure in which each state (active and passive) is adjusted versus each of the land-use classes.

The weighting assignment is done by verifying visually the spatial effects of the weights in the model prototype. The computational time required for each simulation is on the order of a few seconds on a standard PC (1.7 GHz); thus the weight values can be modified interactively until the results fit with the reference land-use map. It is significant that similar weight schemes have been found in simulations for different cities. This is reasonable, considering that in general terms urban land-use

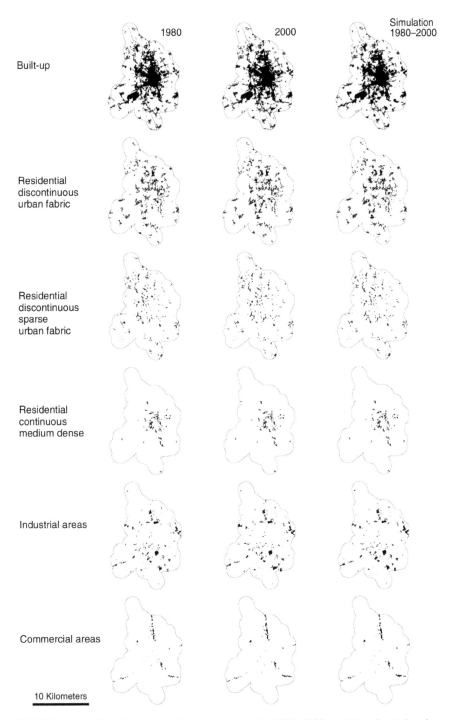

FIGURE 19.1 Udine, built-up and land-use maps for 1980, 2000, and the simulation from 1980 to 2000.

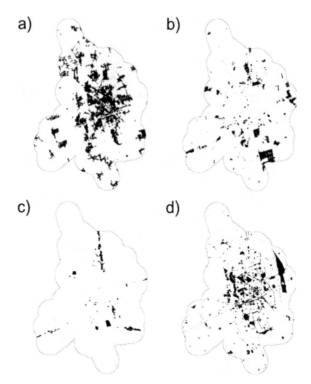

FIGURE 19.2 Zoning regulation data sets for Udine provided from local authorities: (a) residential areas; (b) industrial areas; (c) commercial areas; and (d) public and private services.

classes should respond to similar attraction/repulsion schemes, independently of the city studied. Once all the functions have been calibrated, the model is run again several times in order to verify if the land-use transitions work in a realistic way. Accuracy analyses based on spatial metrics and other procedures such Kappa coefficients (κ) are used in order to produce a fine-tuned version of the simulation.

Another calibrated factor is the random perturbation (see Equation 19.1). It was set at = 0.7 by means of a trial-and-error approach. In general it is set to reproduce the urbanized area of the actual land-use map for 2000. In this case it is also fine-tuned to generate a sufficient number of new seed cells of various land-use classes in new locations such as rural areas (i.e., spontaneous growth), which will subsequently grow into, for example, new industrial, commercial or residential areas.

The preliminary results are a useful demonstration of the extent to which the urban CA can produce realistic simulations using the current master plans or modifying it in order to test several urban development strategies.

19.3 RESULTS: SIMULATION RESULTS TESTING

The model was tested following the same methodology used in other study cases (see [19,20]). The test was carried out using two approaches. A first very intuitive approach was based on a visual comparison between the simulated and actual land-use

TABLE 19.1
Regression Analysis Results from Spatial Metrics for the Actual and the Simulated Land-Use Maps for 2000

Metrics	R-Squared Coefficient
Mean patch area	0.95
Mean patch area CV	0.96
Total edge	0.98
Shape index	0.97
Shape index CV	0.94
Proximity index	0.96
Proximity index CV	0.96
Splitting index	0.99
Simpson's diversity index	Actual city 0.76; simulation 0.75

Note: CV, coefficient of variation.

maps for 2000. The main feature of this analysis is the resemblance between the maps (Figure 19.1). The visual comparison produces a first idea of what the urban CA is able to do. However, statistical tests are needed in order to obtain accuracy values. To this end, several spatial statistical measures of similarity between both maps have been produced. This procedure is based on the use of spatial metrics and regression analyses. It has been shown that procedures using coincidence matrices are not well suited for testing urban CA models [11,22]. The main problem is related to the incapacity of quantifying patterns as such, because coincidence matrices procedures are based on independent comparisons between pairs of cells.

The existence of land-use patterns might be understood holistically at the level of the whole city. Modeling the future of cities should account for the overall appearance of the future city regarding land-use pattern distribution and development. This test is based on the comparison of a set of complementary spatial metrics obtained for the active states in the simulation and in the actual city map. The used metrics quantify landscape composition and/or landscape configuration. More specifically, the spatial metrics at land-use level can be interpreted as fragmentation indices, while metrics at landscape level — in this case only the Simpson's diversity index — considering all land-use classes together, can be interpreted as landscape heterogeneity indices in the overall landscape. The definition of the metrics used can be seen in McGarigal et al. [23]. The metrics have been obtained by analyzing only the active states in both land-use raster maps using the Fragstats software [23]. The metrics from both maps were afterwards compared by regression analyses (see Table 19.1).

The proposed approach is an alternative to cell-by-cell comparison procedures based on contingency tables. By using spatial metrics, it is possible to identify pattern structure differences. This is not possible using methods based on contingency tables and their derived statistics, such as coefficients. This approach offers the advantage of evaluating the overall structure of the landscape, taking into consideration the pattern of each class for each map.

The first result of this analysis is the overall high R-square coefficients obtained for most of the studied metrics, and the similar Simpson's diversity index for both maps. The similar probabilities of both cities showed in the Simpson's diversity index are a consequence of a similar fragmented land-use landscape in both cases. In addition, the R-squared coefficients obtained from the coefficient of variation (CV) of mean patch area, shape index, and proximity index indicate a relatively good fit between the variability of these metrics.

The reasonably good results obtained match with previous studies in the urban CA arena (e.g., [9,11–13,15–17,21,24]). However despite the promising results, it is obvious that both maps are not identical. The spatial metrics have shown that both maps are similar from the point of view of their land-use pattern, which does not mean that the maps are identical, strictly speaking, from a spatial point of view.

19.4 SCENARIO SIMULATION FOR 2020 AND DISCUSSION

The future scenario simulation covers a twenty-year period between 2000 and 2020, and was undertaken using the calibrated model for Udine. In this case, the demands for each land-use class were calculated on the basis of land-use growth trends from recent years. Udine is expected to grow slightly in the coming years, due to population and land-use trends. During the past four decades Udine has grown slightly. These trends can be obviously altered by a number of factors such as changes in the economic climate, which may introduce some important changes in the urban development of the city, producing a greater growth of some activities such as commercial or services.

Some land-use classes are expected to grow following the sparse low-density development style of the city, for example the residential discontinuous land-use class. However, it is likely that there will be an increase of more dense populated areas which might increase the growth of the residential continuous dense land use class. The trends of urban growth have been included in the model for 2020. These trends take into account recent developments of the city. Using urban CA models, an interesting option for planners is to produce several scenarios following several hypotheses (for example, using current trends, increasing, or decreasing it), obtaining in this way different scenarios of the urban future based on several assumptions. The addition of planned transport links or changes in the zoning regulations could also represent an additional strategy for predicting future land-use patterns or for the evaluation of different planning options.

The definition of land-use demands is a key aspect, since these demands will define in an important degree the spatial evolution of the simulated city. All land-use classes are generally foreseen to grow in Udine. Although the urban growth trend included in the model is slightly higher than the current one, the urban land-use classes will not impact highly on the structure of the city. New residential and commercial areas are foreseen to grow mainly in peripheral neighborhoods. On the other hand, another consequence may be the creation of new commercial areas as has been shown by the land-use trends.

The land use demands defined for the year 2020 have been included in the simulation, obtaining in this way the simulated land-use map for that year (Figure 19.3). From a visual point of view, the simulated map maintains the general form of the current city due to the initial conditions of transport network and land-use. The foreseen urban pattern appears to be realistic. However, the form of the city has clearly developed and shows increased built-up nuclei in peripheral areas. Furthermore, it is remarkable that the model has produced iteratively long-term predictions based on several hypotheses for several urban land-use classes, which can be seen as a very useful tool for planning purposes.

In general, the development of the city has occurred as was expected, the bigger built-up areas having grown in most cases continuously from the initial city core. Nuclear patches of new urban areas are distributed around the city core, usually in correspondence with transport links. The transport network, together with the initial land-use map and zoning regulation, represent the major factors for the new developed urban areas.

An obvious problem with future urban simulations is testing their accuracy. Nevertheless, it is clear that simulations with absolute accuracies are not possible, considering the stochasticity of the modeled system and the unforeseeable bifurcations that are frequent in nonlinear systems such as in this case. Despite that, an important aspect is that urban CA models can support "what if" experiments [25], offering the possibility of exploring the future within some degree of accuracy specified in the calibration phase.

A second simulation for 2020 has been developed, representing an "optimistic" hypothesis (i.e., higher growth rates) for two land-use classes: commercial and residential discontinuous urban fabric (see Simulation B in Figure 19.3). In this case, however, the trends take into account recent developments of the city, revealing an increasing urban growth rate for these two land-use classes. This means that the predictions are somewhat greater than the actual trends used in the first simulation for 2020.

19.5 CONCLUDING REMARKS

The results of the simulations for Udine have demonstrated that the urban CA-based model can provide reasonable representations of the future of cities. Added to this, it is worth noting the considerable length of the simulated period of 20 years. On the other hand, the capability of the urban CA to reproduce scattered urban growth patterns is also significant. This is remarkable, considering the aim of the urban simulation prototype. Obviously the inclusion of suitabilities and zoning status data sets yields more reliable results in an adequate time scenario for planning purposes. The inclusion of extensive detailed land-use classes, such as several types of residential classes, together with the fine detail of the data sets, increases the experimental potential of the model, given that it can be used by planners like a simulation

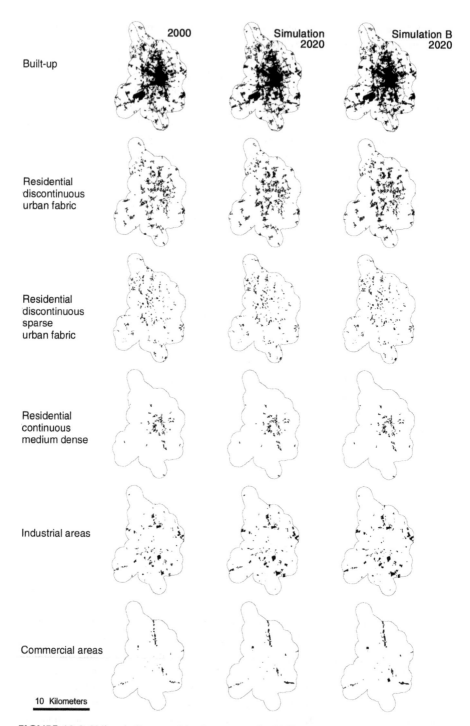

FIGURE 19.3 Udine, built-up and land-use maps for 2000, and two simulations from 2000 to 2020.

box, in which a number of spatial conditions "if … then" can be tested easily in a realistic way within a known degree of accuracy.

The promising results of this urban CA-based model are comparable with previous results in the arena of urban modeling using CA. However, some aspects should be considered for future developments. One of the more relevant weaknesses when future events are modeled is represented by the testing phase. It is obvious that models for future events are difficult to test. An added complication is that, when modeling is made for human or social systems like cities, these systems may be affected by a large series of events, such as economic crisis, changes in planning policies, natural disasters, and other elements that can obviously modify profoundly the aspect and evolution of cities. Furthermore, the stochastic degree and complexity of this kind of system make particularly difficult their simulation and prediction, in particular for long periods of time.

Common sense and rationale are the keywords in the calibration phase of the model. The weighting assignment is the basis of the neighborhood effect factor, which can be considered as the core of the model. These weights can obviously modify the overall result of the model, since the potential for transitions depends highly on their values. The "known" land-use interactions are the basis for their assignment. It would be desirable to focus more research on this area in order to get a better understanding of these interactions. Something similar happens with the stochastic parameter; how can we measure it in real cities? At the moment only previous experience and empirical trial-and-error procedures are available.

On the other hand, the land demands for land-use classes in this prototype are produced outside the model. A more powerful version of this model is being developed and applied to bigger areas and regions. In this case, standard nonspatial models of regional economics and demographics are linked with the urban CA-based model in order to produce a scheme in which changes and trends in population, economics, or environmental aspects will have spatial impacts through the urban CA module. These integrated models will provide planners with more powerful tools for urban and regional scenario generation.

ACKNOWLEDGMENTS

This study is part of a research project funded by the European Commission, Directorate General Joint Research Centre, postdoctoral contract # 17646–2001–03 P1B30 ISP IT, "Monitoring sustainable urban and regional development using spatial dynamics models," awarded to José I. Barredo, and was undertaken as part of the EC's DG-Joint Research Centre MOLAND (Monitoring Land Cover / Use Dynamics) project.

The cellular automata prototype has been developed by Research Institute of Knowledge Systems (RIKS) in the framework of the contract # 14518–1998–11 F1ED ISP NL, awarded by the European Commission, Directorate General Joint Research Centre.

The procurement of the database for the Region Friuli-Venezia Giulia has been financed by the Regional Government of FVG under endorsement n. 24/PT Dec 2000 CCRN, 17250 to the EC-Joint Research Centre.

REFERENCES

1. EEA, *Towards an urban atlas: assessment of spatial data on 25 European cities and urban areas,* Environmental issue report no. 30, European Environment Agency (EEA) and European Commission–DG Joint Research Centre, Office for Official Publications of the European Communities, Luxembourg, 2002, 131 pp.

2. Botequilha Leitao, A. and Ahern, J., Applying landscape ecological concepts and metrics in sustainable landscape planning, *Landscape Urban Plann.,* 59, 65–93, 2002.

3. EEA, *Environmental signals 2002, Benchmarking the millennium,* Environmental assessment report no. 9, European Environment Agency (EEA), Office for Official Publications of the European Communities, Luxembourg, 2002.

4. van Lier, H.N., The role of land use planning in sustainable rural systems, *Landscape Urban Plann.,* 41, 83–91, 1998.

5. Tobler, W., A computer movie simulating urban growth in the Detroit region, *Econ. Geogr.,* 46, 234–240, 1970.

6. Batty, M. and Longley, P., The fractal simulation of urban structure, *Environ. Plann. A,* 18, 1143–1179, 1986.

7. Batty, M. and Longley, P., Fractal-based description of urban form, *Environ. Plann. B,* 14, 123–134, 1987.

8. Cecchini, A. and Viola, F., Eine Stadtbausimulation. *Wissenschaftliche Zeitschrift der Hochschule für Architektur und Bauwesen,* 36 (4), 1990.

9. White, R. and Engelen, G., Cellular automata and fractal urban form: a cellular modelling approach to the evolution of urban land use patterns, *Environ. Plann. A,* 25, 1175–1199, 1993.

10. Itami, R.M., Simulating spatial dynamics: cellular automata theory, *Landscape Urban Plann.,* 30, 27–47, 1994.

11. White, R., Engelen, G., and Uljee, I., The use of constrained cellular automata for high-resolution modelling of urban land use dynamics, *Environ. Plann. B,* 24, 323–343, 1997.

12. Wu, F., SimLand: a prototype to simulate land conversion through the integrated GIS and CA with AHP–derived transition rules, *Int. J. Geogr. Inf. Sci.,* 12, 63–82, 1998.

13. Clarke, K.C. and Gaydos, L., Loose coupling a cellular automata model and GIS: long-term growth prediction for San Francisco and Washington/Baltimore, *Int. J. Geogr. Inf. Sci.,* 12, 699–714, 1998.

14. White, R., Engelen, C., Uljee, I., Lavalle, C., and Erlich, D., Developing an urban land use simulator for European cities, in *Proceedings of the 5th EC–GIS Workshop,* 28–30 June, 1999, Stresa, Italy, European Communities, 179–190, 2000.

15. White, R. and Engelen, G., High-resolution integrated modelling of the spatial dynamics of urban and regional systems, *Comput. Environ. Urban,* 24, 383–400, 2000.

16. Semboloni, F., The growth of an urban cluster into a dynamic self-modifying spatial pattern, *Environ. Plann. B,* 27, 549–564, 2000.

17. Sui, D.Z. and Zeng, H., Modeling the dynamics of landscape structure in Asia's emerging desakota regions: a case study in Shenzhen, *Landscape Urban Plann.,* 53, 37–52, 2001.

18. Couclelis, H., Artificial intelligence in geography: conjectures on the shape of the things to come, *Prof. Geogr.,* 38, 1–11, 1986.

19. Barredo, J.I., Demicheli, L., Lavalle, C., Kasanko, M., and McCormick, N., Modelling future urban scenarios in developing countries: an application case study in Lagos, Nigeria, *Environ. Plann. B,* 32, 65–84, 2004.

20. Barredo, J.I., Kasanko, M., McCormick, N., and Lavalle, C., Modelling dynamic spatial processes: simulation of urban future scenarios through cellular automata, *Landscape Urban Plann.,* 64, 145–160, 2003.
21. Wu, F., Calibration of stochastic cellular automata: the application to rural-urban land conversions, *Int. J. Geogr. Inf. Sci.,* 16, 795–818, 2002.
22. Torrens, P.M. and O'Sullivan, D., Editorial: Cellular automata and urban simulation: where do we go from here, *Environ. Plann. B,* 28, 163–168, 2001.
23. McGarigal, K., Cushman, S.A., Neel, M.C., and Ene, E, FRAGSTATS: Spatial Pattern Analysis Program for Categorical Maps, Computer software program produced by the authors at the University of Massachusetts, Amherst, 2002. Available at: www.umass.edu/landeco/research/fragstats/fragstats.html.
24. White, R., Engelen, G., and Uljee, I., Modelling land use change with linked cellular automata and socio-economic models: a tool for exploring the impact of climate change on the island of St. Lucia, in *Spatial Information for Land Use Management,* Hill, M. and Aspinall, R., Eds., Gordon and Breach, 2000, pp. 189–204.
25. White, R. and Engelen, G., Cellular automata as the basis of integrated dynamic regional modelling, *Environ. Plann. B,* 24, 235–246, 1997.

Part III-B

---

*Learning from Practice:
GIS as a Tool in Planning
Sustainable Development*

Natural and Cultural Heritage

Part III–P

20 The Development of the Cross-Border Region of Hungary and Austria Analyzed with Historical Cadastral and Land Register Data

Susanne Steiner

CONTENTS

20.1 INTRODUCTION

Many land-use planning decisions ultimately are tied to land ownership and parcels. Therefore, the analyses of past land use dynamics cannot be considered only as historic research, but in fact, the development of land and property is an interesting starting point for decisions in recent land management. To serve the demand, for example, of land planners to base their work on research and historic data, a study was conducted to show that past developments in land administration can illustrate future trends. For the study, a border region between Austria and Hungary was chosen, where changing agrarian policy conditions and their effects over the intervening years can be comprehended and compared.

Border regions provide interesting historical details regarding the structure of agricultural land, the enlargement of settlement area, and economic and social situations. An interesting example is the mentioned border region between Hungary and Austria, where changing nationalities of the area due to historical facts influenced the economic development. The last century, with changing political conditions, shaped the region on both sides of the borderline in different ways. Agricultural land as well as building land was affected by political upheavals, and those changes have been registered in land administration documents.

The objective of the study, carried out by a bilateral project team consisting of Hungarian as well as Austrian researchers, was to reconstruct and analyze cadastre-based land use (the term "land use" is referring to cadastral determination, which need not necessarily correspond with real land cover) and ownership changes in the cross-border region of Austria and Hungary and to quantify the influences of external effects like socioeconomic and/or political changes on parcel-based land use and property rights. The investigations in both regions were based on cadastral and land register data that were available in a time series of one hundred years. Based on derived results, some general statements on the rural development trends for these particular border regions could be drawn up.

A geographic information system was used for analyzing and visualizing changes regarding land use and parcel geometry based on a database and on spatial cadastral data.

In this chapter the methodology of acquisition, analyses, and visualization of spatial and nonspatial land administration data is described. It will be shown how the integration of two different land administration data as well as historic and recent data sets into one cross-national database was solved. In addition, a model will be presented which is able to reconstruct a certain point of time in "cadastral history" by imitating the process of land registration. The last section will provide some results concerning land-use, parcel, and ownership changes in the investigation area, and finally the outlook offers some future perspectives on the region based on the analysis results.

20.2 PROJECT AREA

The cross-border region between Austria and Hungary had been an area without hinterland for over 40 years. The so-called iron curtain created an insuperable barrier where social and economic exchange between both regions was no longer possible

FIGURE 20.1 Recent satellite image (Landsat TM) showing the agricultural land at the border area of Hungary and Austria. The white line highlights the boundary between small-scaled patterns of the Austrian territory and the obviously larger fields in Hungary.

[1]. Due to political reasons these two border areas developed in very different ways regarding their rural conditions. Differences are still visible in the landscape; for instance the average size of agricultural patterns in Hungary is much higher than those in Austrian arable land. This characteristic agricultural structure is still recognizable (e.g., on satellite or aerial images; Figure 20.1).

The area of two small adjacent municipalities was chosen for comparing and analyzing ownership and land-use changes (Figure 20.2). They were considered as adequate for the research due to their location near the borderline and due to comparable socioeconomic and demographic conditions. The criteria are, on the one hand, comparable number of inhabitants, similar ownership structures as well as distribution of property and parcel patterns (at least originally in pre-Communistic times). On the other hand, both regions are considered as rural areas from the regional planning point of view, with agricultural land dominating land use, based on similar climatic and soil conditions.

Additional criteria for selecting the project areas *Girm* in Austria and *Harka* in Hungary included the availability of historical and current cadastral maps in terms of complete time series for one century. The investigation area covered building land as well as agricultural-dominated parts.

20.2.1 A Brief Historical Overview

Land administration is defined as "the process of determining, recording and disseminating information about tenure, value and use of land when implementing land

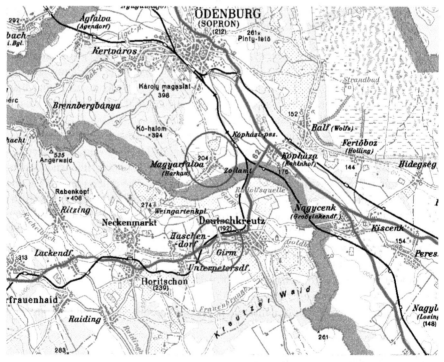

FIGURE 20.2 Location of project areas.

management policies" [2]; in other words, land administration is "the process of administering the complex rights, restrictions and responsibilities pertaining to land" [3].

Land administration generally consists of spatial data, respectively, the cadastre and nonspatial data known as land register. While in the cadastre the parcel information itself concerning location, area, and land use is stored, the land register provides ancillary information (e.g., property rights or debt) related to the parcels and the ownership.

During the era of the Austro-Hungarian Monarchy *(K&K Monarchy)* from 1867 to 1918, the land registration was administered by one common cadastral mapping authority, which basically aimed at creating tax maps of land property. Over time, different political and economic developments (First and Second World War, Communistic era in Hungary from 1949 to 1989) resulted in different agrarian and land administration systems and, combined with that, in different land management systems.

As a consequence, the different systems in land registration led to diversities in registration of spatial parcel-based data and ancillary ownership data. An interesting aspect of this case study was to find out if and in which way political systems and related land management influenced and still influences land use and ownership.

20.2.2 Land Management System in Austria

The cadastre is kept and maintained by the Federal Office of Metrology and Surveying (BEV). Cadastre and Land Register together form the basis for land management in Austria. These two "spatial related data pools" have always been kept separately by two different authorities but with high effort on synchronization of data. The roots of Austrian cadastre mapping trace back to the early nineteenth century, when land surveying had been introduced as taxation basis for the whole monarchy. These ancient cadastre maps are maintained in archives and are still useful for historical investigations of any kind (Figure 20.3).

FIGURE 20.3 Historical cadastral map of the Austrian test site from 1857.

In the 1980s, survey ordinances started digitizing analogue maps within a CAD system [4]. A unique parcel identifier for each estate realized the link between spatial data and land register data that is still kept in a separated database. Recently, the cadastral data is already available as GIS vector format including topology (previously it was simple line data in *dwg* drawing file format), but there has not yet been developed a common geographic database for cadastre and ownership information together. The land register that represents the legal status of all real property is maintained and kept by the local courts of law. Since 1987, the land register has also been available in digital format, but only authorized users such as lawyers have access to these data, in order to protect the data privacy [5].

20.2.3 LAND MANAGEMENT SYSTEM IN HUNGARY

Land and property registration in Hungary has been operational for nearly one and a half centuries, having its origins, as mentioned before, in times of the *K&K Monarchy*.

After all-embracing collectivization processes during the Communistic era, private ownership of agricultural land was substituted by cooperative and state property, leading to huge agrarian production centers and collective farms. These are farms in which a group of farmers pool their land, domestic animals, and agricultural implements, retaining as private property only enough for the members' own sustenance. The profits of the farm are divided among its members, in contrast to cooperative farming, where farmers retain private ownership of the land.

Today's land administration consists of the Department of Lands and Mapping (DLM), Regional Development (MARD), and the Land Office Network (LON). The registration of ownership and immovable properties has been fully computerized since 1991 (with the support of the European Commission under the *PHARE* program). The system is open for the public and is backed by state guarantee. Additional services are the updating of land-use information, classification/evaluation of land, land consolidation piloting, and maintenance of land-related statistics [6].

During the recent economic transition, a major priority of successive governments has been to redistribute land from state ownership and from cooperatives to individuals. This process has been managed for agricultural areas partly by the Land Offices and has placed great demands on the offices, which had to provide information on the past and present ownership status, carry out definitive surveys, subdivide large plots into many small ones, and register more than 2.4 million new owners. "Land privatization affects more than half of the territory of Hungary (5.6 out of 9.3 million hectares). The new parcels created during land privatization are scattered all over the country and this makes it impossible to keep the old cadastral maps up-to-date" [7].

20.3 METHODOLOGY

In this project phase cadastre and land register data from the present land administration systems was acquired and collected. The basis for a comparability of different data sets is the harmonization, which was solved by the integration of data into one database for both countries.

While spatial data (cadastre) was stored within a geographic database, nonspatial data (all additional information mainly about ownership) was registered within a separate relational database. Data flow between these two separate databases was realized through the use of a unique parcel identifier. Based on the resulting comprehensive data records, all land-use and ownership changes that ever occurred for a period of one century could be reconstructed, analyzed, and visualized.

Spatial analyses and the presentation of visual results were performed by a geographic information system. These analyses were topologically based on vectorized (digitized) cadastral maps, where each parcel represents a spatial object having a unique identifier.

20.3.1 Previous Investigations and Results

Preliminary studies had been carried out that were considered as prototypes for the study. The core aims of this pilot project were basically the same (analyzing land use and parcel changes as well as ownership changes), but focusing on a much smaller investigation area and covering only the center and parts of agricultural land of both municipalities. The results and experiences of these previous studies turned out as basis for further investigations. Due to limited temporal and financial resources, the pre-study mentioned was restricted to just fifty years as a research period.

20.4 DATA ACQUISITION

Data collection of land register data was carried out at the responsible local and regional public authorities in Hungary and Austria. Historical data could be tracked out in several archives and was transferred into digital format (database entries of ownership data, digitizing of cadastre data) while the current data sets were already digitally available. Data acquisition represented the most time-consuming phase of the project. For this reason, a well-planned database for fast and semiautomated data entry was an important precondition.

20.4.1 Nonspatial Data

Nonspatial data in our case have a direct relation to a parcel and therefore to location, but are not considered as "real" spatial data, in the sense of data that can be linked to locations in geographic space. Therefore the inclusion into the project's database and the later linkage to the spatial data had to be well figured out. In this case nonspatial data comprise all ancillary information related to the single estate regarding property rights, distribution of property among owners, and number of owners per parcel.

20.4.2 Conceptual Database Design

The project database consists of the main entities "ownership," "parcel," and "land use," defined as a one-to-many relationship (Figure 20.4). Basically, the relation between ownership and parcel in Austria and Hungary is a one-to-many relationship:

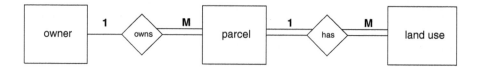

FIGURE 20.4 Entity relationship model of project database.

one owner holds one or more parcels. The particular regulation in the Austrian cadastre and land register though knows so-called "storage numbers," which are identification numbers for one or more owners. These owners can be married people, brothers and sisters, or agrarian communities. Therefore, the exact relationship should be described as a many-to-many relationship. Since this fact makes modeling much more complicated, the project consortium agreed on the simplified version of a one-to-many relation: one ownership holds one or more parcels. This system was valid for the Hungarian system until about 1950 as well. After the assumption of the Communistic system the conditions turned to a "modified" one-to-many relationship: one collective held a lot of parcels. Recently the registration system in Hungary was again changed into a "real" one-to-many relationship: one owner (not ownership) holds one or more parcels. Since for one parcel often exists more than one land use type (e.g., building and adjacent garden on one parcel), also the relation between land use and parcel was organized as a one-to-many relationship. Further tables act as look-up tables in the database. Figure 20.4 shows the entity relationship model.

For the conceptual database design these considerations were important basic issues. The most challenging part of the project was the harmonization of two national land administration systems and the modeling of the "cadastral and land register reality."

Detailed work had been invested in the conceptual design of the relational database, especially for the data acquisition. The database had to guarantee an easy, fast, and semiautomated entry of thousands of records. Moreover, redundancies and incorrect entries should be avoided. The database had to deal with historical as well as with recent data by simultaneously compensating different format, language, and measuring unit systems of diverse systems (recent metric system and historic "*Klafter (Öl)*"system).

A relational database organizes data in tables. Each table is identified by a unique table name and is organized in rows and columns. Since data are often stored in several tables, those tables can be joined or referenced to each other by common columns (relational fields). These columns often contain identification numbers that act at the same time as primary key [8]. The primary key here is a combination of parcel-ID (including sub-ID resulting from parcels that had been split up previously) and the identification number of each cadastral district, which represents a unique identifier for each test area. The composition of two or more columns guarantees the uniqueness of the primary key. Based on this key, the records are joined to the spatial objects, in particular, parcels in the GIS.

20.4.3 SPATIAL DATA

Spatial data (or geographic, topographic, or geo-data) are data which describe phenomena directly or indirectly associated with location. Topography can be defined as configuration of a surface, including its relief and the position of natural and human-made features. In addition, geographic features (here: parcels) are characterized by "topology," which regards the spatial relationship of objects stored with respect to one another.

The spatial data for the project were derived from historical cadastral maps by digitizing. After the preprocessing steps of scanning and geo-referencing the analogue maps, each year of interest was digitized, including updates and geometry changes of parcels. Digitizing basis was dependent on the availability of cadastral maps (see Figure 20.3). Cadastral data was generated in both countries with a commonly used desktop GIS. Current cadastre maps are already digitally available.

A new approach in storing geographic data related to nonspatial data is the geodatabase model from ESRI® (Environmental Systems Research Institute, Inc., Redlands, CA). Especially for purposes like the present project, it would be feasible because it facilitates digitizing and editing by implementing topology rules, and it is able to validate features by previously defined integrity rules. The geodatabase model supports an object-oriented vector-data model where entities are represented as objects. An object is a collection of data elements and operations that together are considered as a single entity, which can adopt properties, behavior, and relationships [9]. It is a data model where geo-data (and nonspatial data) from several data sources can be unified into one general system. Editing rules, especially splitting or merging policies, are interesting functionalities for the purposes of digitizing parcels.

Due to the limited project time, the advantages of this editing and data storage approach could not be tested in its entirety, but it can be suggested as a convenient method for further research in the field of cadastral investigations.

20.4.4 MODELING THE CADASTRAL REALITY

In general terms, a model is a representation of reality. In this project it was intended to represent and imitate the reality of land registration in a simplified model in order to reproduce a status quo of property and land for a requested analyses period.

Combined with the conceptual database design, the most challenging part of the data acquisition phase was therefore the modeling of the real-world land registry transactions into simplified functions. These "functions" represent the process of registration regarding changes of ownership, land use, as well as parcel size. The first registration of a parcel or owner, the reunion or cancellation of a parcel, the adjustment of area, and many more are examples of such "functions."

By anatomizing each course of registration into its basic elements, the modeling of the real world was simplified, and basic modules were established. The combination of various atomic components resulted in realistic functions, which were part of the database model. Together with the date of change, the feasible function for each action in the land register was stored as an entry in a special table of the database. If more steps (respectively, modules) for one "land registration" were

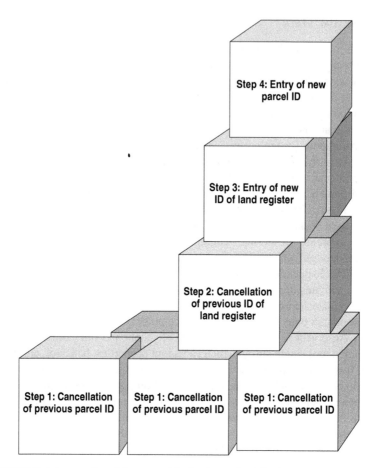

FIGURE 20.5 Scheme of modular assembly for the collectivization process in Hungary.

necessary, more entries were stored, whereupon the exact time flow had to be considered to maintain the referential integrity rules of the database.

Figure 20.5 shows the scheme of modular assembly for the collectivization process in Hungary; in this example the action consists of four basic modules (four steps).

The function "collectivization of parcels," for example, which was applied for the Hungarian test area, consists of four basic modules: "cancellation of old (previous) parcel," "cancellation of land register ID," "new entry of new land register ID," and "new entry of new parcel" (in that order). The modules, according to this example, exactly represent the procedure of the real-world land registration. Each function is predefined by an identification number (type of change) and further parameters.

20.4.5 PARCEL-RELATED AND OWNERSHIP CHANGES

The starting point for modeling the "cadastre reality" was the knowledge of basic changes and processes that can occur on land use, parcel geometry, and ownership.

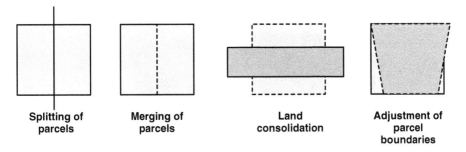

Splitting of parcels **Merging of parcels** **Land consolidation** **Adjustment of parcel boundaries**

FIGURE 20.6 Schematic presentation of possible geometry changes.

The following list of possible changes regarding cadastre and land register were relevant for the project areas:

- Change of parcel geometry (e.g., merging or splitting of parcels)
- Land consolidation (basically similar to splitting or merging parcels)
- Collectivization in Hungary (comparable to merging parcels)
- Privatization process in Hungary (comparable to division of parcels)
- Adjustment of parcel boundaries based on recent surveying

The schematic presentation of possible geometry changes is shown in Figure 20.6.

An example of a typical land use change is the transformation from arable land to residential area. Ownership changes can be distinguished between the change of property rights due to assignment, inheritance, purchase, expropriation, etc. and the changing number of owners.

20.4.6 DATABASE PROCESSING

In order to process database entries, a small program was written in Visual Basic, which is convenient for small self-made applications. The advantage of this language is the well-working interoperability with the database management system used.

The program performs an automated processing of all database entries for each year, including all changes for each parcel that had been recorded. The results are stored in three tables per year (owner, parcel, land use) and copied into an extra database (resulting database). These tables, which can be referred to each other according to the database model (Figure 20.4), represent the status quo of land use and ownership conditions of any requested year or date. The original entries remain untouched in this process in order to avoid possible loss of information.

Finally, table joins in the GIS organize the linkage between the nonspatial data from the database tables and the spatial objects (parcels). These joins formed the basis for analyses, queries, and visualization in the GIS.

20.5 RESULTS

Based on the integrated spatial and nonspatial data sets and combinations from those, analyses were conducted regarding changes of ownership, land use, and parcel

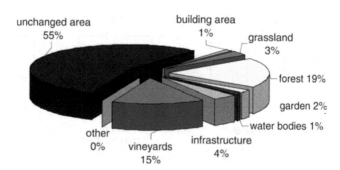

FIGURE 20.7 Land use changes in the Austrian test site; most of the affected arable land changed into forested land (19%).

geometry. The last two issues were mainly analyzed using a GIS, since in this case the geometric basis was significant. For ownership changes it was sufficient to execute queries directly within the database. The visualization of the results was performed within the GIS. The following examples give a brief overview on the results of the investigation.

20.5.1 LAND-USE CHANGES

Land-use change means a transition from one type of land use to another for various reasons. Within one hundred years, half of the area (55% of the analyzed parcels) remained unchanged in the Austrian test site *Girm* while a considerable percentage of area turned into forests but also into vineyards (Figure 20.7).

The conversion of agricultural land into forested land is a phenomenon that can be observed very often in rural and border regions in Austria: an area of low productive conditions is given up or afforested, and a favored area with good conditions gets intensified or is released for the increasing residential area [10].

Land-use changes in the Hungarian test area *Harka* can be mainly traced back to collectivization processes between 1960 and 1990, where the agricultural land was cropped according to centralized plans. The parcel structure changed from small fields into large agricultural production areas. In the phase of restitution and reprivatization from 1995 up to now, the parcel size decreased again, and related to that, the variation in land use types increased.

20.5.2 CHANGES IN PARCEL GEOMETRY

Based on extensive land consolidations, the geometry of the larger part of the agricultural land in the Austrian test area was reshaped between 1981 and 1986 (Figure 20.8).

As Figure 20.8 shows, land consolidation was necessary because of the narrow, long fields in the center of the village making difficult an efficient cultivation of land. In the same period the large-scale landed property of a big land owner in the north of the territory was portioned and bequeathed. Apart from those main encroachments, only marginal parcel changes were registered within the one hundred years of the research period.

FIGURE 20.8 Parcel changes in the Austrian test site between 1981 and 1986: only the gray colored parcels remained untouched (right).

Even more affected by geometrical changes was the Hungarian test area, where the collectivization process from 1960 to 1990 caused reorganizations of nearly all analyzed parcels and again from 1995 to 2000, where the whole area was "shuffled" due to reprivatization of the previous collective land.

20.5.3 OWNERSHIP CHANGES

Chart 20.1 and Chart 20.2 illustrate the changes of ownership in the tested municipalities of Austria and Hungary summarized and visualized as five-year steps.

In the case of *Girm* there is one characteristic low point (1945) in the chart reflecting the years of World War II where the real estate market was nearly broken down. The number of ownership changes is increasing from then to a peak in the years around 1970, a reason for the high percentage could not be statistically fixed though it is supposed that the figures can be traced back to the general booming tendency in this period. A high number of property changes are noticeable also for the years around 1985 where the related effects of land consolidation (exchange of parcels, replacement, etc.) were documented in land registration (Chart 20.1).

The municipality of *Harka* reflects perfectly the political conditions in Hungary (Chart 20.2). There are just a few ownership changes until 1940, also with a low point in 1945. The highest number of property transfers can be noticed in the years around 1960, when the collectivization was carried out and parcels of private ownership became one collective agricultural land, the *"Nyugatmajori Állami Gazdaság"*

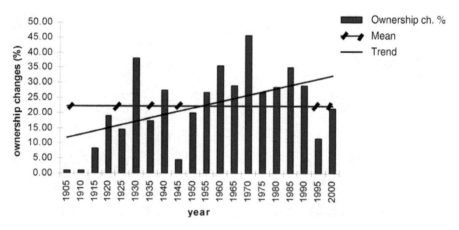

CHART 20.1 Ownership changes in the Austrian test area (in percentage).

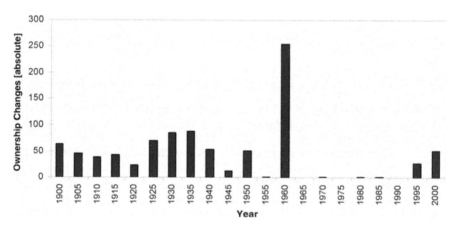

CHART 20.2 Ownership changes in the Hungarian test area (in absolute figures).

("Nyugat major state collective"). Between 1960 and 1991 nearly no ownership changes are registered; there seems to be a hold up in the real estate market due to the political conditions. From 1995 to 2000 the beginning of the reprivatization process and the restitutions of the former collectivized land are illustrated by constantly increasing figures.

20.6 PROBLEMS ENCOUNTERED

Originally the idea of the present project was to conduct analyses back to the middle of the nineteenth century, where land register was founded in the *K&K Monarchy* and the earliest cadastral maps were established. Unfortunately several "missing links" like lost historical cadastral maps or land registration documents due to changing political conditions and social turbulence made it impossible to follow the red thread.

Changes in the numbering system of parcels without registering the old ones and changing or lost ownership identification numbers in Hungarian land administration systems represented further obstacles in reproducing historical facts. The consequence was to restrict the investigation period to the last century.

20.7 OUTLOOK

Based on the results of this study, possible future scenarios of land use, agriculture, rural development and property rights in the border region can be estimated considering Hungarian's access to the European Union in May 2004.

With the termination of the Structure Funds programming period of the European Union at the end of 2006, the "objective 1" funding stops for the Austrian border region, and instead Hungary will benefit from this funding. Effects on land use (e.g., decrease of arable land due to abandonment, afforestation, or transformation into building land) and distribution of property (increase of dynamism of tenure) can be expected. In Hungary, based on the increased traffic of tenure and the reprivatization process the average size of arable land per private or cooperative farm is presumed to decrease. After the process of compensation and restitution, since 1992 agricultural cooperative societies of a new type have been founded (corporations, limited companies) [11]. These cooperative farms have economic and production advantages compared to the small-scale agriculture in the Austrian border area. This could lead to an intensification of cultivation, since soil conditions in West Hungary are optimal. The land market of Hungary has not yet been fully liberalized with the accession to the European Union, but nevertheless the trans-boundary traffic of land in the Hungarian border area to Austria is very high. Farms run by Austrians on Hungarian territory are estimated around 500 and 700 (± 20%) for the year 1999 [11]. In the meantime the number is supposed to have increased.

With the constantly increasing road traffic in middle Europe, the extension and completion of the transnational road networks can be expected. This assumption would definitely evoke direct and indirect consequences and changes on the border region of Austria/Hungary, combined with impacts on land use, ownership structure, and traffic of real estate.

20.8 CONCLUSIONS

The present project demonstrated the methodology of data collection of two border municipalities of Austria and Hungary. The areas were eligible for the research due to their comparable basic conditions. The research approach was a historic one, and the results are based on historic and recent land administration data. If historic cadastre maps are available, they represent a suitable data source for investigations related to time series. They provide a good insight into the past land use and ownership distribution of an area. It must be mentioned though, that land use recording in the recent cadastre does not necessarily correspond to actual land use in terms of land cover (e.g. derived from classification of satellite images), due to different definition for "land use" in land administration.

It was demonstrated that it is possible to solve the integration of land administration data from different national systems and data covering one century by providing a dedicated relational database. In order to analyze the development of land
use, property rights, and parcel changes for a border region, the time component
played a basic role within the database design and the project as a whole.

The results show that external effects like varying political and/or socioeconomic
factors can influence the development of rural and/or agricultural areas. The visualized issues regarding the changes of parcel geometry and ownership demonstrate
in an impressive way that political conditions, especially in Hungary (Communistic
era with collectivization of agricultural land from 1960, reprivatization and restitution recently) shaped the land, and these impacts are reflected by the cadastre.

ACKNOWLEDGMENTS

The present study was conducted within the project "ELNA2," which was financed
by the Austrian-Hungarian initiative *"Aktion Österreich Ungarn"* (project number
51ÖU5) in order to encourage and support bilateral projects and to improve scientific
cooperation between Austria and Hungary [12]. The project was realized in collaboration with the Department of Geoinformatics in Székesfehérvár, University of
West-Hungary and the Institute of Surveying, Remote Sensing and Land Information,
BOKU University of Natural Resources and Applied Life Sciences, Vienna.

We are grateful to all Hungarian and Austrian authorities for providing relevant
project data without any administrative obstacles. Special thanks go to the Austrian
and Hungarian students for the time-consuming collection of historic and recent
land register data and the processing of the derived data.

REFERENCES

1. Seger, M. and Belusczky, P., Bruchlinie Eiserner Vorhang: Regionalentwicklung im
   österreichisch-ungarischen Grenzraum. Breakline Iron Curtain: Regional development at the Austrian-Hungarian border area, P. Belusczky, Ed., Böhlau, 1993.
2. FIG, The Bathurst Declaration on Land Administration for Sustainable Development,
   Conference paper, United Nations-FIG, Washington, DC, 1999.
3. SDI, in www.sli.unimelb.edu.au/research/SDI_research/research/background.htm,
   2003.
4. Heine, E., Mansberger, R., and Ernst, J., Landadministration und Dateninfrastruktur
   — Nationale Problemstellungen und europäische Entwicklungen In: Chesi, G.,
   Weinold, T. (Hrsg.): 12. *Internationale Geodätische Woche*, Obergurgl, 2003.
5. Feil, E., Grundbuchsgesetz: Ein Kurzkommentar für die Praxis. Land register law: A
   Comment for Practice, Linde Verlag, 1998.
6. Remetey-Fülopp, G., Land Administration in Hungary. Institutional support for the
   implementation of the agricultural, rural development and land tenure policies, FAO
   SEUR, Budapest, 2001.
7. DLM, UN-ECE Documentation of Land Administration Projects. Workshop: Customers–Cooperation–Services, Department of Lands and Mapping, Hungary, Vienna,
   2003.

8. Sauer, H., *Relationale Datenbanken. Relational Databases*, Addison-Wesley, Reading, MA, 2002.
9. Zeiler, M., *Modeling our World. The ESRI Guide to Geodatabase Design*, E. Press, Ed., Environmental Systems Research Institute, Inc., Redlands, CA, 1999.
10. Steiner S., Räumliche Analyse der Landschaftsentwicklung in der Grenzregion Mühlviertel-Böhmerwald anhand von Fernerkundungsdaten. Spatial Analyses of Landscape Development in the Cross-Border Region of Mühlviertel and Böhmerwald. Strobl, Blaschke, Griesebner (Hrsg.) *AGIT — Angewandte Geographische Informationsverarbeitung XIV*, Salzburg, 2001, p. 471.
11. Greif F., *Die Grenzgebiete Österreichs und seiner östlichen Nachbarn. Border areas of Austria and its eastern neighbors*, Schriftenreihe, Nr. 91, BAWI—Federal Institute of Agricultural Economics, Vienna, 2001.
12. AÖU, in http://www.omaa.elte.hu/, 2003.

21 Computer-Aided Reflexivity and Data Management in Archaeology

Anthony Beck and Assaad Seif

CONTENTS

21.1 INTRODUCTION

Archaeology is approaching a sea change. Computer hardware and software have reached an unprecedented level of sophistication, yet many archaeologists continue to uncritically collect raw data in traditional ways, eschewing the benefits of computerization. However, recent theoretical developments demand a range of dynamic interpretations and could provide a springboard for computerization. This chapter considers the pitfalls and potentials of digital archaeology, examining data collected at different scales and the impact of recording systems. Finally, issues of data dissemination and interpretation are examined. It is envisaged that a ground-up reappraisal of archaeological collection, storage, and analysis will increase the potential of the resource, liberating analytical frameworks.

21.2 ARCHAEOLOGICAL DATA, ANALYSIS, THEORY, AND TECHNOLOGY

Archaeological data are possibly some of the most complex and diverse information sets studied within any discipline. Data of archaeological relevance encompass virtually all areas of nature and culture. The use of technology to manage archaeological resources has led to the phenomenon of information explosion. A survey of users' needs [1] highlights the increased production of and demand for digital data. This was helped by the increased usage of computerized storage and analysis systems such as relational database management systems (RDBMS), computer aided design (CAD), geographical information systems (GIS), multimedia software and, most significantly, "office" applications. However, although digital techniques predominate, structured approaches to data integration that allow the archaeological record to be fully articulated are rare.

All raw archaeological data need classification and interpretation, given the modern complexity of classification systems. Computerized systems offer the only environment in which to perform multicriteria and other complex analyses. Indeed it can be argued that the reconstruction of past environments and society within a spatial context can only be fully articulated and analyzed using GIS. As such, GIS have been heralded, since the early 1990s, as the pivotal analytical tool for archaeological data [2]. These environments are now referred to as archaeological data management (or information) systems (ADMS or AIS). At this juncture it would seem appropriate to define "data" and "information." Data are the raw material collected, for example, during excavation. Information, on the other hand, is derived from data through some form of analysis. As such, information transforms data to varying degrees and has an intellectual value defined by analytical goals (see Figure 21.1).

Current interest in the complex relationships between theory and practice, particularly on the point of reflexivity, has reopened discussions on the application of

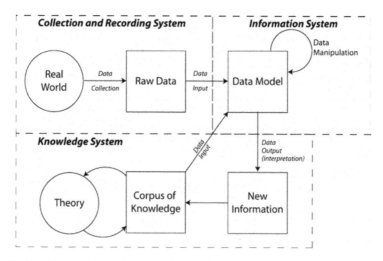

FIGURE 21.1 The transformation of archaeological data between collection and synthesis.

technology [3,4]. Reflexivity is a component of the post-processual theoretical movement. Post-processualists focus on context and argue that knowledge comes from a dialogue between the subject (archaeologist) and object (the archaeological record). This dialogue can occur at any scale. Reflexivity refers to multiple iterations of this dialogue using different subjects and objects.

Reflexive approaches to field recording and interpretation, by their very nature, require rapid feedback of complex data in a digestible format (information synthesis on the fly), a process that is best conducted through an AIS. One of the other major barriers to reflexive archaeology is the reliance on a paper-based record in the field. Paper-based records, in comparison to a digital facsimile, are notoriously difficult to access. Although many experiments have been conducted in "total digital excavation" (for example DigIT, www.english-heritage.org.uk), there has been little general uptake in these practical advances. It has been argued by Beck and Beck [5] that the reemergence of theoretical excavation is a facilitator for field-based computer applications. However, there is an inherent contradiction in many interpretations of reflexivity: more information is required for immediate feedback (reflexive), yet there is distrust in the very technology which best facilitates this feedback (hierarchical and specialist, hence *un*-reflexive). The role of technology within these discussions has been misunderstood. Technology itself is a facilitator; however, the implementation and use of this technology is subject to organizational pressures [6,7].

The majority of current archaeological computer applications attempt to shoehorn "off the shelf" recording systems into a computerized format without considering how this data could or should be analyzed (the transformation of data into information). This is akin to creating unintelligent electronic information repositories (or catalogs) rather than using computers as tools designed specifically for analysis and synthesis.

Although more data are now stored in digital format, accessing these data sets is difficult. Coherent digital archiving and dissemination is still in its infancy. This is an inherent barrier to virtually all archaeological research and enquiry. Previous archaeological studies are not easily available in digital form, and where they are, many do not provide the "raw" data that are of most import. The authors believe that the broad disillusionment and lack of uptake with computerized systems is partially due to an organizational lack of maturity and vision when implementing the technology and uncertainty about cost implications. Rarely have organizations gone back to first principles and reappraised their data collection and analysis systems in light of the benefits of computerization. Furthermore, digital deposition is, to many, an expensive luxury that has yet to prove its reuse value.

Without wishing to advocate an inflexible framework, we will attempt to outline an environment where high-quality data can be collected, integrated, generalized, and disseminated for analysis at any scale. The outlined environment would provide a research framework that will allow the integration and generalization of archaeological data whatever the scale of collection. The ability to access multiple high quality data sets should unify some of the disparate agendas of archaeological theory and practice which, for too long, have existed in isolation and are exacerbated in contract scenarios [4,8].

21.3 ARCHAEOLOGICAL DATA AND ANALYSIS: AN OVERVIEW

Most nonarchaeologists think of the subject in terms of research excavation: that something *interesting* has been found and *must* be excavated to establish its date and nature. However, most archaeological works are now carried out within the development process. Assessments of archaeological potential are made using a variety of noninvasive techniques (i.e., aerial photography, geophysical survey, topographic mapping, and historical research) in order to gain an understanding of the underlying archaeological resource *without* excavation. This information is used by the curatorial body to establish whether development may take place and, if so, what level of archaeological mitigation is required [9].

Research bodies also use a broad range of techniques, the difference being that researchers have more freedom to pursue theoretical agendas. Aerial photographs and satellite imagery allow interpretations to be made within an archaeological landscape [10,11]. Anomalies identified from these sources may then be targeted for ground survey (i.e., geophysics, fieldwalking, and topographic survey), which can often better define the form, structure, relationships, and date of the resource [12–14]. Historical records (mapping, documents, etc.) and artifact and ecofact analyses provide depth and detail to any archaeological interpretation.

Excavation is only likely to take place after the examination of such alternative data sources. Archaeological excavation is a nonreproducible destructive event. Consequently, the site archive is the most important aspect of the excavation process because this represents the raw information source from which future archaeologists will form their interpretations. Therefore, when excavation does occur, a whole battery of techniques and specialists are employed. For example a modern excavation, in addition to its excavation team, will normally have a number of surveyors, planners, soil scientists, and artifact and ecofact specialists (for a more extensive discussion see [15]).

Archaeological excavation is a very labor-intensive and costly exercise. In the United Kingdom the document Management of Archaeological Projects (MAP) [2,16] outlines a technique for project management, dividing fieldwork and post-excavation into two temporally discrete activities. In this framework, fieldwork is a data collection phase with limited record checking, processing, or analysis. Post-excavation is the period when artifacts and ecofacts are processed, assessed, and analyzed, plans digitized, and context sheets entered into a database. More importantly, time is given over to reflection of the "integrated" archaeological whole. This framework was developed to facilitate successful project management. However, the fragmentation of the collection and processing framework has led to reduced understanding during excavation itself. In essence, a system has evolved that does not provide adequate feedback during the excavation process when it is most critical [5]. This approach has been questioned, in practice, by contractual excavations conducted under the aegis of Framework Archaeology and research excavations under the direction of Ian Hodder at Çatalhöyük, Turkey. Both have applied a more integrated approach to archaeological project management in order to fulfill their "reflexive" research agendas [3,17–19].

Hence, archaeological data are derived from a number of different sources and can be collected at a number of different scales. The analysis of these data sets can occur at any level, dependent upon the type of problem that the archaeologist is studying. For example, aerial photographic interpreters can analyze settlement and hinterland evidence over a landscape but, due to the limitations of the technique, are not able to undertake sophisticated analyses about these individual entities. Geophysics could produce more refined data, leading to a greater understanding of local or regional processes. Finally, excavation provides a wealth of information on local social, political, economic, and natural conditions over time. Traditionally, scale has been utilized as a convenient tool to partition the landscape (i.e., site — micro, inter-site — meso, and regional/landscape — macro) in order to help facilitate the production of syntheses or narratives. More recently research has focused on the relationships between scales of synthesis, data scales, and data resolutions [20–22]. Integration of the aerial photographic, geophysical, and excavation data discussed above will aid interpretation at any of the scales by providing them with different contexts. For example, an aerial photographer may want to determine the relative date of two intercutting boundary ditches. This would be nearly impossible from aerial photographic evidence alone. However, if records of excavated components of these ditches were integrated into the data set, then determining the relative date of the landscape features could be easily constructed.

The key to understanding these information resources is to find appropriate mechanisms to collect, integrate, analyze, generalize, and synthesize the archaeological record. Computerization is the obvious solution to this problem.

21.3.1 A BRIEF HISTORY OF ARCHAEOLOGICAL INFORMATICS

> Computers not only change the way we do things, but more importantly, they change the way we think about what we do and why we do it [23, p. xiii]

Archaeological research methods are inevitably related to the technical skills of the practitioners and the general technological level of society. Ancient Greeks and Romans speculated about the past, demonstrating that anthropological and archaeological studies are not a recent phenomenon [24, p. 19]. Since the introduction of modern archaeological techniques (widely attributed to General Pitt Rivers who excavated at Cranbourne Chase during the 1880s) archaeologists have extended their theoretical, methodological, and interpretative pantheon by borrowing techniques from allied disciplines (the hard sciences, social sciences and humanities). The incorporation of these techniques has caused a number of shifts in archaeological theory, method, and interpretation.

This was exemplified by the paradigm shift caused by radiocarbon dating in the mid-1950s: the introduction of scientific dating techniques destroyed many contemporary theoretical models and opened the door to the theoretical movement that would become known as processual archaeology. Processual archaeologists borrowed extensively from the hard sciences and attempted to apply scientific rigor and hypothesis testing when collecting and interpreting archaeological data.

The 1960s saw the birth of archaeological computing. Archaeologists employed a range of quantitative, statistical, and applied mathematical techniques on large mainframe systems. Given the scarcity and expense of computing resources at this time, access was only provided to a small sector of the archaeological community.

This expansion continued in the 1970s, when computerized classification systems proliferated and the first relational database models were introduced. The first symposium on the applications of computers in archaeology was held in this decade.

During the 1980s personal computers became widespread. Archaeological applications continued apace, with a number of archaeologists starting to use GIS. In the 1990s both software and hardware platforms started to mature, and archaeological GIS applications proliferated. More emphasis was placed on data and systems interoperability. The World Wide Web was given wide user appeal by the development of html. The first decade of the twenty-first century has seen the maturing of archaeological computer applications. Less emphasis is placed on the specifics of the technology, and greater emphasis is shown toward data quality, multivocality, the impact of data analysis on theory, and dissemination using the web [5,19,23,25,26]. Virtual applications have also been embraced, providing archaeologists with a number of significant tools to conceptualize and interpret archaeological landscapes and sites [27,28].

21.3.2 GIS AND ARCHAEOLOGY

In the past decade GIS and associated technologies have taken a preeminent position in archaeological computing applications. It is expected that GIS should fulfill a wide range of user needs (from basic cartographic plotting to high-end spatial analyses) and should fit seamlessly into an organization's infrastructure. However, in reality successful implementation is not guaranteed. The management of technological change is not just about providing people with tools or software and giving them free rein. Successful implementation of technology involves dealing with characteristics of the organizational culture, people, and perceived requirements and outcomes [29, p. 50; 30,31].

In common with a number of other GIS user groups, archaeologists and archaeological organizations have traversed a number of identifiable steps in their implementation and end-user adoption of GIS technology. As discussed by Heywood et al. [31], GIS applications tend to fall into the following categories: pioneering, opportunistic, and routine. Each category implicitly defines the skill base and market strategy (i.e., the cost/risk position) of the users. However, these positions are not static; an application that was once pioneering can, over time, become routine through a technological trickle-down effect. The first archaeological GIS applications were pioneering; some archaeologists even wrote their own GIS software. These pioneers are technologically sophisticated, are prepared to bear the cost and risk of development, and have often fulfilled the role of "GIS champion." The opportunistic users are relatively more risk averse and take the more robust developments of the pioneers and provide them to a broader user base. This is the preferred strategy of technologically savvy archaeological units. Routine users are cost and risk averse and probably have a lower technological skill base. These users encompass the rest of the archaeological community.

In this process, it must be remembered that GIS technology is continually developing. GIS software has moved a long way from single-user professional workstations, although these still exist. Software is now available for different hardware platforms (including handheld devices) and is accessible through the Internet as thin clients (i.e., the client machine does not require GIS software to be installed locally; see for example [32]). This proliferation of different GIS types has helped the dissemination of the technology, particularly to routine users. For example, Microsoft's Map Point® (http://mappoint.msn.com) is a network analysis GIS application that can be used to locate maps or derive driving directions (using street address or postcodes). The majority of users do not know that they are using GIS software and, furthermore, do not need to know. The simplicity of the user interface belies the complexity of the underlying GIS application. This proliferation has in part been supported by user demand, which underlines the shift of the GIS industry from the private (local government and education) to the commercial sector [33].

In summary, archaeological GIS applications are increasing in sophistication, and more archaeologists are using GIS for management and research purposes. Not only do archaeologists analyze complex data sets in order to understand aspects of human inhabitation over time, they are increasingly applying different theoretical perspectives to these analyses. The key to understanding these enormous information resources is to find appropriate mechanisms to collect, integrate, analyze, generalize, and synthesize the archaeological record.

21.4 RECORDING SYSTEMS

A recording system is the crux of any analytical framework. The system is only as good as the questions one can answer with it. As ever, the adage "garbage in garbage out" still holds. In the United Kingdom and elsewhere, bastardized versions of the Museum of London Archaeological Services (MoLAS) recording system are favored [34]. The MoLAS system developed in the 1970s provides a standardized framework within which archaeological data sets could be compared. However, the theoretical environment has moved forward, and the MoLAS system is now incompatible with modern theoretical and analytical models. The pseudo-objectivity of the record, coupled with a reductionist approach, produced a system through which it is difficult, if not impossible, to get results with archaeological value that match current theoretical agendas. For example, in an environment where people can consider sites in terms of Schiffer's [35] principles of formation and deformation, it would seem appropriate to expect formation and deformation attributes to be recorded in a structured manner. However, they are not; the information is nearly always present but hidden in a combination of fields (or normally only in free text).

If the research goals of a project are not answered by the recording system, then the recording system is inappropriate. If our scales and directions of analysis are well understood, then recording systems can be tailored accordingly. Formation-driven multicriteria analysis is employed in the Archaeological Services of the University of Durham (ASUD) recording system [36] and appears to provide a significant advance in one aspect of field recording. This is one of the few recording systems that explicitly addresses the problem of extracting information from data.

Designing recording systems that will always provide appropriate data to address ever changing research agendas will be impossible. However, since the 1970s, it has been rare for archaeologists to debate recording systems at all [25, although see 36].

This situation must change; approaches need to be developed which can extend the flexibility of databases for recording so that the data can be analyzed with rigor. Furthermore, appreciation of the multidimensional and contextual nature of archaeological data must be encouraged [19]. Different modes of theory, analysis, and collection require different levels of recording and/or synthesis. Furthermore, the theoretical constructs underpinning different interpretations and the "intellectual audit trail" delineating how one came to those interpretations should also be recorded. This will provide a greater understanding of the transformation of data into information and how this can be used to construct multiple narratives.

A new approach has to be oriented toward rethinking the standards and criteria used in the recording process so that multiple data sources can be integrated. A major issue will be content and structure: the recording system must be able to address all common archaeological scenarios and be flexible enough to record data in unforeseen circumstances. This will address the basic level of data recording; however, the system should also allow relationships between different data collected at different scales to be expressed (for example, linking aerial photographic data with excavation data). This will allow researchers to access the different "depths" within the database in a framework analogous to data mining. Thus, researchers can find correlations or patterns at different levels of granularity.

Furthermore, comparable thesauri (data dictionaries or simple lookup tables) must be established. For example artifact specialists classify objects on a number of criteria. Specialists may use different criteria to describe the same artifact. Hence, one specialist may refer to tile as "ceramic building material" and another may refer to it simply as tile. A thesaurus will be able to link these initially disparate interpretations into a framework whereby they can be analyzed together. CIDOC [37] is drawing up standards to facilitate this kind of integration on an international scale.

Finally, the linearity of recording is starting to be reappraised. Photography has long been an important component of archaeological recording, as the photographic image elegantly conveys complex information. Increasingly, archaeological projects are employing digital audio and video to augment photography. These tools produce more immersive interpretative environments. Not only do they allow researchers access to more information, they may also capture the more ephemeral aspects of the archaeological record that are normally underrepresented. For example, tagged digital video files can contain information on the field archaeologists' first impressions about the features they are excavating. This information may be invaluable during subsequent interpretations.

Hence, the archaeological data collection process employs a complex arrangement of digital and analogue techniques to describe the resource. It is important that this data is articulated at an early stage so that an excavation can be effectively managed and research aspects prioritized. There is no point discovering in post-excavation that the most interesting archaeological features did not receive enough attention during excavation, because in all likelihood, by this time, the whole area has been redeveloped. Rigorous and accurate data feedback allows the archaeologists to be discriminatory during the data collection phase: elements of the resource that are not understood can be excavated in preference to those that are and hypotheses can be tested.

21.4.1 DATA GENERALIZATION

Data generalization is an essential component of archaeological analysis. Archaeological processes and mechanisms of collection are scale dependent and require changing degrees of detail for analysis. Hence, there is a need for variable levels of abstraction for different modeling and analytical purposes [38, p. 3]. Excavation data are normally synthesized for intersite and landscape analysis. However, the general rule has been to synthesize these data sets with limited long-term reference to the raw data sets. Hence, over time, these data sets lose their precise analytical value. Ideally these data should be integrated into a coordinated spatial and related attribute data set that can be automatically generalized rather than multiple data sets that need maintenance and refreshing.

Database generalization has been an overlooked area within all the spatial disciplines. The cartographers' challenge is to create meaningful maps at different scales (the ratio between the size of an object on a map and its real size) and resolutions (the smallest object that can be represented). This same analogy is being applied within the GIS and Computer aided mapping (CAM) industries to manage automated and semiautomated generalization of spatial data [39]. Until recently the software to enable cost-effective spatial data generalization has been either nonexistent or prohibitively expensive. However, modern software can, at least theoretically, automatically generalize spatial and nonspatial data sets. Encoding a generalization index or algorithm that is applied to the spatial component and its nonspatial attribute information is, potentially, one technique of successful generalization [38]. The use of object-orientated data models such as ESRI's geodatabase® (Environmental Systems Research Institute Inc., Redlands, CA) could be pivotal [40, pp. 3–5 gives an overview of proposed generalization functions for ESRI, 41]. Ideally this information should be encoded within metadata [42]. These metadata indices should be created for all potential scales of analysis (both spatially and nonspatially). For example, for intersite analysis the metadata index would automatically generalize raw data (spatial and nonspatial) to its context groupings (i.e., enclosures, structures, etc.).

An approach of this type could still retain any level of complexity and could be employed at a variety of information levels, automatically reducing the data to fit into any scale of analysis without influencing the primacy of the raw data.

21.5 DATA INTEGRATION AND DISSEMINATION

Archaeological data is collected at a variety of different scales. For example, remote sensing, particularly aerial, geophysical, and satellite imagery, provides a huge corpus of raw and interpreted data at the site, regional, and landscape levels. Improvements in remote sensing will continue to impact on archaeological survey, prospecting, and management techniques. Systems such as LiDAR [for descriptions see 43], high resolution satellite imagery (i.e., Ikonos), and most importantly, high-resolution geo-referenced aerial hyperspectral data will have a huge impact on archaeology. English Heritage's National Mapping Programme is currently creating a benchmark for aerial photography throughout the United Kingdom [44]. Although this is important, as discussed earlier it would be more useful if this information set were

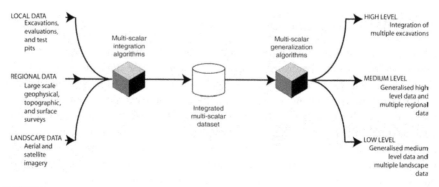

FIGURE 21.2 Data schema for multiscalar data integration.

integrated with excavation records to reflect a "lower level" (in terms of scale) of data collection. This would allow the contextualization of both the aerial and excavation records. This raises an important point about theoretical applications of methodology. Many reflexive approaches are only used at one level of data collection: excavation. Surely it would be equally appropriate to integrate information from different scales into reflexive practices.

In order to integrate information from multiple scales into the data model described above, metadata recording is essential. Once scales of collection and analysis have been identified, it must be incumbent on projects to record information for the data set at its basic scale of collection, all scales of generalization, and its linkages to "higher level" data. Figure 21.2 details a data schema for this integration process. It must be emphasized that the algorithms (in the black boxes) may be difficult to determine. However, unless we think in this way, archaeological data can never be fully articulated.

Once an integrated data set has been produced, a coherent dissemination structure is required. Other geographical science industries have consolidated their provision of spatial information through a few web-based geographic data portals (e.g., ESRI's geography network www.geographynetwork.com). These portals tend to be collaborative, multiparticipant systems allowing discriminatory access to geographical data. The majority of leading GIS and remote sensing software vendors include functionality to access data sets over such distributed networks. Higher-speed communication systems (such as broadband) and improved compression algorithms mean that larger file sizes can be easily accessed. For example the use of Enhanced Compressed Wavelet (ECW) technology by ER Mapper Inc.® has facilitated the deployment of terabytes of data over the web (e.g., a single 25 cm-resolution mosaiced aerial photograph of the whole of Denmark is available at www.kortal.dk). It is, however, acknowledged that at present this technology is predominantly available in Europe and America.

It is not difficult to envisage how such mediums can be used to supply a variety of raw and manipulated data themes (such as land use, hydrology, geophysical, aerial, and categorized multispectral imagery) on demand. Granted, there are structural problems with this, such as what geographic projection to use, how to push currently archived information into the framework, and how to maintain such a highly structured archaeological theme. However, given current trends in computing technology, this should not be considered a long-term issue, although organizationally it could be more problematic. Given the current structure it is difficult to determine where such a body or organization would fit and who would pay for its maintenance. It would be hoped that an organization like the Archaeological Data Service (ADS) would be a facilitator in such an arrangement, although it would be preferable if the organization had a more international outlook.

However, it would be easier for individual projects, particularly those supported by academic departments, to make their data available over the web using thin client GIS interfaces. As discussed previously, these interfaces do not require specialist software at the user end and are a cost-effective way to provide GIS functionality to a large number of people. Thus, researchers could be empowered by having direct online access to primary archaeological data or their synthetic derivatives, archived remotely sensed imagery, geo-referenced historic mapping, environmental themes (geology, hydrology, etc.), and virtually any other data set that would aid their enquiry. More importantly, during excavations fieldworkers would be able to contextualize their results within an integrated landscape of information, allowing multiple and diverse interpretations to occur.

21.6 CONCLUSIONS

Academic research has a tendency to turn toward the easiest and most publishable issues that are on the agenda at a given time — rather than turning toward the most relevant and urgent issues (because those are more complex) [38 p. 13].

This chapter has attempted to address a future of technical and computer-based archaeology and its relationship to current theoretical and methodological approaches to data collection and interpretation. The authors would not be so confident as to predict that all these approaches will happen, although increased use of technology during fieldwork is inevitable [15]. However, the issue here is not based on the technology (technology will continue to develop irrespective of archaeology) but based on what is done with it. Archaeology's underpinning rationale is to collect data that have analytical value (i.e., data that can easily be transformed into information). If reflexive, multivocal interpretative frameworks are to be realized, then high-quality data collection systems employing an analyzable recording system are essential. These should encourage new and exciting ways of considering the execution and interpretation of archaeology at the "trowel's edge." Approaches like those proposed by ASUD, although not answering the reflexive problem, do allow rapid and meaningful computer-based analysis at a fixed scale. Enhancing the recording processes to take advantage of technological improvements should be prioritized.

Real developments will come through improvements in global positioning and communication technology, product convergence, and cost reduction. The creation of a single integrated gadget that has photographic, digitizing/planning, database, GIS, and communication functionality at a realistic price and speed is not far away. Some archaeologists do use cheap handheld personal digital assistants (PDAs) to record a "digital" context sheet in a relational database. All contextual data is stored on a relational database in the PDA, which can be directly "synchronized" with a desktop database in Microsoft Access® [45]. In June 2004 the PDA and software cost c. $100. Furthermore, if the guidelines to recording outlined above are implemented, then the record has been produced in a manner that allows simple generalization, analysis, and synthesis. In the future this will provide "data democratization," because each "gadget" will always carry the most up-to-date record through which the field archaeologist can browse, cross-reference, record, and analyze [46]. It must be stressed that, by their very nature, these tools must be easy to use. Data collection and analysis should be open to everyone; data management is the specialism.

The ability to then integrate these information sets into a coherent scale-free data set is an ambitious plan, but one that would provide integrated data of unprecedented research value. The challenge for archaeology is to develop frameworks within which we can capitalize on our data diversity by integrating data into a coherent whole.

Enhancing this archive with multimedia data provides an interpreter with an alternative viewpoint. Furthermore, it makes the significant point that the recording process was conducted by people. Too often archaeological archives have their human side removed for the sake of objectivity.

This would provide a multifaceted, multiscalar data and information system through which users can navigate. This experience can be made more or less detailed depending upon user requirements. It is hoped that such systems could help archaeologists to articulate the record in fresh and exciting ways, liberating them from dogmatic interpretation procedures.

One criticism of computer-based archaeology is that it is elitist [4]. At the moment this is true, but that is not necessarily the fault of the practitioners and definitely not a fault of the technology. Universities should prepare and train students in the use of digital technologies and the analysis of digital data [see 47]. Paradoxically, fieldwork and data manipulation are practically nonexistent in most university syllabi. However, as Müller's quote above demonstrates, are academic archaeologists the people who should manage this research? In our view it is essential that the gaps between contractual and academic archaeologists should be bridged [for examples of this in Sweden, see 48].

Archaeological data and data collection methodologies have been underrepresented in recent debates. It is almost as if the problems of data collection were solved in the 1970s. Future theoretical and methodological issues can only be addressed through the corpus of data we collect and archive. GIS technology is one of the leading tools for analyzing this data. Organizations such as the ADS are trying to ensure that digital data will be available over the long term. It is incumbent upon the archaeological community to make sure that this data has a high analytical and reuse potential. This will require increased liaison and dissemination of information both within and between contractual, academic, and curatorial archaeologists. A coordinated data and analysis system will produce profound benefits including improving theory, practice and curation. Given that the world's cultural heritage is under significant threat, particularly from development, it is imperative that our data systems provide accurate information to support the management and sustainability of this fragile resource. Future generations of archaeologists will not thank us for our mismanagement of our primary archaeological legacy: the data.

ACKNOWLEDGMENTS

This chapter owes a great deal to many people. We would particularly like to thank the European Space Agency, EurISY (particularly Mary and Valerie), the University of Durham, Natural Environmental Research Council (NERC) (Beck) for their financial assistance [research award GT04/1999/TS/0053], and the editor.

REFERENCES

1. Condron, F. et al., Strategies for digital data: findings and recommendations from digital data in archaeology: a survey of user needs, Archaeology Data Service University of York. York, UK, 1999.
2. Allen, K., Green, S., and Zubrow, E., Eds., *Interpreting Space: GIS and Archaeology*, Taylor & Francis, London, 1990.
3. Beck, A.R., Intellectual excavation and dynamic information management systems, in *On the Theory and Practice of Archaeological Computing*, Lock, G. and Brown, K., Eds., Oxbow: Oxford, 2000, pp. 73–88.
4. Chadwick, A., What have the post-processualists ever done for us? Towards an integration of theory and practice and a radical field archaeology, in Proceedings of the 2001 Interpreting Stratigraphy Conference, Roskams, S. and Beck, C.M., Eds., in prep.
5. Beck, A.R. and Beck, M., Computing, theory and practice: establishing the agenda in contract archaeology, in *Interpreting Stratigraphy Site Evaluation, Recording Procedures and Stratigraphic Analysis*, Roskams, S., Ed., Tempus Reparatum, 2001, pp. 173–182.
6. Campbell, H. and Masser, I., Implementing GIS: the organisational dimension, in *Association of Geographic Information*, AGI, 1993.
7. Reeve, D. and Petch, J., *GIS Organisations and People*, Taylor & Francis, London, 1999.
8. Chadwick, A., Taking English archaeology into the next millennium: a personal review of the state of the art, Assemblage, 2000(5), http://www.shef.ac.uk/assem/5/chad.html.
9. Hunter, J. and Ralston, I., Eds., *Archaeological Resource Management in the UK: An Introduction*, Sutton Publishing, Stroud, 1993
10. Scollar, I., *Archaeological Prospecting and Remote Sensing. Topics in Remote Sensing*, 2, Cambridge University Press, Cambridge, UK, 1990.
11. Wilson, D.R., *Air Photo Interpretation for Archaeologists*, Tempus Publishing, Stroud, 2000.
12. Gaffney, C. and J. Gater, *Revealing the Buried Past: Geophysics for Archaeologists*, Tempus Publishing, Stroud, 2003.
13. Banning, E.B., *Archaeological Survey. Manuals in Archaeological Method Theory and Technique*, M. Schiffer, Series Ed., Kluwer Academic, New York, 2002.
14. Brothwell, D.R. and Pollard, A.M., *Handbook of Archaeological Sciences*, Wiley, Chichester, 2001.
15. Roskams, S., *Excavation*, Cambridge University Press, Cambridge, UK, 2001.
16. English Heritage, Management of Archaeological Projects, HBMC, London, 1991.
17. Andrews, G., Barrett, J., and Lewis, J., Interpretation not record. *Antiquity*, 74, 525–530, 2000.

18. Hodder, I., *On the Surface: Catalhoyuk 1993–1995*, British Institute of Archaeology Ankara, London, 1996.

19. Hodder, I., *The Archaeological Process: an Introduction*, Blackwell, Oxford, 1999.

20. Rossignol, J. and Wandsnider, L., Eds., *Space, Time, and Archaeological Landscapes. Interdisciplinary Contributions to Archaeology*, Plenum Press, New York, 1992.

21. Ramenofsky, A.F. and Steffen, A., Eds., *Unit Issues in Archaeology: Measuring Time, Space, and Material. Foundations of Archaeological Inquiry*, University of Utah Press, Salt Lake City, 1998.

22. Allen, K., Consideration of scale in modelling settlement patterns using GIS, in *Practical Applications of GIS for Archaeologists: a Predictive Modelling Toolkit*, Wescott, K.L. and Brandon, R.J., Eds., Taylor & Francis, London, 2000.

23. Lock, G.R., *Using Computers in Archaeology*, Routledge, London, 2003.

24. Greene, K., *Archaeology: an Introduction*, Batsford, London, 1982.

25. Lucas, G., *Critical Approaches to Fieldwork: Contemporary and Historical Archaeological Practice*, Routledge, London, 2001.

26. Lock, G. and Brown, K., Eds., *On the Theory and Practice of Archaeological Computing*, Oxbow, Oxford, UK, 2000.

27. Gillings, M. and Goodrick, G., Sensuous and Reflexive GIS: Exploring Visualisation and VRML, Internet Archaeology, 1998, http://intarch.ac.uk/journal/issue1/gillings_index.html.

28. Forte, M. and Williams, P.R., *The reconstruction of archaeological landscapes through digital technologies: Proceedings of the 1st Italy–United States Workshop, Boston, Massachusetts, USA, November 1–3, 2001.* BAR International Series; 1151. Oxford: Archaeopress, for British Archaeological Reports, 2003.

29. Huxhold, W.E. and Levinsohn, A.G., *Managing Geographic Information System Projects. Spatial Information Systems*, Oxford: Oxford University Press, New York, 1995.

30. Chan, T.O. and Williamson, I.P., The different identities of GIS and GIS diffusion, *Int. J. Geogr. Inf. Sci.*, 13(3), 267–281, 1999.

31. Heywood, D.I., Cornelius, S., and Carver, S., *An Introduction to Geographical Information Systems*, Prentice Hall, New York, 2002.

32. Peng, Z. and Tsou, M., *Internet GIS*, John Wiley and Sons, New Jersey, 2003.

33. Longley, P. et al., *Geographic Information Systems and Science*, Wiley, New York, 2001.

34. Westman, A., *Archaeological Site Manual*, 3rd ed., Museum of London, London, 1994.

35. Schiffer, M.B., *Formation Processes of the Archaeological Record*, University of New Mexico Press, Albuquerque, 1987.

36. Adams, M., The optician's trick: an approach to recording excavation using an iconic formation process recognition system, in *Interpreting Stratigraphy Site Evaluation, Recording Procedures and Stratigraphic Analysis*, Roskams, S., Ed., Tempus Reparatum, Oxford, UK, 2001, p. 91–102.

37. Crofts, N., Doerr, M., and Gill, T., The CIDOC Conceptual Reference Model: A Standard for Communicating Cultural Content, CIDOC, 2003.

38. Müller, J.C., Lagrange, J.P,. and Weibel, R., Eds., *GIS and Generalization: Methodology and Practice. GISdata*, 1, Taylor & Francis, London, 1995.

39. Morehouse, S., GIS-based map compilation and generalization, in *GIS and Generalization: Methodology and Practice*, Muller, J.C., Lagrange, J.P., and Weibel, R., Eds., Taylor & Francis, Compiegne, France, 1995, p. 21–30.

40. Lee, D., Automation of Map Generalization., ESRI, 1996.

41. MacDonald, A., *Building a Geodatabase: GIS by ESRI*, Environmental Systems Research Institute, Redlands, CA, 2001.

42. Wise, A. and Miller, P., Why metadata matters. Internet Archaeology, 1997(1) http://intarch.ac.uk/journal/issue2/wise_toc.html.

43. Bewley, R. and Raczkowski, W., Eds., *Aerial Archaeology: Developing Future Practice*. Life and Behavioural Sciences. Vol. 337., IOS Press, Oxford, UK, 2002, 376.

44. Bewley, R., Aerial photography for archaeology, in *Archaeological Resource Management in the UK: An Introduction*, Hunter, J. and Ralston, I., Eds., Sutton Publishing, Stroud, 1993.

45. Beck, A.R., Getting your hand dirty: in the field with the Handspring PDA and ThinkDB. *Archaeol. Comput. Newsl.*, 59, 11–15, 2002.

46. Ryan, N.S., More, D.R., and Pascoe, J., Enhanced reality fieldwork: the context aware archaeological assistant, in *Computer Applications in Archaeology, 1997*, Gaffney, V., Exon, S., and van Leusen, M., Eds., BAR, Oxford, UK, 1998.

47. Wheatley, D. and Gillings, M., *Spatial Technology and Archaeology: the Archeaological Applications of GIS*, Taylor & Francis, New York, 2002.

48. Berggren, Å., Between structure and individual — reflexive approaches to archaeological fieldwork in Malmö, Sweden, in *Proceedings of the 2001 Interpreting Stratigraphy Conference*, S. Roskams and M. Beck, Eds., in prep.

Part III-C

Learning from Practice: GIS as a Tool in Planning Sustainable Development

Society and Environment

22 A Geographical Approach to Community Safety: A U.K. Perspective

Jonathan Corcoran and
Bernadette Bowen Thomson

CONTENTS

22.1 INTRODUCTION

Crime and disorder are inevitable realities of society, affecting all of the populace either directly or indirectly. Their formal control has traditionally been the responsibility of the police. Increasingly, recent years have seen the control of crime and disorder, in England and Wales, charged to a range of both nationally and locally governed agencies. The requirement to minimize community problems through tackling crime and disorder issues was formalized in the Crime and Disorder Act [1]. The Act formally introduces the creation of multiagency Crime and Disorder Reduction Partnerships (CDRPs) within each local authority area. A legal obligation was placed on these CDRPs, particularly the local authority and police, to work in tandem to develop, publish, and implement three-year strategies to tackle crime and disorder. The production of an informed crime and disorder strategy relies heavily upon an in-depth local community safety audit, which provides a snapshot of crime and disorder-related issues, a further stipulation of the Act. Each audit, consisting of multiagency data and community consultation, attempts to encapsulate the community dynamics within a given area. In addition, the Act stipulates the necessity to work with other key agencies, including the health authority (Sections 5–7, Crime and Disorder Act 1998 [1]), while Guidance recommends the expansion of the partnership to business and voluntary sectors.

Section 17 of the 1998 Crime and Disorder Act extends the scope of responsibility for controlling crime and disorder. It places a statutory obligation on local authorities and the police to consider crime and disorder implications in all its functions [1]. Part of this legislation (Section 115) enables partners to share previously internalized data for crime and disorder reduction purposes. If these agencies are to embrace the principles within the Act, then the production of a holistic strategy is essential. Such a strategy would enable a variety of agencies to use their expertise for crime and disorder reduction and prevention purposes and for the benefit of the community, thus realizing increased community safety. Key to achieving their missions is the ability to assimilate an understanding of criminal dynamics, which are inherently complex. Geographical tools have the potential to provide invaluable insight into these dynamics.

It has been shown that crime and disorder recorded by the police constitute only a partial descriptor of community issues [2]. Therefore, to understand the dynamics and requirements of a region, there is the need to consult additional data, sourced from a range of organizations at the local level [3]. On this basis, local partnerships have been promoted to guide and facilitate the data collation, aggregation, and analysis process. Hough and Tilley [4] outline six guiding principles that support the requirement for local partnerships:

- The police alone cannot control crime and disorder
- No single agency can control crime and disorder
- Agencies with a contribution to reducing crime and disorder need to work in partnership
- Evidence-based problem solving approaches promise the most effective approach to reducing crime and disorder
- Problems of crime and disorder are complex, and there are therefore no panaceas
- Crime and disorder problems need to be understood in their local contexts and strategies need thus to be locally tailored

Hough and Tilley [4, p.1]. With permission.

In the remainder of this chapter, the importance of geography for crime and disorder analysis and nature of community safety is discussed. This is followed by a discussion of the design, development, and implementation of Holistic approach to strategic crime and disorder evaluation (HASCADE), a geographical approach to strategic crime and disorder analysis.

22.2 THE IMPORTANCE OF GEOGRAPHY

The mapping of crime has a long history as a tool for understanding crime's spatial distributions. It can be traced back as far as the nineteenth century in France [5,6], where mapping was first utilized to visualize and analyze crime information. Crime data and the computational tools that are used for their collection and analysis have, over recent years, grown in importance. Academics and practitioners have seen value in their potential to analyze crime and disorder issues.

A central theme in the geo-analysis of crime and disorder data is the quest to better understand their dynamics, which in turn can be applied to formulate targeted responses. The U.K. Home Office advocates a geographically orientated approach to crime analysis. This is reflected by the marked growth of computerized mapping by U.K. police forces [7], the trend set to continue. However, a report of the auditing process [8] revealed that less than half (42%) made use of a GIS.

22.2.1 WHAT IS COMMUNITY SAFETY?

Community safety is a recent concept, the definition of which has amassed much debate. Since the Morgan Report [9], the term community safety has witnessed increasing popularity in Britain. The Morgan Report (para. 3.7), considers community safety "as being concerned with people, communities and organizations including families, victims and at risk groups, as well as attempting to reduce particular types of crime and the fear of crime. Community safety should be seen as the legitimate concern of all in the local community" (cited [10, p.6]). Community safety is recognized as comprising situational and social characteristics. The situational characteristics of community safety include crime prevention. Crime prevention, in its simplest form, indicates a situation whereby crimes would have occurred if they had not been prevented [11]. In general, crime prevention techniques can be applied to a variety of approaches that aim to reduce the likelihood of an individual or group encountering crime events. Social characteristics refer to the socioeconomic and cultural aspects of people's lives; thus individuals and groups should be able "to pursue, and obtain fullest benefits from, their social and economic lives without fear or hindrance from crime and disorder" [12]. Partnership working is fundamental to community safety because it recognizes that community safety is not the sole responsibility of the police. The change in policy focus toward partnership working at the community level in the United Kingdom implies that an advantage will be achieved if broader multiagency and multifaceted approaches are applied. Walklate identifies that "a genuine desire for policy to work for change needs above all to be cognizant of the importance of the local context in which that policy is set. This desire needs to work with rather than against the historical and socioeconomic circumstances which structure that local context" [13, p.62]. As such, diligence in ensuring that inclusive approaches are implemented and that these approaches are appropriate to the community should have primacy.

HASCADE attempts to inform short, medium, and long-term strategy through the examination of data at the community level. This information endeavors to provide insight into the crime, disorder, and potential vulnerabilities that include socioeconomic factors present within such geographical areas. It seems, then, that the term community safety renders itself more easily toward applying holistic approaches, thus potentially increasing engagement from partners, community, and agency. In relation to this chapter, the term community safety will be used, recognizing that such a term includes crime prevention and that its definition can extend beyond the realms of crime and disorder.

22.2.2 CURRENT APPROACHES TO COMMUNITY SAFETY

Current approaches to achieving community safety in England and Wales often reflect traditional crime prevention concepts, commonly involving applied situational crime prevention (SCP) techniques [14]. The application of such techniques has positively impacted upon crime and disorder reduction in communities, often achieving a rapid effect. Felson and Clarke [15] note numerous examples where the application of targeted "opportunity-reducing measures" has produced effective outcomes. Such measures include reductions in check frauds occurring in Sweden, through the introduction of new identification measures, and the establishment of CCTV cameras in Surrey University car parks that resulted in reductions in crime.

The HASCADE model is receptive to the important contribution of crime prevention, but it also endeavors to inform strategy development centered upon consideration of wider, holistic, community safety issues, particularly those based around social exclusion. Crime and disorder strategies informed solely by analyzing the spatial distribution of crime and disorder events, provide only a partial view of community issues. Typically this can involve the analysis of police crime and disorder data, to identify hotspots (areas exhibiting disproportionately high levels of crime and disorder). The results from this exercise are then used to design strategy to combat crime and disorder within the identified locales. However, such use of crime and disorder data is likely to increase the risk of only responding to community safety through the application of primarily situational methods (for example, the use of locks to deter burglaries within identified burglary hotspots). In addition, these are likely to result in imbalanced strategy that may not identify key facets of a community's needs. Such imbalanced approaches increase the risk of exclusion, while potentially reducing trust within communities (for a more detailed discussion of trust and exclusion see [16,17]).

HASCADE introduces a joined-up strategic framework. Its holistic nature encourages multifaceted methods for improving community safety, as opposed to applying a singular methodology. The use of geographically referenced multiagency data is key to informing this holistic approach.

22.3 THE HASCADE APPROACH TO COMMUNITY SAFETY

The relevance of incorporating multiagency data within the crime and disorder audits reinforces the underlying principles of partnership working that is promoted throughout the Crime and Disorder Act. In addition, all guidance relating to the crime and disorder audits has advocated the use of multiagency data sets [4], but little information exists yet on how these multiple data sets should be incorporated and analyzed.

22.3.1 DATA REQUIREMENTS AND ISSUES

The importance of a geographically orientated approach has already been stated. However, modeling the geography of crime in a way applicable to CDRP objectives requires an alternative approach to that demanded by police operations. Modeling techniques to direct, monitor, and evaluate community initiatives demands the adoption of a holistic approach, in which a range of local information is analyzed in an appropriate manner.

The ability to visualize the precise locations of events has been welcomed and promoted by the government [18]. As such, microlevel analysis has become of particular interest for those implementing SCP programs. Microlevel analysis can prove successful in such programs, because the objective is to uncover the specifics of a locale. Furthermore, it offers an explanation of a locale's propensity toward observed events (for example, a series of houses within a neighborhood particularly subject to burglary). SCP techniques should not constitute the entire audit analysis, because an imbalanced strategy, with a tendency to short-term gains, would result. Such a strategy would discount the social, cultural, and community characteristics of an area. A primarily SCP focus impedes the development of a strategy that strives to address community safety in the short, medium, and long term. Moreover, in the context of the audit, the use of such techniques places large demands upon each partner to provide full address information from which the data can be geocoded to the fine scale typically demanded by SCP programs. In addition, the role of the audit is to provide an overview of a whole local authority area; thus fine resolution analysis is arguably not the primary objective. Therefore, the audit should put in place a series of analyses that are capable of identifying the broad issues. At this stage a micro level analysis could take place to isolate the specific issues (for example, vulnerable houses and common modus operandi) to ensure a correct application of preventative measures (such as a lock-fitting scheme and security advice), while simultaneously directing attention on the broader social issues that impact upon the community.

During the development of HASCADE, several constraining political, technical, and administrative issues were identified. These issues are discussed below in turn.

22.3.1.1 Technical Issues

A fundamental issue at the commencement of auditing was the identification of key technical personnel within each partner agency, possessing the necessary technical skills and knowledge pertaining to their data systems. This formed a vital stage, following which questions regarding data (for example, descriptions of coding protocols) could be posed and replied to efficiently.

22.3.1.2 Security Issues

Despite the caveat (Section 115) in the 1998 Crime and Disorder Act facilitating the sharing of previously internalized data, many concerns for partners still remained. This was partially resolved through the implementation of data security measures in addition to the specification of what could be disseminated. Clearly stated dissemination

TABLE 22.1
Minimum Data Provision Proforma

Attribute	Content
Temporal Reference	Date (dd/mm/yyyy)
Spatial Reference	Full postcode
Incident or Event description	Numeric code or standardized text
File formats	MS Excel, MS Access, Delimited text

protocols were enforced, whereby no information was to be published (cartographic or text based) beyond the confines of the CDRP without formal approval by all partners. On a technical level, it was agreed that a single machine would be utilized for all data analysis and presentation, the access to which was strictly controlled.

22.3.1.3 Data issues

Many partners had concerns regarding the sharing of data. The concern was in perceived contravention of the 1998 Data Protection Act, despite the clear caveat stipulated within Section 115 of the Crime and Disorder Act facilitating their use. To appease such concerns, a data wish list proforma was designed to adhere to the general, while providing the necessary information from which targeted mapping and statistical output could be generated. Using the identified technical contacts, all partners completed this, and a final standard was agreed, with each agency providing as a minimum the contents summarized in Table 22.1.

In addition, the confirmation of any problems, inconsistencies, and known errors were established prior to the analysis, which in the main involved changes in counting rules and coding protocols. This established known, but not necessarily published, information concerning data reliability. The result of this process was to either omit or amend their use within the audit. This could then be attached to the audit, not necessarily as a formal appendix, but as a reference from which decisions to include or omit certain facets of data could be supported.

One of the most problematic issues was achieving a common temporal coverage across all contributing agencies. Typical obstacles to accomplishing this were modifications to software systems that rendered data prior to particular dates difficult to access. In addition, the requested data spanned alterations in collection protocols and, thus, introduced potentially immeasurable inconsistencies.

22.4 THE HASCADE MODEL

The HASCADE model (Figure 22.1) uses both spatial and statistical techniques to provide an insight into the dynamics of crime, disorder, and vulnerability across the partnership region.

The result of the data collection process (following the resolution of the aforementioned issues), achieved the collection of eight partners' data spanning a twelve-month

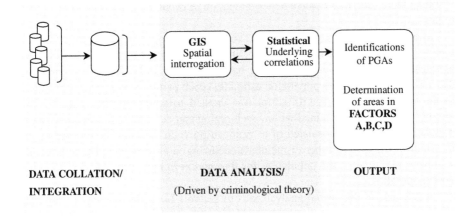

FIGURE 22.1 The HASCADE model.

TABLE 22.2
Meanings Attributed to Data

Crime and Vulnerability Indicators		Associated Datasets
Crime and Disorder		Police Incident Figures, Benefit Fraud, Arson
Vulnerability indicators	Economic poverty	Council tax benefit claimants
	Peer and family criminality	Police reprimand, final warning, sentence — youth offending team (YOT); supervised and unsupervised data — probation
	Lack of educational attachment and future risk of low educational attainment	School exclusions
	Risk of social exclusion	All of the above, looked after children and benefit fraud

period. Each data set was then attributed a meaning, classified as either a crime and disorder data set or a vulnerability data set (Table 22.2).

The following sections detail the spatial and statistical techniques that were employed to generate the final output.

22.4.1 SPATIAL METHODS

Using the spatial reference provided by all partners, data was geocoded utilizing OS Code-Point® [19] to assign x,y coordinates at a postcode unit level. Once geo-referenced, all data were inspected at point level and overlain with street and boundary information to provide context. In many cases the volume of mapped incidents created visualization problems at the point level, where multiple incident localities appeared as a single occurrence (each point was simply positioned one above another and, thus, appeared incorrectly as a single incident). Therefore aggregate and density mapping were used to provide a better indication of event intensity across the CDRP region.

In order to provide a greater context to the underlying population geography to which the various events were related, it was necessary to generate a series of aggregate maps. Using a boundary set based upon the 1991 Census enumeration districts (containing 1999 population estimates) each partner's data were aggregated to the new framework. A GIS script was created to automate the calculation of number of incidents contained within each region and derivation of rate based upon population. The script consisted of a "point in polygon" test for the partner's data to calculate the total number of incidents occurring in each region. The total count was then used against the population for that region to derive the incident rate per 1000 population.

A key part of the spatial analysis was to derive a boundary network to provide the closest representation of each partner's data. This boundary network then formed the foundation from which statistical analysis could be conducted. Creating and validating the boundary network first involved examining the event distributions (using hotspot mapping — as this best describes event distribution) from each agency's data. Where there was an identified lack of coterminosity, the boundary network could be modified to provide a closer fit to the agency data. Typically, a modification included aggregating two or more regions together. In certain circumstances, however, the imposed boundaries could either under- or overfit areas of high event volume. Because a perfect match could not be achieved, it was deemed acceptable where the boundaries were generally representative of all data, (Figure 22.2).

One limitation of this process was the input boundary network constraining the minimum size of areas for which population data was available. Thus, if agency data underfitted a region, this could not be redefined to a subdivision, because incident rates could not be calculated. A second limitation was the decision criteria used to assess whether areas were representative or required modification. For this stage visual inspection rather than any quantitative techniques were employed because it was recognized that high-level precision could not be achieved due to the nature of the agency's data. A visual comparison was therefore considered sufficient.

At this stage it was possible to commence a primary identification of priority geographical areas (PGAs) on the strength of point, hotspot, and aggregate mapping across all data sets. Aggregate outputs offered an indication of vulnerability at a broad neighborhood scale, while point and hotspot maps identify more specific subneighborhood localities internal to these regions.

22.4.2 STATISTICAL ANALYSIS

Spatial analysis using GIS techniques offered a tool by which visualization and aggregation was conducted. In many audits, outputs from this stage are taken no further. Therefore one was able to overlay and visualize, but unable to quantify interactions between various layers of information.

Statistical analysis targeted the correlation between the various data sets to reinforce relationships identified through the spatial analyses. Establishing significant statistical linkages, together with identification of PGAs, provided the foundation from which

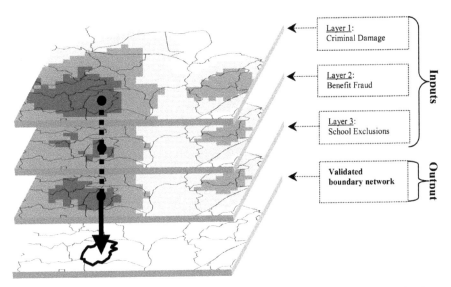

FIGURE 22.2 Method of validating the boundary network. Identification of areas with similar event intensities was achieved using the GIS to overlay each agency's data (for example, see dotted line between layers). The degree to which hotspots matched the boundary network could be assessed and modifications made if the majority of layers lacked coterminosity. (From Corcoran, J. and Bowen Thomson, B., *Br. J. Commun. Justice*, 2 (1), 45, 2003. With permission.)

partnership strategy was developed through building a fundamental comprehension of community safety processes.

Pearson correlation coefficients [20] were used as the basis from which significant relationships were established. Using the rates for each data set, for each small area, significant correlations were flagged and used as the basis to establish key dependencies between the various data areas (the term small area refers to a single areal unit of the validated boundary network). Thus, for a neighborhood area identified as a vulnerable locale for events such as school exclusions and youth offending, spatial inspection can be further queried through statistical linkages, in addition to suggestions of further potential associations that may exist. The result of this hypothetical scenario would be to identify a target set of agencies required to jointly direct interventions in the specified area. In addition, such output and evidence produced through the implementation of this framework reinforces the necessity of the partnership, supporting the current drive toward joined-up government working practices.

22.4.3 Results from HASCADE

The final part of the analysis was to produce a final output in which areas from the validated boundary were classified to reflect each of the partners' data. To achieve this categorization, a count system was designed. Using a natural break method, each partner's data was first mapped using five classes. For each data set, those areas that were classified in the top two categories were tabulated (Figure 22.3).

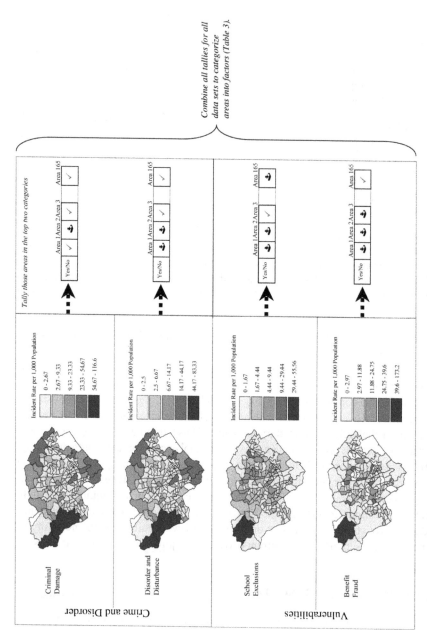

FIGURE 22.3 The HASCADE count system (using artificially generated data).

TABLE 22.3
Classification of Small Areas into Factors

		Area 1	Area 2	Area 3	Area 4	Area 165
Crime & Disorder	Burglary		1	1		1
(C&D)	Disorder	1		1		
	Violence	1		1		
	Robbery	1		1		
	Criminal damage	1		1		
Vulnerabilities	Benefit fraud	1	1		1	1
(Vul)	Probation Unsupervised	1	1		1	
	Probation Supervised	1	1		1	
	School exclusions		1			
	YOT Final warning	1			1	
	YOT Police reprimand	1			1	
	YOT Sentence				1	
Total (C&D)		4	1	5	0	1
Total (Vul)		5	4	0	6	1
FACTOR		A	B	C	D	—

The tabulation of all cartographic outputs (representing each partner's data) was then used to distinguish areas of disproportionately high crime, disorder, and vulnerability from those exhibiting lesser levels. Analyzing the results of the tabulation indicated several different area types, from which four categories were defined (Table 22.3).

Each of the four distinct categories (termed Factors A, B, C, and D) reflects a blend of community issues within identified areas where a different combination of community safety responses is required (Table 22.4).

The result of the factorization process resulted in the production of a single cartographic output (Figure 22.4) that is supported by the statistical analysis. This final output was designed to minimize the use of numerous representations (for example, graphs, tables, and multiple maps for each partner agency), providing a concise indication of crime, disorder and vulnerabilities across the CDRP region, from which strategy could be derived.

22.5 DISCUSSION

HASCADE provides partners from a variety of agencies with a method of describing community dynamics within an area, based on the prevalence of crime, disorder, or community vulnerabilities. The factorization of an area enables practitioners to consider how their expertise could be distributed across the larger region. Guidance for CDRPs notes the importance of referring to multiagency information to inform strategy. The partnership Auditing Toolkit [21] advocates the use of GIS and its incorporation into crime and disorder audits, but its emphasis is upon the use of such systems for microlevel analysis of crime and disorder incidents, thus risking

TABLE 22.4
Description of Geographical Areas and Type of Response

	FACTOR A	FACTOR B	FACTOR C	FACTOR D
Description of Geographical Areas	Several categories of crimes or disorder PLUS community vulnerability	One category of crime or disorder PLUS several types of community vulnerabilities	Several categories of crimes or disorder BUT NO community vulnerabilities	No crime and disorder, BUT SEVERAL types of community vulnerabilities
Type of Response	Situational crime prevention, recourse to criminal justice system, plus prevention of future criminality	Situational crime prevention, recourse to criminal justice system plus prevention of future criminality	Recourse to criminal justice system plus situational crime prevention methods	Promotion of community safety and prevention of future criminality

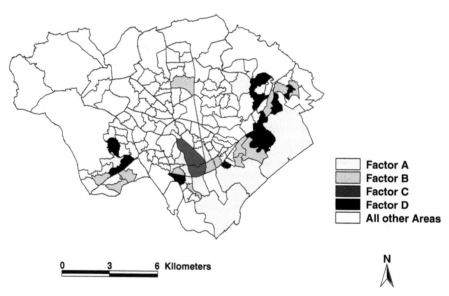

FIGURE 22.4 Factor map showing PGAs (From Corcoran, J. and Bowen Thomson, B., *Br. J. Commun. Justice*, 2 (1), 47, 2003. With permission.)

imbalanced strategy development, primarily concerned with SCP approaches. It has been suggested that the application of such approaches may result in reductions in interpersonal trust or trusting individuals being considered irresponsible. "Instead, much of what remains of our trust will reside in locks and alarms, CCTV, environmental design, and regulation, rather than in some conception of others as trustworthy" [22, p.41]. The inclusive approach of HASCADE does not, however, rely upon such microlevel analysis. Such analysis is likely to deter partner involvement, as not all partners are able to produce information in the detail required (for example, high-resolution full-address information), nor would their service be motivated to respond by purely situational means. HASCADE attempts to reduce the risk of such an imbalanced approach to community safety strategy by informing decisions based upon various vulnerabilities, for example, school exclusion data, alongside crime and disorder, which can then be combined to inform the development of a more holistic strategy. Additionally, the holistic nature of HASCADE aims to provide partners with an insight into the community dynamics of an area, thus informing multifaceted approaches to community safety based upon each agency's expertise. The extension of HASCADE beyond the mapping of crime and disorder incidents facilitates the development of holistic processes. Subsequently, such processes are not restricted to criminality-based service provision. Thus, multiple agencies are better able to identify how they can support community safety, while adhering to the principle objectives of their own service delivery.

HASCADE identified a number of factors that aimed to describe areas where the prevalence of crime, disorder, or community vulnerability was disproportionately high. These areas could thus be considered as PGAs. Examination of an area identified as a Factor A, for example, would consist of a relatively high prevalence of crimes and disorders coupled with a relatively high prevalence of community vulnerabilities. A multiagency partnership approach to community safety could involve a cluster of techniques incorporating SCP approaches (for example, a lock-fitting scheme for burglary reduction) recourse to the criminal justice system (for example, a prison sentence), and the prevention of future criminality (for example, youth diversionary schemes such as sport and vocational provisions). Examining the data layers that contribute to its factorization would provide various agencies with information that could inform how service provision could be collectively targeted, thus indicating intralocale relations.

In addition to intralocale relations, the factorization scheme has the potential to indicate interlocale relations. Our analysis identifies possible relationships between different factor areas that are adjacent to one another (Figure 22.5). Areas exhibiting a variety of high crime and disorder types and vulnerabilities (Factor A) are located next to areas exhibiting disproportionately high levels of vulnerabilities and relatively low levels of crime and disorder (Factor D). Areas exhibiting high levels of vulnerabilities may include disproportionately high levels of offenders, young people who have been excluded from school, or benefit claimants. Figure 22.5 is possibly indicative of interrelations between Factor A and D areas.

Overall, the division of areas into Factors A, B, C, and D, has the potential to inform strategic decisions regarding resource allocation from partner agencies across the partnership area. The positioning of factor areas and, where appropriate, the interlocale relations between them require a variety of community safety approaches. Simply targeting the high crime and disorder areas may not prove the most effective way of achieving community safety. Instead, agencies may wish to expand relevant schemes to the adjacent Factor D areas.

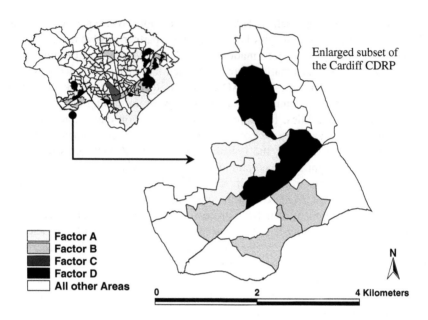

FIGURE 22.5 Enlarged factor map.

Guidance on reducing crime within areas mostly focuses upon multiagency intralocale relations [23,24]. HASCADE identifies high crime areas and a number of underlying vulnerabilities, but it also attempts to inform partners of potential multiagency interlocale relations. Research shows that offenders do not travel large distances to commit crimes; Wiles and Costello [25] found that the average distance traveled to commit crimes, by offenders, was 1.93 miles from their home. In relation to crime and disorder partnership areas, 1.93 miles represents a relatively large area, possibly extending an administrative boundary or many other imposed boundaries. The significance of inter- and intralocale relations when developing community safety strategy is more pertinent, because it supports holistic community safety approaches, thus enabling partners to play an active role in community safety, increasing partner participation.

To follow are two examples of how HASCADE was applied to inform strategic decisions, as well as the application of operational measures to support community safety objectives. At the time of writing, both case studies are ongoing.

The application of HASCADE has informed the targeting of a number of projects. One such project is the Safer Cardiff Bus Project; a mobile community provision. One of the primary functions of this bus project is the provision of activities for young people, particularly young people "at risk," within areas that were classified as Factor A, B, and D. The final aspect of this project involved the identification of specific locales and timings to ensure efficient and effective service delivery. To achieve this, further analysis of relevant data relating to issues considered to be pertinent to the target group were required. For each data set hotspot mapping was generated. Using this microlevel analysis, coupled with local knowledge and perception, allowed informed operational decision-making regarding the most appropriate location of the bus over given times. The outcome of this microlevel analysis revealed areas previously known to agencies working with the target group. Most importantly, however, additional, previously unknown, areas were identified that potentially could benefit from the service.

Similarly, HASCADE was used to identify geographical environments for the application of a Homesafe project (a lock-fitting service). The areas identified consisted of disproportionately higher levels of crime and disorder, particularly burglary dwelling and disproportionately higher levels of vulnerabilities than elsewhere within the partnership region. As the areas used within HASCADE are relatively small (i.e., approximately equivalent to the U.K. Census enumeration districts), Homesafe was able to provide its free service to households, up to a 10% saturation point within the priority area (as recommended by the project director).

The aforementioned examples illustrate the flexibility of HASCADE, as it is able to inform strategic decision-making and operational implementation. Additionally, HASCADE is organic in nature. The model can be easily adapted to include additional data layers, which collectively aim to further the explanation of community dynamics.

22.6 FUTURE DEVELOPMENTS

HASCADE represents the first stage in the development of a geographically grounded community safety methodology. Future developments of HASCADE are numerous and can be categorized under four broad categories, which are discussed in turn in the remainder of this chapter.

22.6.1 An Integrated Deployable Solution

Future work should develop HASCADE into a solution that is deployable to CDRPs whereby the scalability of the model can be more fully evaluated. To date, HASCADE relies upon the use of multiple packages and a largely manual process for the transfer of data between such applications. The focus for future development should be on the creation of a single application capable of carrying out the necessary mapping, aggregation, and statistical functions. Of particular note is the method by which the validated boundary network is created. The current process (based upon visual inspection) should be updated by a quantification of the visual technique (for example, calculating a measure of fit from which the best boundary configuration is selected), to ensure a more robust and replicable procedure.

Finally, embedded within this application should be the capability of providing expert advice to the user (for example, on the appropriate use of agency data combinations — the suitability of using benefit fraud data alongside data from probation and police incidents and what this combination is potentially illustrating).

22.6.2 Increased Data Sets

In spite of legislation facilitating the exchange of previously internalized information for crime and disorder reduction purposes (Section 115, Crime and Disorder Act,

1998), some agencies may be reluctant to provide personalized information for multiagency use. The HASCADE model attempts to alleviate such concerns using a postcode as the minimum spatial reference required. Furthermore, data outputs from HASCADE are aggregated based upon postcode centroids. Consequently, confidence regarding the initial provision of information is increased for agencies that may be hesitant about sharing personalized information.

Future developments for HASCADE would include expanding the crime, disorder, and vulnerability data layers currently incorporated in the model. The minimum data requirements increase the potential for agencies, particularly smaller, or not-for-profit agencies, to be included in the process and actively involved in the planning and delivery of community safety.

REFERENCES

1. The-Stationery-Office, Crime and Disorder Act, Home Office, London, 1998.
2. Shepherd, J., Shapland, M., and Scully, C., Recording by the police of violent offences; an accident and emergency department perspective, *Med. Sci. Law*, 29, 251–257, 1989.
3. Graham, J., Ekblom, P., and Pease, K., Reducing offending: an assessment of research evidence on the ways of dealing with offending behaviour, Research study 187, Home Office, London, 1998.
4. Hough, M. and Tilley, N., Auditing crime and disorder — Guidance for local partnerships, Home Office, London, 1998.
5. Guerry, A.M., Essai sur la statistique morale de la France. *Westminster Rev.*, 18, 357, 1833.
6. Quètelet, A., *A Treatise on Man*, Chambers, Edinburgh, 1842.
7. Corcoran, J. and Ware, J.A., Helping with enquiries, in *Geo:Connexion*. 2002, p. 36–40.
8. Phillips, C., Considine, M., and Lewis, R., A review of audits and strategies produced by crime and disorder partnerships in 1999, Home Office, London, 2000.
9. Morgan, J., Safer Communities: The Local Delivery of Crime Prevention Through the Partnership Approach, Home Office — Standing Conference of Crime Prevention, London, 1991.
10. Feenan, D., Community Safety: Partnerships and Local Government, Criminal Justice Review Group Research, Report 13, Stationary Office, Belfast, 2000.
11. Duff, R.A. and Marshall, S.E., Benefits, Burdens and responsibilities: some ethical dimensions of situational crime prevention, in *Ethical and Social Perspectives on Situational Crime Prevention*, Von Hirsch, A., Garland, D. and Wakefield, A., Eds., HART Publishing, Oxford, UK, 2000.
12. Ekblom, P., Community Safety and the Reduction and Prevention of Crime — A Conceptual Framework for Training and the Development of a Professional Discipline, Home Office Research and Statistics Directorate, London, 1998, http://www.homeoffice.gov.uk/docs/cstrng5.html.
13. Walklate, S., Trust and the problem of community, in *Crime, Risk and Insecurity*, Hope, T. and Sparks, R., Eds., Routledge, London, 2000.
14. Clarke, R.V., Situational crime prevention, in *Building a Safer Society*, Tonry, M. and Farrington, D., Eds., University of Chicago Press, Chicago, 1995.

15. Felson, M. and Clarke, R.V., Opportunity makes the thief: practical theory for crime prevention, Police Research Series Paper 98, Home Office, London, 1998.
16. Zedner, L., The pursuit of security, in *Crime, Risk and Insecurity*, Hope, T. and Sparks, R., Eds., Routledge, London, 2000.
17. Wakefield, A., Situational crime prevention in mass private property, in *Ethical and Social Perspectives on Situational Crime Prevention*, Von Hirsch, A., Garland, D. and Wakefield, A., Eds., HART Publishing, Oxford, UK, 2000.
18. Home-Office, Review of Crime Statistics: a Discussion Document. 2000, http://www.homeoffice.gov.uk/rds/pdfs04/review.pdf.
19. Ordnance-Survey, Code-Point® Data. Southampton, UK, 2001.
20. Galton, F., Co-relations and their measurement, chiefly from anthropometric data. *Proc. R. Soc. Lond.*, 45, 135–145, 1888.
21. Crime-Reduction-Website, Crime Reduction Toolkits, Partnership Working, 2003, http://www.crimereduction.gov.uk/toolkits/p00.htm.
22. Kleinig, J., The burdens of situational crime prevention, in *Ethical and Social Perspectives on Situational Crime Prevention*, Von Hirsch, A., Garland, D. and Wakefield, A., Eds., HART Publishing, Oxford, UK, 2000.
23. Crime-Concern, Reducing Neighborhood Crime a Manual for Action, Crime Concern, Wiltshire, 1998.
24. Crime-Concern, Reducing Neighborhood Crime, Crime-Concern, Wiltshire, 2000, http://www.crimeconcern.org.uk/pubs/pspredneighbourhoodcrime.pdf
25. Wiles, P. and Costello, A., Road to Nowhere: The Evidence for Travelling Criminals, Home Office Research Study No. 207, Home Office: London, 2000.
26. Corcoran, J. and Bowen Thomson, B., New insights into community safety: An application of the HASCADE model, *Br. J. Commun. Justice*, 2 (1), 37–50, 2003.

23 GIS Application to Support Water Infrastructures Facilities Localization in Particularly Valuable Environmental Areas: The Eolian Islands Case Study

Giuseppe Cremona and Luisella Ciancarella

CONTENTS

23.1 INTRODUCTION

This chapter describes an original experience in GIS application to support an environmental prefeasibility study of Public Facilities as defined by the Italian Outline Law on Public Works (L 109/94 and later modifications and additions). The Law introduces the feasibility study as a key step in the decisional process on public investments, having to provide all the elements enabling the administration to make informed decisions before planning start-up. The environmental prefeasibility is a component of the study that assumes specific relevance for their natural heritage in particularly valuable areas such as the Eolian Islands.

The volcanic Eolian Archipelago is in fact the only Italian naturalistic site belonging to the UNESCO World Heritage List, as an "outstanding record of volcanic island-building and destruction, and ongoing volcanic phenomena."

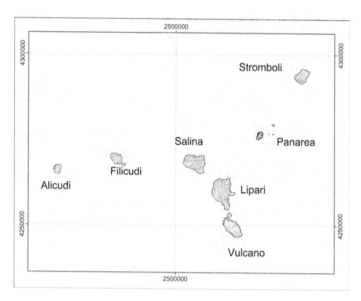

FIGURE 23.1 Eolian Archipelago.

This volcanologic peculiarity joins important naturalistic and landscape conno-
tations and significant portions of the islands are included in the European Natura
2000 Network as Community Importance Sites and Special Protection Zones. Furthermore, Regional Natural Reserves have been established in all the islands aiming
at the conservation of the natural heritage.

The Eolian Archipelago is composed of seven islands standing in the southeast
Tyrrhenian Sea about 40 km off the Sicilian coast. Actually, there are ten volcanoes
building over the deep-sea plain, forming a ring of which only seven emerge from
water. In decreasing size order, the largest islands are Lipari, Salina, and Vulcano,
followed, with their smaller extent, by Stromboli, Filicudi, Alicudi, and Panarea
(Figure 23.1). Tourism is a vital part of the archipelago economy and, as in many
other small Mediterranean islands, the planning and management of limited environmental resources is a special challenge [1]. In this context the total lack of natural
water sources and a water supply system based largely on tankers has become an
ever-more-pressing issue, together with the need to treat wastewater in a sustainable
manner. A sustainable management of water resources is, on the other hand, "an
important determinant in the location of a tourism enterprise or ensuring the viability
of existing operations" [2], and a specific "competitiveness factor," too [3], especially
in small islands where the deterioration of fragile ecosystems is the principal cause
of destination's decline.

The Italian Ministry of Environment, in the frame of the "Sustainable Development of Italian Minor Islands" objective of a more comprehensive Program Agreement, entrusted to ENEA (Italian National Agency for New Technologies, Energy
and Environment) a feasibility study to find out the infrastructures required for sustainable and integrated management of Eolian Islands water resources, to be realized
through project financing. The goal was, therefore, to remove the inefficiencies of the

present supply system, to increase the qualitative standard of the water facilities, and to identify new and up-to-date infrastructures that can locally produce drinking water and provide the collection and treatment of waste water, providing all the elements to enable the administration to find a promoter.

These integrated infrastructures certainly contribute to the environmental sustainability of the Eolian Islands, but a careful assessment of their environmental compatibility is essential, in order not to compromise further this particularly fragile territory.

For this reason the environmental prefeasibility, beyond a preliminary description of the state of the environment, assumed the macro-localization aspects as the main point of view through which to analyze the environmental implications of the proposed solutions.

In the frame of works to be performed for the integrated management of Eolian Islands water resources, the final goal was to achieve a geographic data management and analysis able to effectively support management options and location alternatives, considering both planning regulations and constraints and potential impacts on biodiversity and environmental matrixes' quality. Therefore a spatial analysis application has been developed using GIS, which has made it possible to map the macro-localization potentiality of water infrastructures for each Eolian Island belonging to Lipari Municipality.

This application, which the promoter could further update, has laid the foundation for the assessment of works consistency to the land planning framework and also of the potential impacts and compensation measures to be foreseen.

23.2 PLANNING IN THE EOLIAN ISLANDS

The normative rules established in the Eolian Islands to preserve natural and cultural heritage have not been supported by proactive management able to exploit the sustainable use of the environment, both protecting the basic resources of the Eolian economy, and identifying innovative opportunities for excessive seasonal tourism.

The lack of accurate and well-balanced land and urban planning is indeed one of the reasons for the unsuccessful control and minimization of the negative impacts caused by the increasing mass tourism which has raised the local community's well-being but, in the long run, does not assure further improvements or natural resource preservation. Vested interests, which delay development and approval of the urban plans, and the lack of management tools for the natural reserves, when not suspended, have raised illegal building phenomena with little attention to natural resource conservation and sustainable use.

Against this background, the Territorial Landscape Plan of the Eolian Islands [4], approved in February 2001 according to the National ACT 431/85, meets with strong opposition by local authorities, who filed an appeal to the Regional Administrative Court (TAR) of Sicily, which involved the Constitutional Court (which already passed a sentence in favor of the Plan). While waiting for the TAR judgment, the Territorial Landscape Plan is fully in force.

The Plan approach is centered on Eolian volcanism as the "decisive factor of the anthropic settlement in the Eolie and then of all cultural heritage related to the

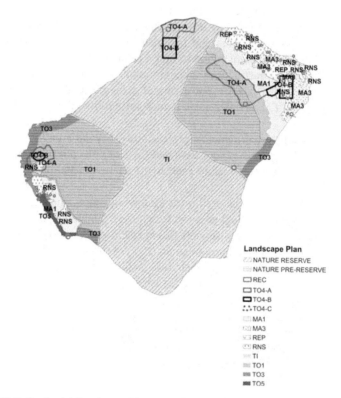

FIGURE 23.2 Territorial Landscape Plan of Stromboli Island.

human process." The main Plan category, called "Landscape Forming Cultural Heritage," is therefore represented by morphologic-volcanic-tectonic territorial systems, by wide patterns of the morphologic-volcanic-tectonic landscape, by significant volcanic elements, and by geomorphologic posteruptive territorial systems.

The other important category, called "Landscape Featuring Cultural Heritage," is represented by vegetative, faunal, anthropic, and historical heritage and archaeological heritage systems (Figure 23.2).

Starting from the identification, hierarchization, and localization of these categories as territorial context, normative rules are applied regulating the compatible activities, the compatible activities for recovery interventions only, and the incompatible activities (i.e., not pertaining to the given territorial context).

Specific rules concern "works significantly transforming the territory" (Law Art. 39, 40–41, Title IV), including technological infrastructures such as desalinators, water conditioners, and reservoirs, requiring an environmental and landscape impact assessment inclusive of compositional and formal factors and infrastructure executive details, with reference to historical and environmental features (geological, ecological, botanical, faunal) dealing with the territorial context involved.

The localization of these infrastructures is allowed in the zones concerned with transformation processes (normative rules: RIO, MO1, MO2, TR., etc.) but also elsewhere, providing that these infrastructures are objectively essential and that it is objectively impossible to locate them in the previous zones (Art. 47).

23.3 LAND USE AND VEGETATION IN THE EOLIAN ISLANDS

Land use and acknowledgement of vegetation species with a high degree of wildlife and landscape value are outstanding topics in order to describe the state of the environment, and in order to detect the prospective macro-localization solutions.

In 1992, the Municipality of Lipari implemented an agronomic and forest survey of the six Eolian islands to support the development of the Urban Plan. [5]

The high degree of detail of the survey (1:10,000 scale) makes it a valuable tool in the assessment of any kind of intervention, because it allows reading recent outcomes of evolutionary processes that have conditioned and modified land use in a meaningful way. It also allows evaluation of the consistency (which appears residual already at the time of the study) and distribution of agricultural grounds, for prospective strategies of reclaimed water reuse (Figure 23.3).

Table 23.1 shows a concise distribution of each island's surface according to the most significant classes of land use and vegetation types. It underlines the residual farming ground surfaces opposed to the growing weight of the Mediterranean maquis, a vegetative association composed of numerous (often endemic) high wild-life value species.

23.4 THE GIS APPLICATION

Starting from the two thematic bases mentioned earlier, a macro-localization suit-ability map of water infrastructures has been drawn for each Eolian island, except Salina Island due to its incomplete data set information (at present only the Territorial Landscape Plan is available). The first step had been particularly work intensive because it was necessary to develop a geo-database with the acquired data, which were not geo-referenced and, in the case of the landscape plan, were CAD data. [6,7]

The second step was to classify the spatial zoning of the landscape plan, through a close analysis of normative rules, on the basis of location compatibility of new infrastructures and facilities (Table 23.2).

This second step has then allowed creation of a generalized classification in terms of macro-localization suitability and its mapping (Figure 23.4).

In the same way, starting from the Agricultural and Forest Study of the six Eolian Islands, a classification of land use and vegetation has been created, classifying the localization suitability for the infrastructures and installations potentially impacting on the environment. The criterion, in this case, is to adopt an inverse scale, compared to the natural and landscape value scale used in the study, for each single class and for each island. In brief, a class with a high natural level has been linked to a low or null localization suitability (i.e., maquis or garrigue associations where the prev-alent vegetation is characterized by endemic species with particular importance) [8].

In contrast, a class was connected to high or medium localization suitability value if it was an urban settlement or a deteriorated vegetation area. The only

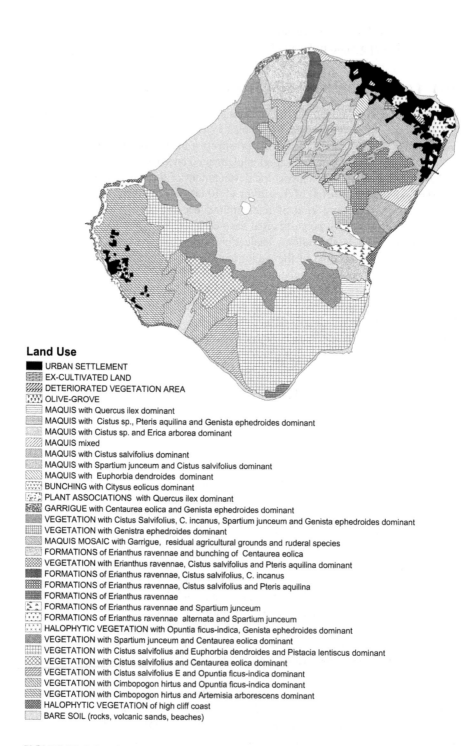

Land Use

- URBAN SETTLEMENT
- EX-CULTIVATED LAND
- DETERIORATED VEGETATION AREA
- OLIVE-GROVE
- MAQUIS with Quercus ilex dominant
- MAQUIS with Cistus sp., Pteris aquilina and Genista ephedroides dominant
- MAQUIS with Cistus sp. and Erica arborea dominant
- MAQUIS mixed
- MAQUIS with Cistus salvifolius dominant
- MAQUIS with Spartium junceum and Cistus salvifolius dominant
- MAQUIS with Euphorbia dendroides dominant
- BUNCHING with Citysus eolicus dominant
- PLANT ASSOCIATIONS with Quercus ilex dominant
- GARRIGUE with Centaurea eolica and Genista ephedroides dominant
- VEGETATION with Cistus Salvifolius, C. incanus, Spartium junceum and Genista ephedroides dominant
- VEGETATION with Genistra ephedroides dominant
- MAQUIS MOSAIC with Garrigue, residual agricultural grounds and ruderal species
- FORMATIONS of Erianthus ravennae and bunching of Centaurea eolica
- VEGETATION with Erianthus ravennae, Cistus salvifolius and Pteris aquilina dominant
- FORMATIONS of Erianthus ravennae, Cistus salvifolius, C. incanus
- FORMATIONS of Erianthus ravennae, Cistus salvifolius and Pteris aquilina
- FORMATIONS of Erianthus ravennae
- FORMATIONS of Erianthus ravennae and Spartium junceum
- FORMATIONS of Erianthus ravennae alternata and Spartium junceum
- HALOPHYTIC VEGETATION with Opuntia ficus-indica, Genista ephedroides dominant
- VEGETATION with Spartium junceum and Centaurea eolica dominant
- VEGETATION with Cistus salvifolius and Euphorbia dendroides and Pistacia lentiscus dominant
- VEGETATION with Cistus salvifolius and Centaurea eolica dominant
- VEGETATION with Cistus salvifolius E and Opuntia ficus-indica dominant
- VEGETATION with Cimbopogon hirtus and Opuntia ficus-indica dominant
- VEGETATION with Cimbopogon hirtus and Artemisia arborescens dominant
- HALOPHYTIC VEGETATION of high cliff coast
- BARE SOIL (rocks, volcanic sands, beaches)

FIGURE 23.3 Land use and vegetation of Stromboli Island.

TABLE 23.1
Prevalent Land Use Classes and Vegetation Sorts in Eolian Islands

ISLAND	Surface Area (Hectares)	Farming Grounds (%)	Maquis (%)	Garrigue (%)	Areas With High Level of Anthropization[a] (%)
Lipari	3749.7	15.5	33.1	21.0	6.6
Vulcano	2097.6	1.0	37.8	19.0	7.7
Stromboli	1266.4	0.2	10.5	6.0	4.4
Filicudi	932.7	1.0	39.9	45.7	2.1
Alicudi	508.2	—	38.0	36.9	1.9
Panarea	336.8	1.1	35.5	21.4	10.3

[a] Surfaces include enclosed green areas and small farmlands.

exception to criteria previously seen has been to assign a medium potentiality (and not high, as derived from the low naturalistic and landscape value) to the classes connected to local, residual, and typical culture (vineyard and olive-grove) for their relevance in typical local productions and for their potential recovery of environmentally sustainable productive traditions.

With this method a second thematic layer has been created for the six Eolian Islands in terms of macro-localization suitability of new water infrastructures on the basis of results of the Agricultural and Forest Study (Figure 23.5).

Starting from the previous thematic classifications, through spatial overlay procedures driven by logical criteria on the attributes of geographic features, a macro-localization suitability map of water infrastructures has been made for each island belonging to Lipari Municipality (Figure 23.6). The first step was to produce new polygon coverages deriving from the spatial overlay of the coverages first created for the Territorial Landscape Plan and for the Agricultural and Forest Study. At this stage, while the high number of polygons derived from the precision and positional accuracy of geometric characteristics of input maps rather than real spatial variability, there already was a strong matching asset between the results associated to the landscape plan and the Agricultural and Forest Study, confirming the naturalistic values within the Plan itself. However, because the Plan determinations are affected by safeguard requirements at different levels, the various potentialities of localization associated with different classes of land use and a wide series of environmental and cultural heritage have also allowed grading, under natural profile, the localization potentiality for the areas forbidden in the Landscape Plan.

This opportunity appears of particular importance when the localization alternatives could be impracticable in higher potentiality areas (possibility, as previously highlighted, also foreseen by the Plan itself for objectively essential facilities) and it should be analyzed in more depth for a full consideration of biodiversity patterns in the Eolian Archipelago.

The second step was to create a classification of the resulting polygons, representing the evidence drawn and at the same time useful to read the final theme created. So, the criteria have been to keep the minimum and maximum of suitability,

TABLE 23.2
Classification of Eolian Islands Landscape Plan Spatial Zoning

Rules	Description	Infrastructures/Facilities	Suitability
TV	Art. 10 Volcanic protection (Ti+To+Ts)	No compatibility	Null
TI	Art. 11 Integral protection of the natural ecologic system	No compatibility	Null
TO1	Art. 13 Protection oriented to agricultural and productive areas	No compatibility	Null
TO2	Art. 14 Protection oriented to ludic activities	No compatibility	Null
TO3	Art. 15 Protection oriented to termal, talasso-termal, therapeutic, and recreational fruition and to social and public usefulness fruition of the sea	No compatibility	Null
TO4[a]	Art. 16 Protection oriented to the archeological and archeotermal landscape exploitation	Prevailing more restrictive rules provided by the competent departments of Sovrintendenza ai Beni Archeologici, according to national Law 490/99, in these areas can be exclusively carried out trasformations compatible with the protection level and activity tipology related to the territorial context involved	Null
TO5	Art. 17 Protection oriented to the environment restoration	No compatibility	Null
TS1	Art. 18 Special protection 1: Vulcano Terme di Levante, Acque Calde	No compatibility	Null
TS2	Art. 19 Special protection 2: Pilato III	No compatibility	Null
TS3	Art. 20 Special protection 3: Papesca—Porticello, Acquacalda	No compatibility	Null
RP1	Art. 21 Urban settlements to be reorganized through landscape specified and ruled parts	No compatibility until new urban plans approval	Average with urban plan reference
RCS	Art. 22 Urban historical centers recovery		Average with urban plan reference

RNS	Art. 23 Urban generating nuclei recovery	No compatibility until new Urban Plans approval	Average with urban plan reference
REP	Art. 24 Urban propagation recovery according to the historical pathway patterns	No compatibility until new Urban Plans approval	Average with urban plan reference
ZM1	Art. 25 Mining Zone 1 with plan normative rules	No compatibility	Null
ZM2	Art. 26 Mining Zone 2 with plan normative rules	No compatibility	Null
MA1	Art. 27 Preservation of rural landscape arrangement in areas enclosed between Vulcanic Protection zones (TV) and zones at different level of antropization	No compatibility	Low with urban plan reference and landscape compatibilization
MA2	Art. 28 Preservation of urban landscape (ex-buffer zones not suitable for further building)	No compatibility	Low with urban plan reference
MA3	Art. 29 Preservation of not suitable for further building areas with high landscape value and strategic functions	No compatibility	Null
RIO	Art. 30 Landscape rearrangement through Detailed Plans	No compatibility until new urban plans approval	High with urban plan reference
MO1	Art. 31 Compatible modifications of agricultural landscape (agritourism)	No compatibility until new urban plans approval	Low with urban plan reference and landscape compatibilization
MO2	Art. 32 Compatible modifications of extra-urban and peri-urban landscape	No compatibility until new urban plans approval	High with urban plan reference and landscape compatibilization
TR	Art. 33 Compatible transformation (Urban Plan)	No compatibility until new urban plans approval	High with urban plan reference
RES	Art. 34 Areal landscape restoration with indication of elements to compatibilize (DP1-DP2)	No compatibility	Low with urban plan reference and landscape compatibilization
REC	Art. 36 Conservative building recovery oriented to cultural-productive re-use	New transformations are exclusively compatible with the protection level and activity typology related to the territorial context where the buildings and feasibilities are plugged-in	Null

TABLE 23.2
Classification of Eolian Islands Landscape Plan Spatial Zoning (continued)

Rules	Description	Infrastructures/Facilities	Suitability
DP7	Art. 35 Point landscape restoration with indication of typological elements to compatibilize (amenity perceptive DP)	No compatibility	Null
DP6	Art. 35 Point landscape restoration with indication of typological elements to compatibilize (infrastructural DP)	No compatibility	Null
DP5	Art. 35 Point landscape restoration with indication of typological elements to compatibilize (environment polluting DP)	No compatibility	Null
DP4	Art. 35 Point landscape restoration with indication of typological elements to compatibilize (ablative structural DP)	No compatibility	Null
DP3	Art. 35 Point landscape restoration with indication of typological elements to compatibilize (additive structural DP)	No compatibility	Null

[a] This normative rule has been further disaggregated in the Figure 23.2 as TO4-A, archeological areas; TO4-B, archeological potentiality areas; TO4-C, delimited archeological areas; TO4-D, isolated emergencies; and TO4-E, archeo-termal areas.

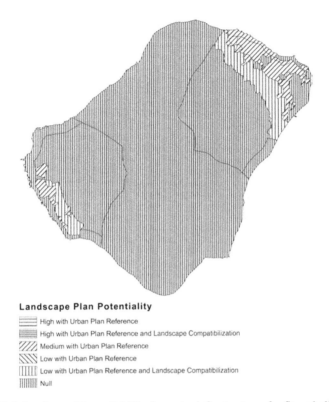

Landscape Plan Potentiality

☰ High with Urban Plan Reference
☰ High with Urban Plan Reference and Landscape Compatibilization
▨ Medium with Urban Plan Reference
▧ Low with Urban Plan Reference
▥ Low with Urban Plan Reference and Landscape Compatibilization
▥ Null

FIGURE 23.4 Landscape Plan suitability for water infrastructures for Stromboli Island.

according to the higher convergences (positive or negative) of the values assigned to various areas by the two input instruments, and to graduate the potentiality of lesser convergence areas, giving a higher weight to the Landscape Plan.

23.5 CONCLUSIONS

The procedure developed and the results achieved are, certainly, a preparatory step for a concrete localization hypothesis that must allow for other variables, both under a technology/infrastructure profile and an environmental profile. The positioning of the existing water networks on the final maps could also be a new subsequent step in a location analysis that will be taken into account by the promoter in a proper scale.

The evaluation of the real free surfaces suitable for water infrastructure localization inside areas with higher suitability (i.e., through a buffering analysis of the existing buildings) gives much important information both for a closer localization analysis and for compensative actions to provide for impact minimization.

The procedure is also a first step of a decision-making support tool based on a more formalized multicriteria suitability analysis [9]. An ex-ante assessment of environmental compatibility during first stages of public works planning stages is necessarily of a running type and assumes great relevance also as informative support to start up a participatory process. This appears particularly advisable in contexts,

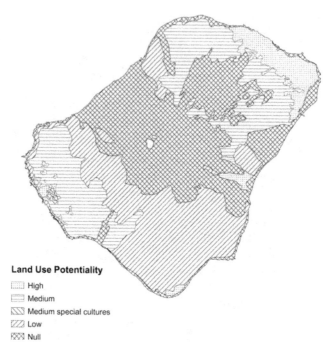

FIGURE 23.5 Land use suitability for water infrastructures for Stromboli Island.

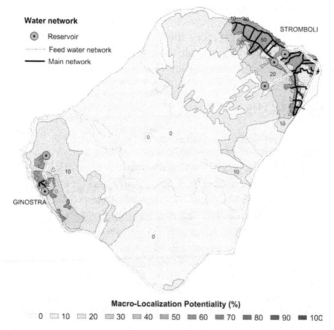

FIGURE 23.6 Macro-location suitability of water infrastructure for Stromboli Island.

such as the Eolian Islands, where conflicting views about development paths persist and transparent information could help in the settlement of potentially diverging interests.

REFERENCES

1. Pigram, J.J., Water resources management in island environments: the challenge of tourism development, *Tourism* 49 (3), 267, 2001.
2. United Nations Environment Programme, *A Manual for Water and Waste Management: What the Tourism Industry Can Do to Improve Its Performance*, UNEP, 2003.
3. Mihalic, T., Environmental management of a tourist destination. A factor of tourism competitiveness, *Tourism Manage.*, 21, 65, 2000.
4. Regione Sicilia—Assessorato Beni Culturali ed Ambientali e Pubblica Istruzione, "Piano Territoriale Paesistico dell'arcipelago delle Isole Eolie," Scientific Coordinator Prof. Vincenzo Cabianca, 2001.
5. Amministrazione Comunale di Lipari, "Studio Agricolo Forestale delle Isole del territorio comunale," Scientific Supervisor Prof. G. Asciuto, 1993.
6. Compagnia Generale Ripreseaeree (CGR), Digital color orthophoto at nominal scale 1:10.000, Program IT2000, Volo 1998–1999.
7. Regione Sicilia—Assessorato Territorio e Ambiente, Carta Tecnica Regionale in scala 1:10000, formato raster in coordinate piane riferite al sistema nazionale Gauss-Boaga, 1986–1987.
8. Lo Cascio, P. and Navarra, E., Guida Naturalistica alle Isole Eolie, L'EPOS, Palermo, 1997.
9. Baja, S., Chapman, D.M. and Dragovich, D., Using GIS-based continuous methods for assessing agricultural land-use potential in sloping areas, *Environ. Plann. B*, 29, 3, 2002.

24 Influence of Data Quality on Solar Radiation Modeling

Tomaž Podobnikar, Krištof Oštir, and Klemen Zakšek

CONTENTS

24.1 INTRODUCTION

The Sun is the main energy source for all living beings on Earth, and it enables photosynthesis and life. It provides natural influence on the Earth's atmosphere and climate. Solar energy affects all physical, chemical, and biological processes in the terrestrial ecosystem. One of its products is also oil, the energy source of our economy, which is (compared to the length of our lives) nonrenewable. On the other hand, "pure" solar radiation is almost unlimited and is considered to be the energy of the future.

The duration of solar radiation and the energy that reaches the ground thus represent a very important piece of spatial information. Different aspects of solar radiation research are significant for meteorologists, foresters, agronomists, geographers, and other practitioners. Solar radiation is commonly measured with pyranometers, which measure the incoming shortwave radiation at a constant angle in the shape of a hemisphere oriented upwards. Recently it has also been measured from satellites orbiting the Earth. For each location on the Earth's surface, an energy budget calculation is made using hourly visible radiation information obtained from Japan's Geostationary Meteorological Satellite GMS-5, with 6-km data resolution. The U.S. SORCE satellite recently started to provide measurements of incoming radiation. One can also obtain solar data from NASA data sets, which provide solar energy data with 1° resolution over the globe.

Solar radiation measurements provide data with huge spatial dispersion and possible gross errors. If we desire more precise data, available for any period, we have to model it. Solar radiation energy depends mostly on the incidence angle, which is defined by astronomical and surface parameters. Simple modules for computing solar radiation, which require only surface data and the zenith angle of the Sun, are available in some geographical information systems (GIS) programs. More sophisticated models and software also take into consideration the astronomical, surface, and meteorological influences [1].

In the late 1970s research regarding solar radiation modeling in Slovenia began with the study "Spatial Dispersion of Solar Potential in Slovenia" [2]. Later, a map of solar radiation in Slovenia was produced for the Slovenian Geographical Atlas [3]. The study presented in this chapter has been developed aiming at creating a more accurate model using supplementary and higher quality input data. The current model has been improved, and the significance of particular input data regarding the expected quality of the results was evaluated. Special attention was paid to the preparation of a digital elevation model (DEM) using advanced methods of radar interferometry [4] and data fusion [5].

The model was tested and implemented for all of Slovenia. Surface data was calculated from the digital elevation model with the resolution of 25 m (DEM 25).

24.2 SOLAR RADIATION MODEL

In general, the solar radiation energy is influenced by astronomical, surface, and meteorological parameters. Many general GIS and dedicated applications include tools for solar radiation modeling. The main difference between them lies in the meteorological part, in which equations are mostly empirically determined. Therefore, a variation within the results can be expected when using different models.

Most of the commercial GIS packages do not have modules for solar energy computation, or have only basic solar programs with simplified models, and one has to program its own functions and models. The programs are rather good for shadow determination and incidence-angle calculation. On the other hand, astronomical and meteorological parts are simplified. In most cases the astronomical part considers that the Earth travels around the Sun in a circle and not in an ellipse. The meteorological model is usually replaced by a constant, which represents the percentage of unabsorbed radiation. The reason for such poor meteorological modeling lies in the difficulty of gathering accurate data. In contrast, there are usually no problems with the availability of surface, such as digital elevation model (DEM), and astronomical data.

Complex meteorological models take into consideration more parameters, such as temperature, relative humidity, and atmospheric pressure [6]. The most sophisticated models also include transmission coefficients based on atmospheric composition (vapor water, ozone). Even snow depth can be important, because it can be used to estimate albedo, which can have an important role in diffuse radiation. However, some sophisticated models take too long processing time, or they are specialized for different tasks, like Meteonorm's calculation of global solar energy that does not consider surface data. If one wanted only general information on solar energy on a specified area, it would be better to use a less complex model. The use of the most complex models is therefore suitable only for smaller areas and over a short period. The general model, on the other hand, should be sophisticated enough, yet at the same time optimized so it could be used also for larger areas and a long period.

24.2.1 THEORETICAL BACKGROUND

Solar energy drives the Earth's climate. Slight variations in solar radiance could offset or increase global warming. The 23.5° angle between the Earth's spin axis and its orbit around the Sun results in the seasonal cycle, causing the length of day and the sunlight angle to vary during the year. As a result, summer is much warmer than winter, and the polar regions are dramatically colder than the tropics. The Earth's orbit around the Sun is an ellipse; thus it is slightly closer in early January than in July. This results in about 7% more solar radiation reaching the Earth in January than in July. Both of these seasonal effects, the tilt of the Earth's axis, and the orbital distance to the Sun, are stable and predictable. The energy of quasi-global radiation is influenced by the astronomical, surface, and meteorological parameters.

24.2.1.1 Astronomical Parameters

An incidence angle of the Sun can be calculated if the normal vector to the surface tangent plane and a vector in the direction of the Sun were previously determined. Both vectors were defined in the global geocentric Cartesian coordinate system. Input data enables the calculation of the geographical coordinates, ellipsoidal heights, slopes, and expositions. The normalized normal vector $\vec{n}$ to the surface is given by:

$$\vec{n} = \begin{bmatrix} -\cos\lambda & -\sin\lambda & 0 \\ -\sin\lambda & \cos\lambda & 0 \\ 0 & 0 & 1 \end{bmatrix} \cdot \begin{bmatrix} \sin\varphi & 0 & -\cos\varphi \\ 0 & 1 & 0 \\ \cos\varphi & 0 & \sin\varphi \end{bmatrix} \cdot \begin{bmatrix} \sin\eta\cdot\cos\alpha \\ \sin\eta\cdot\sin\alpha \\ \cos\eta \end{bmatrix}$$

where φ is geographical (ellipsoidal) latitude, λ is geographical (ellipsoidal) longitude, η is slope, and α is exposition.

The vector of sunrays can be defined if the virtual Sun motion is simulated. An observer on Earth can discover that the position of the Sun constantly changes. Therefore, the vector of sunrays is a function of time and the Sun's declination. The normalized vector of sunrays $\vec{s}$ is given by (valid for Central European Time [CET]):

$$\vec{s} = \begin{bmatrix} \cos\delta\cdot\sin\beta \\ \cos\delta\cdot\cos\beta \\ \sin\delta \end{bmatrix}$$

where δ is Sun declination, t is time (CET; in hours), and β is the time angle defined by:

$$\beta = \frac{\pi\cdot\left(t-7^h\right)}{12^h}$$

The angle between the normal vector to the surface and the vector of sunrays can be computed with a scalar product. If both vectors are normalized, then the value of the scalar product equals the cosine of the angle between the selected vectors.

24.2.1.2 Surface Parameters

As the most important surface parameter, geomorphology influences the effective possible duration of solar radiation, that is the period of sunshine in clear weather, subtracted by the period of shade due to relief obstacles. The surface exposed to the Sun receives not merely the diffused radiation, like the surface in the shadow, but also the energy of direct radiation. It is therefore important to know which part of the surface is in the shadow. Geomorphology also influences the solar radiation through slope and exposition (incidence angle).

There are two types of shadows: hill shade and cast shade. The hill shade occurs when the surface is oriented away from the light source. The hill shade can be easily determined if the incidence angle is previously calculated. In contrast, cast shade determination is more complex. It occurs only on surfaces with an exposition in the direction of the Sun when a geomorphological obstacle between the Sun and the surface is present (Figure 24.1).

The surface could be exposed to the Sun when the incidence angle is smaller than 90° (α_1). If sunrays are almost parallel to the surface (incidence angle is close to 90°), then the surface is only partially insulated due to vegetation, micro-geomorphology and man-made obstacles (α_2). If the incidence angle is over 90°, the surface is in the shadow (α_3) hill shade. Due to geomorphological obstacles, some parts of the surface could be in the shadow even if the incidence angle is smaller than 90° (α_4) cast shade. Solar radiation can be direct or diffuse. The surfaces in the shadows receive only energy from diffused radiation, which is normally at least twice as small as the energy of direct solar radiation.

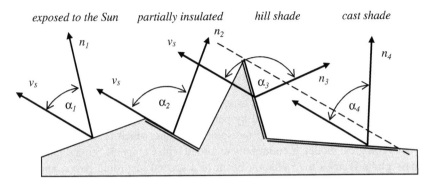

FIGURE 24.1 Incidence angle and shadows of solar radiation with regard to geomorphology.

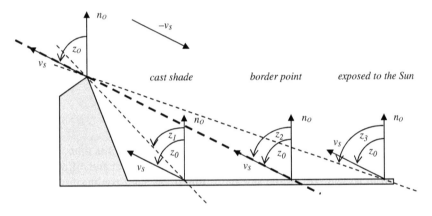

FIGURE 24.2 Cast shade determination.

The first step in the cast shade determination is finding the top of the obstacle. The obstacle's highest point is the point at which the hill shade occurs first. This is where the incidence angle becomes smaller than 90°. After finding a potential obstacle (Figure 24.2) the algorithm moves through DEM in the direction closest to the azimuth of sunrays. The azimuths to neighboring cells are calculated and then compared with the azimuth of the sunrays.

If the zenith distance from the standing point to the top of the obstacle (z_1) is smaller than the zenith distance of the Sun, then the surface is in the cast shade. If the zenith distance to the peak of the obstacle (z_3) is greater than the zenith distance of the Sun, then the surface is exposed to the Sun. If the standing point is on the border between the illuminated area and shadow, then the zenith distance to the top of the obstacle (z_2) equals the zenith distance of the Sun. Thus, the key to determining the cast shade is in finding the border:

$$\frac{\pi}{2} - \arctan\left(\frac{h_0 - h}{d}\right) = \arccos\left(\vec{s} \cdot \vec{n}_E\right)$$

where h_0 is obstacle height, h is standing point height, d is horizontal distance between top and standing point, $\vec{s}$ is vector of the sunrays, and $\vec{n}_E$ is normal vector to Earth's ellipsoid.

24.2.1.3 Meteorological Parameters

Clouds and fog can be understood as a filter, which increases the dispersal in the atmosphere. The climate is studied by parameters obtained by measures at meteorological stations. The duration of the solar radiation, which is most important for the meteorological model, is commonly recorded with Campbell-Stokes heliographs. A glass globe focuses the radiation beam to a special recording paper, and as the Sun moves, a trace is burned onto the paper. When no sunshine is sensed, no records occur. Climate data is important for prediction over a longer period. On the other hand, weather data is important in the analyses of a specific shorter period.

The quasi-global radiation energy is the amount of energy that is received by a random inclined surface in a specific time interval. The daily energy of quasi-global radiation E_{qr} is theoretically given by time integration of quasi-global radiation surface density from sunrise to sunset by:

$$E_{qr} = \int_{\text{sunrise}}^{\text{sunset}} j_{qr} \cdot dt$$

where j_{qr} is the surface density of quasi-global radiation, and t is time.

Surface density of quasi-global radiation is a sum of direct and diffuse radiation coming from a clear part of the sky and diffused radiation coming from a cloudy part of the sky. Equation for j_{ko} was derived by Hočevar [2]:

$$j_{qr} = I_o \cdot \rho^2 \cdot q_a^m \cdot \left(q_s^m \cdot (\vec{n} \cdot \vec{s}) \cdot D + 0,5 \cdot \cos^2 \frac{n}{2} \cdot (1 - q_s^m) \cdot \cos^{4/3} z \cdot (D + (1 - D) \cdot C) \right)$$

where I_0 is a solar constant, ρ is Earth-Sun distance in astronomical units, q_a are transmission coefficients relative to absorption, q_s are transmission coefficients relative to dispersion, D is relative duration of solar radiation, $\vec{n}$ is surface normal vector, $\vec{s}$ is Sun direction, C depends on the type of clouds and the zenith angle, n is slope of the surface, z is zenith distance of the Sun, and m is optical mass.

24.2.2 Application Model of Solar Radiation

The described model was implemented and tested for all of Slovenia (area of 20,273 km²). A grid cell of 25 by 25 m was selected due to the availability of DEM and adequate processing speed. The calculation of quasi-global radiation energy was numerically simplified. Hourly values were calculated for the mean day within a ten-day period. Daily values were obtained as a sum of hourly energies and multiplied by ten, and 10-day energies were calculated. The annual quasi-global radiation energy was calculated as the sum of energy over all decades. Three types of data were used in the model: astronomical data (Sun declination, Earth–Sun distance), surface data (DEM, geoid model), meteorological data (duration of solar radiation, transmission coefficients relative to absorption and dispersion).

24.2.2.1 Astronomical Data

Sun declination (coordinate of celestial equatorial system) was used to determine the vector of sunrays. Declinations were determined to one hundredth of a minute. Earth–Sun distance was calculated as input data in the meteorological part of the model. The required values were obtained from the astronomy almanac.

24.2.2.2 Surface Data

The DEM 25 (with grid cell of 25 by 25 m) was used for surface elevations. This has been produced with enhanced modeling of data fusion of DEMs, including InSAR DEM 25 produced with radar interferometry and other global and local DEMs available for Slovenia [4,5].

The shape of the Earth was approximated as a rotational ellipsoid; therefore the coordinates had to be transformed from the national coordinate system to geographical coordinates. Geoid undulations were used to transform orthometric to ellipsoidal heights [7]. Slopes and expositions were then computed from the DEM modified with a geoid. A 5-km buffer zone around Slovenia was used to efficiently compute the shadows on low positions. All calculations have been performed over an area covering the entire country including a buffer zone of 10 km. At the end the results were clipped along the Slovene border.

24.2.2.3 Meteorological Data

The meteorological model included data about the duration of the solar radiation and transmission coefficients relative to absorption and dispersion [2]. Data for meteorological stations were obtained from the Meteorological Office of the Environmental Agency of Slovenia. The measurements cover a 30-year period between 1961 and 1990. A characteristic area surrounding each station was determined in order to produce 24 climate regions. The meteorological model used in modeling is complex but, due to processing issues, still rather simplified. It distinguishes among direct and diffuse solar radiation and considers the following influences: zenith angle of the Sun, duration of solar radiation, type of clouds, and transmission coefficients.

24.2.2.4 Results of the Modeling

The result of the radiation modeling is a data set of annual energy of quasi-global radiation for Slovenia. The mean value of the quasi-global radiation energy for the entire country equals 4020 MJm^{-2} with a standard deviation of 520 MJm^{-2}. According to our model the most illuminated surface in Slovenia is next to the village of Sočerga (5360 MJm^{-2}) and the least illuminated surface is under the northern wall of Mount Triglav (840 MJm^{-2}). The coastal region in the Southwest receives the most solar energy and the Alpine region in the Northwest the least (Figure 24.3).

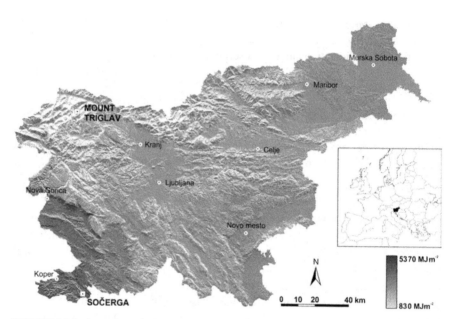

FIGURE 24.3 Annual quasi-global radiation energy for Slovenia; minimum at Mount Triglav and maximum at Sočerga.

The results clearly show that the energy of quasi-global radiation is mainly affected by surface data geomorphology. The influence of the surface on the solar energy radiation was therefore further evaluated. The north-facing slopes receive significantly less quasi-global radiation energy (3600 MJm^{-2}) than the south-facing slopes (4400 MJm^{-2}). The quasi-global radiation energy on the south side also has a much smaller standard deviation (260 MJm^{-2}) than the energy on the north exposures (620 MJm^{-2}). The slopes facing east or west do not differentiate to a great extent. The surface geomorphology also depends on the surface roughness and slope [5]. Plains receive the most quasi-global radiation energy in Slovenia (mean value 4200 MJm^{-2}, standard deviation 130 MJm^{-2}) and mountains the least (mean value 3560 MJm^{-2}, standard deviation 880 MJm^{-2}).

Slovenia receives the largest amount of solar energy in July (mean value 580 MJm^{-2}) and the smallest in December (mean value 70 MJm^{-2}). As expected, the quasi-global radiation energy changes the fastest around both equinoxes and the slowest around both solstices.

24.3 QUALITY EVALUATION OF THE MODEL

In order to determine which parameters have the highest effect on the results and their quality, we performed a detailed evaluation of the applied model. As previously described, numerous parameters that might influence solar energy are not taken into consideration. Heights of the surface obstacles are not known and the Earth curvature is neglected in shadow determination. These two influences might lead to local errors, but when considering larger areas they can be considered not significant. The meteorological part and the astronomical part of the model could also be improved.

As already noted there are three main groups of parameters used in the model: astronomical, surface, and meteorological. Precise astronomical parameters are important only together with high-quality surface data set (DEM). If surface data would be coarse, even very general astronomical parameters would be sufficient. If the application area is small, the same astronomical parameters could be used for the entire area. In contrast, in large areas that are significantly spreading in the north–south direction, different astronomical parameters are required. It is clear that surface parameters do not dominate in large areas, but they are important in small areas with a rough surface. If, for example, one were interested in larger areas (several 100 km^2), a selected DEM 25 or even coarser would not greatly affect the results. However, if the conditions are evaluated on smaller local areas (just a few cells) then more precise DEM is needed. The size of the area is a relative term in evaluating meteorological circumstances, because the meteorological condition also depends on the geomorphology. If the surface is rough (typical for Slovenia), the weather and climate can change very rapidly over small distances. Thus, one must consider Slovenia as a large area.

Usually methods for modeling error propagation are used for evaluating the quality of the application [8]. This is not an easy task in our application, because there are a large number of variables with complex mathematical operations used in the model. The quality of the application results was estimated through a simulation of the most probable error of the important input data sets (variables). After considering all of the influences one can make the final evaluation of the quality of the application.

24.3.1 INFLUENCE OF ASTRONOMICAL DATA

Astronomical parameters were evaluated at the beginning. The standard deviation of the Sun declination equals one hundredth of an angle minute. Even when the Sun declination was changed for the value of its standard deviation, the mean value of quasi-global radiation energy hardly changed at all (difference was only 0.01%). Therefore, we can conclude that astronomical parameters have the smallest influence on quasi-global radiation energy.

24.3.2 INFLUENCE OF SURFACE

For the solar radiation modeling, a high-quality DEM (which is the most important data set for solar energy simulation) was used. The first step of the high-quality DEM modeling was InSAR DEM 25, which was produced with radar interferometry from ERS-1 and ERS-2 radar imagery of the European Space Agency. The model has a spatial resolution of 25 m and a vertical accuracy of 1.9 m in the planes, 5.2 m in the hills, and 13.8 m in the mountains, with an average of 5 m for all of Slovenia. InSAR DEM 25 was one of the first successful applications of radar interferometry imagery for large areas [4].

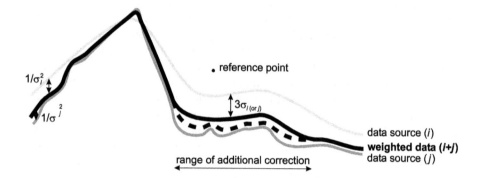

FIGURE 24.4 Interpolation of DEM 25 with the method of weighted sum of data sources with geomorphologic enhancement.

The DEM used, marked as DEM 25 was produced through a fusion of some other different quality data sets. For DEM 25 modeling, numerous spatial databases which are available nowadays were used: InSAR DEM 25, another photogrammetrically produced DEM 25, DEM 100, contour lines in scales 1:25,000 and 1:5000, the hydrology network, building database, various geodetic points, local data sets, etc. Numerous data sets were provided by the Surveying and Mapping Authority of the Republic of Slovenia. The quantity of data sets carries a potential for an enhanced production of DEM with data fusion that is of better quality than the one currently obtainable. Two methods for the integration of different data sets were proposed. The first, simultaneous data interpolation, is based on the simultaneous interpolation of all useful spatial data sources, which is supported by a proper prediction function according to the weights. The second, weighted sum of data with geomorphologic enhancement, sequentially combines individual data sets according to their weights and is maintained by geomorphologic enhancement. In both methods the weights depend on the output of the used spatial data set's quality control. With the second method we produced a DEM 25 in which heights H_{i+j} (considering weights of individual data sets w_i and w_j) were calculated as:

$$H_{i+j} = \frac{w_i H_i + w_j H_j}{w_i + w_j} = \frac{\sigma_j^2}{\sigma_i^2 + \sigma_j^2} H_i + \frac{\sigma_i^2}{\sigma_i^2 + \sigma_j^2} H_j$$

regarding variances (calculated from the evaluated error) σ_i^2 and σ_j^2. The quality of DEM 25 was evaluated with a number of developed statistical and visual methods. Its vertical accuracy is 0.9 m in the planes, 3.3 m in the hills, and 7.7 m in the mountains, presenting an average of 3.2 m for all of Slovenia with a mean error of 0.6 m [5] (Figure 24.4).

The quality of DEM 25 regarding the solar radiation model was evaluated with the simulation of the surface error. The simulated autocorrelated mean error of this surface's height is 0.6 m with a standard deviation of 3.2 m, which is equal to the average height accuracy of DEM. This simulated surface was added to the DEM 25 test area and then compared to the mean value of quasi-global radiation energy before and after the error simulation. As expected, the mean values did not greatly differentiate (0.05%). However, the energy in single cells could differ more. The error of DEM 25 has a strong influence on the results in single cells; however, in the global perspective it is insignificant.

The application of solar radiation was compared with the previously applied model [3]. Since the previously used solar data set was not available, only both DEMs were compared. The DEM 25 used in our case ensures better spatial resolution and heights precision than the DEM 100 used previously. As expected, DEM 100 has lower extremes and consequentially a lower standard deviation than DEM 25 (DEM 100 is also smoother). In order to compute the difference in heights and slopes, DEM 100 was resampled to 25-meter resolution so it could truly be compared with DEM 25. The mean difference in height equals 1.8 m with a standard deviation of 8.9 m, and the mean difference in slopes equals 2.5° with a standard deviation of 6.2°. The correlation for heights is high (0.99), but for the slope it is poorer (0.83). The dissimilarity and covariance for heights are acceptable, while on the other hand, they are unsatisfactory for slopes, which means that the poor slope data obtained from DEM 100 affect the quality of the incidence angle.

It should also be mentioned that the consideration of the ellipsoid instead of the Earth's geoid did not have much effect on the quality of the solar radiation model, either for small areas or for all of Slovenia. Even considering the entire world, the deviations are not large enough to greatly influence the solar energy simulation.

24.3.3 INFLUENCE OF METEOROLOGICAL DATA

Before evaluating the effect of the meteorological parameters, we should differentiate climate from weather. Weather data are required for short terms, because weather is just a temporary state in the atmosphere. On the other hand, climate can be defined as the average weather over a period of years. Therefore, climate data are used for long terms. Meteorological data used in the application were gathered over a 30-year period. Besides, the similarity of meteorological conditions in large areas was assumed in this model, which means that the climate was modeled in the application.

Weather, which can change rapidly over small distances, could also be modeled. For example, there could be fog in one small valley but in the next valley sun could shine. We have to be aware, that the Sun is at least twice as weak in cloudy weather when compared to clear skies. Thus, it is very important to have precise meteorological data, for example, images taken by different weather satellites.

Unfortunately, no information on climate data quality was available for the evaluation. Therefore, the last decimal place of data was considered as the standard deviation measure. A matrix with normally distributed values with a standard deviation calculated as stated above was produced. This random matrix was added to the meteorological data matrix and after we compared the mean values before and after error evaluation. It was established that the values differentiate by approximately 1%.

24.4 SUMMARY AND CONCLUSIONS

This chapter discusses a methodology for quasi-global radiation energy modeling. A theoretical model is elaborated and implemented for all of Slovenia (over an area of more than 20,000 km²). In the process, the virtual Sun motion over the surface represented with DEM was simulated. Meteorological data and a shadow determination algorithm were developed and integrated into the final model. The applied methodology was also tested and compared with previous studies, both in Slovenia and elsewhere in the world.

In the model astronomical, surface, and meteorological data were used. Astronomical parameters (Sun declination and Earth-Sun distance) were obtained from the astronomy almanac. The DEM 25, produced with data fusion from various sources, was used as a source of elevations. The meteorological part of the model included data as regards the duration of solar radiation as well as transmission coefficients relative to absorption and dispersion, provided by averaging measurements over a period of 30 years.

The main result of the modeling is an annual quasi-global radiation energy data set for all of Slovenia. The data set has the same resolution as the input DEM, meaning that for every square measuring 25 by 25 m the annual amount of solar energy is known. It was determined that the mean value of solar energy equals 4020 MJm⁻² with a standard deviation of 520 MJm⁻². As expected the coastal region receives the most solar energy (maximum 5360 MJm⁻²) and the Alpine region the least (minimum 840 MJm⁻²). North-facing slopes receive significantly less energy (3600 MJm⁻²) than the south-facing slopes (4400 MJm⁻²). There is almost no difference between east- and west-facing slopes. It was also proven that the geomorphology strongly influences the received energy, with plains receiving the most energy (mean value 4200 MJm⁻²) and mountains the least (3560 MJm⁻²). Slovenia receives the greatest amount of solar energy in July (580 MJm⁻²) and the smallest in December (70 MJm⁻²).

In the described application the most important parameters were considered and included into the model. Unfortunately, they were very heterogeneous. The quality of astronomical and surface data was high and did not influence the results. In contrast, surface data (DEM 25 and the geoid model) had the strongest influence on the quasi-global radiation energy due to the mostly rough relief in Slovenia. The meteorological part of the model was simplified. It is considered to be the weakest part of the application and the processing could be improved with better meteorological data and a better model. Meteorological data (duration of solar radiation and transmission coefficients) determines the regional conditions.

Following the modeling of solar radiation, quality control was performed. The precision of the annual solar radiation energy is approximately 1% for all of Slovenia. The lowest effect on the average value of quasi-global radiation energy is presented by astronomical data, followed by meteorological and surface data. Furthermore, the quality of the application is almost unaffected by the astronomical and surface data. The greatest influence on quality is represented by poor meteorological data. Therefore, better surface and astronomical data would not make a substantial difference in the final quality, if higher quality meteorological data are not used (Figure

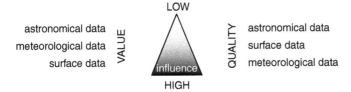

FIGURE 24.5 Influence of astronomical, surface, and meteorological data on the average value of the quasi-global radiation energy in Slovenia and on the quality of the application.

24.5). The quality of the described modeling (long-term for all of Slovenia) is acceptable and comparable with other sources.

Considering the effects of the individual parameters, one can conclude that the model provides good results for longer periods on larger areas. Because the meteorological part of this model is rather simplified, it is inappropriate for shorter periods. Thus, an improved meteorological model with better meteorological data would represent a great improvement to the results. Furthermore, the shadow algorithm could also be improved, but the changes would only slightly influence the results. Another significant improvement would be represented by the use of vegetation data, since forests cover over half of the study area.

The presented study demonstrated the capabilities of solar energy modeling. It has also proven that input data quality is very important for the reliability of the results. Even a very sophisticated model with low-quality data cannot produce better results than a simplified model using high-quality data. It is therefore important to improve the existing sources as shown in this case study by a data fusion process to produce high-quality DEM.

REFERENCES

1. Schaab, G., Modellierung und Visualisierung der räumlichen und zeitlichen Variabilität der Einstrahlungsstärke mittels eines Geo Informationssystems/ Modelling and visualization of the spatial and temporal variability of the irradiation strength by means of a geoinformation system, *Kartographische Bausteine,* Dresden, 2000.
2. Hočevar, A. et al., *Razporeditev potenciala Sončeve energije v Sloveniji /Spatial Dispersion of Solar Potential in Slovenia,* Ljubljana, (RSS), 495 pp., 1980.
3. Gabrovec, M., Solar radiation and the diverse relief of Slovenia, *Geografski zbornik/ Acta geographica,* 36, 47, 1996.
4. Oštir, K., Analysis of the influence of radar interferogram combination on digital elevation and movement models accuracy, Ph.D. Dissertation, University of Ljubljana, Ljubljana, 2000.
5. Podobnikar, T., Various data sources of different quality for DTM production, *Proceedings of 5th AGILE Conference on Geographic Information Science,* Palma, 517, 2002.
6. Bowling, L., Technical note: solar radiation, 1999, http://www.hydro.washington.edu/Lettenmaier/Models/VIC/Technical_Notes/NOTES_radiation.html.

7. Ambrožič, T. et al., Approximation of local geoid surface by artificial neural network, in *Proceedings of the FIG Commission 5 Seminar* Geodesy and Surveying in the Future — The Importance of Heights, Lilje, M., Ed., FIG, Gävle, Sweden, 1999, p. 273.

8. Heuvelink, G., *Error Propagation in Environmental Modelling with GIS,* Taylor & Francis, London, 1998.

Part III-D

Learning from Practice:
GIS as a Tool in Planning
Sustainable Development

Public Participation

25 GIS Support for Empowering Marginalized Communities: The Cherokee Nation Case Study

Laura Harjo

CONTENTS

25.1 INTRODUCTION

This chapter discusses several uses of GIS and its role in community empowerment at the Cherokee Nation GeoData Center, which is an entity within the Cherokee Nation tribal government. The role of the GeoData Center is to provide spatial analyses and data analyses to assist in decision support, policy issues, and tribal planning efforts.

The philosophy behind the Cherokee Nation GeoData Center is to reinforce traditional values specific to Indian culture, particularly that of Cherokee citizens, thereby advocating for and empowering them. The GIS technology is housed in the GeoData Center. Previous GIS projects have integrated traditional values specific to Indian culture, a guiding principle is that the work must advance the lives of Cherokee citizens. Many of the traditional values have been recognized since time immemorial. For example, placing a high importance upon one's social network ensures that one's actions are measured and will have positive implications on one's social network, including the nuclear family, extended family, and surrounding physical environment. The natural environment influences and gives order to a people's culture [1].

GIS has become a tool of empowerment for Indigenous peoples across the world. Indigenous peoples are strategically using the technology of GIS to empower and advocate for their causes. They are using it internally to empower their people and externally, with governmental entities, to advocate for their people. Empowerment and advocacy on behalf of Indigenous peoples is what ensures survival and prosperity. Efforts of measured, methodical, and substantial research are forging a path for Indigenous people worldwide in profound ways. Common threads can be drawn between Tribal peoples of the United States and Indigenous peoples of the world that are living as marginalized peoples. Indigenous peoples from all over the world are using GIS to advance their peoples. For example, the Maori of New Zealand are gathering spatial information for land claims, at the same time gathering traditional knowledge and teaching their elders how to use GIS [2]; the Seri Tribe in Sonora, Mexico, are delineating areas for sustainable resource practices [3]. This chapter outlines methods Cherokee Nation currently uses to sustain its citizens and its tribal government. This in turn leads to its survival and prosperity, coupling the tools of modern science with traditional Indigenous values.

Advocacy for Indigenous peoples is construed as standing up for Indigenous peoples in opposition to a governmental entity or any other entity that is a threat or detriment to the survival of a people or to its traditional way of life. There have been a string of occurrences within the United States against Tribal peoples. Many stories have not found their way to national interest; however, they are appalling. A couple of the worst transgressions of the twentieth century against Tribal peoples in the Oklahoma region were the placement of Indian children in boarding schools during the late 1800s to mid-1900s, where they were subjected to beatings and molestations and punished for speaking their native languages and, during the 1970s, coerced sterilization of Indian women of childbearing age at Indian Health facilities [4]. This is only the proverbial tip of the iceberg of the political climate that Tribal peoples live in within the state of Oklahoma. The tribal history contains a string of injustices; however, the pressing contemporary struggles include legislative battles and agreements at the state and federal legislative level. These are struggles that can be met with the technology of GIS.

In the contemporary United States there are multitudes of tribes clamoring for survival. The struggle is urgent, many tribes are small groups, and once the last person dies, they are gone forever. Prior to contact there were over a million Tribal individuals with distinct cultures, languages, and governing systems. Much of the Indigenous population was annihilated after European colonization. Many died from new diseases brought to the shores. The Indian population was reduced by approximately 70%, from one million at the time of Columbus, to 300,000 in 1900 [5]. The culprits of this dramatic reduction in population were mainly war and disease.

The policy of the United States government toward American Indian tribes has been a tumultuous journey. The federal government dealt separately with the various Indian tribes until 1871, when the U.S. Congress enacted a law that prohibited further treaty-making with tribes.

Historically, the United States government has dealt with Indian tribes as sovereign nations; today tribes are considered domestic sovereigns [6]. Federal Indian policy has had several eras up to the present day. In order to gain a better understanding of philosophies and principles of American Indian tribes, it is necessary to understand their history.

In Stephen L. Pevar's *The Rights of Indians and Tribes* [5], the reader is taken through seven eras of federal Indian policy up to current Indian policy. A brief discourse on United States Indian policy illustrates the underpinnings from which the concept of Indian law was borne as well as an understanding of contemporary struggles American Indians face today, some of which include lasting consequences for Tribes from failed Indian policy.

The first era was from 1492 to1787: Tribal Independence [7]. Tribes were independent nations. They provided settlers with assistance in what is now America. During the periods of war, various European nations sought their support. In the French and Indian War, tribes allied with the British, the British and American Indians proved successful. As a result the King of England made a proclamation to limit the taking of Indian lands by colonists [8]. However, neither this proclamation nor laws, nor treaties still to come in the future would make a difference.

Proclamations, laws, treaties were rarely enforced to the benefit of Indians. Indian land was taken, and tribal peoples were moved off their land to make way for colonists. Another turning point was the American Revolution, which took place between the colonists and the British, with much of this war fought on the aboriginal territories of Indians. Indians perceived this war as a war among outsiders; however colonists would burn villages to engage their participation in the war [9]. Tribes were independent entities that new arrivals in America looked toward for assistance in surviving, assistance in fighting wars. As the new arrivals settled in so did their insatiable want for Indian land.

The era of 1787–1828 is called Agreements between Equals. This began the treaty period between tribes and the United States. During this time there were several land cessions on the part of American Indians. The United States benefited the most any time there was a land cession deal brokered with Indian tribes. At the end of this era a dark period for the Indian tribes of the southeastern region of the country came. The Cherokee Tribe among others was subjected to atrocious human rights violations.

1828–1887 was the era of Relocation of the Indians. During this time, what once was a covert policy now became an overt mission, to remove Indians from homelands desired by white settlers. Land was desired either for farming potential, gold, or extrication of other resources from the land. The influx of settlers was pushing resources to an unsustainable condition. This is a problem that is pervasive in the United States even today: unsustainable practices. This era saw tribes located in the Southeast subjected to a forced march called the Trail of Tears, which began in the southeast and ended in the state of Oklahoma. This forced march included Cherokee Nation.

In 1887 the Dawes Act was passed, which allowed the government to take and allot the communally held tribal land base in Oklahoma. This land base was demarcated as the new Cherokee Nation after land cessions of aboriginal homeland during the removal period. The United States government split tribal land bases into aliquot parts and assigned parcels of land to tribal members. After tribal members were assigned property, the United States government then deemed the remaining property as surplus. This surplus property was then opened up for sale to non-Indians. The intent of this land tenure policy was to assimilate Indians into white culture by virtue of converting Americans Indians into land-owning farmers. The boarding schools curriculum where Indian children were sent was steeped in principles of training Indians to become farmers and service providers.

The period of 1934–1953 was coined Indian Reorganization, this was a period of readjustment of the existing Indian policy, and it began during the Great Depression. A need for Indian land began to dwindle, due to lack of financial resources by non-Indians at the time as a result of the Great Depression. A critical report was released during this time, the Merriam Report, which outlined the poor status of Indians in the United States. This began a wave of change in Indian policy. It should be noted that the United States had and still takes a paternal stance with Indian tribes.

In 1934 policies were put in place that allowed tribes to reorganize with the blessing of the United States. This was a time of nation building since the tribes were allowed to reorganize.

In modern times the situation of Tribal peoples in the United States parallels that of other racial minorities. Data trends frequently indicate socioeconomic and health disparities among minority groups, American Indians included. The existence of American Indians in the contemporary United States takes its place along with other marginalized peoples. Marginalized peoples in the country for the most part include racial minorities, Blacks, Latinos, as well as white lower-income ranks. The common thread is that, from a socioeconomic standpoint, they all seem to suffer from the same ills: poverty, health disparities such as diabetes, heart disease, cancer, and a general lack of wealth, defined as assets such as homes and financial portfolios. Racial discrimination is a pervasive factor in the United States; it is insidious in that sometimes is not called what it really is. This is demonstrated by the amendments to the Constitution, the Bill of Rights that guarantees suffrage for all peoples; however people of color still fight for social justice even today.

The use of GIS technology and mapping is rising to meet the challenge of social justice for Cherokee citizens. They have been relegated to a position in which they are on the fringes of people's consciousness; in other words, they have been marginalized. Socioeconomic variables can be factored into this. Native people usually suffer the highest of all social ills (i.e., poverty rates, unemployment, risk factors

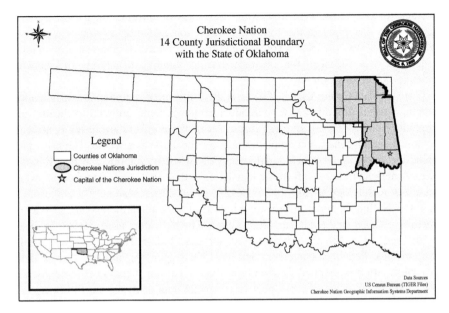

FIGURE 25.1 Map of Cherokee Nation jurisdiction.

for substance abuse). GIS has been used to measure trends in these socioeconomic variables.

25.2 GEOPOLITICAL CONTEXT OF CHEROKEE NATION

The Cherokee Nation is located in the northeastern corner of the state of Oklahoma. The Tribal jurisdictional area is comprised of a 14-county area, which is approximately 7,000 square miles, although only about 45,000 acres of the tribal jurisdiction are actually owned by Cherokee Nation (Figure 25.1).

The geopolitical context of Cherokee Nation within the United States is that of a dependent sovereign [10]. There are approximately 500 federally recognized tribes in the United States. The Cherokee Nation Jurisdictional Area services a population of over 80,000 citizens with a land base of 6,945 square miles in northeastern Oklahoma. The service population is defined as the number of Cherokees who live within the 14-county jurisdiction; the number of Cherokees worldwide is over 240,000 citizens. Cherokee Nation operates a tribal government, comprised of an executive branch, judicial branch, and legislative branch.

25.3 CHEROKEE NATION GEODATA CENTER

This program is considered a resource provider within the tribe; it researches, gathers, and analyzes information to enable informed and strategic decision-making. The GeoData Center is housed under the major division of Information Systems and serves as a GIS shop to all programs within Cherokee Nation. The GeoData Center is often requested to perform data and spatial analyses and produce descriptive statistics for programs. Typically, when this form of information and analysis is requested, it is applied in the following ways: substantiating a legal argument or position that allows the tribe a voice in a particular matter, or demonstrating a level of need such as the need for adequate diabetes funding to address the diabetes epidemic in Indian Country.

The GeoData Center uses its GIS to address five areas: community participation mapping, Tribal empowerment, individual empowerment, community health and community empowerment through maps. All of these efforts are in the context of social justice, advocacy, and empowerment for the Tribe as a whole. The principles behind the mapping also include education about socioeconomic and health issues facing American Indians, informing legislators at the tribal, state, and federal levels, and use of technology usually only available or used by researchers at universities. Cherokee Nation mapping efforts are a hybrid of GIS, scientific and social science research methods, and local tribal knowledge. The indicators used in most analyses are derived from tribally administered surveys, Indian Health Service RPMS (Resource and Patient Management Services), State agencies, and U.S. Census data. This information is mapped frequently at three areal units: Indian community, zip code, and county. From the perspective of spatial data analysis, the following is a brief list of methods used: spatial data visualization, analysis of clusters and trends, creation of socioeconomic indices using quartiles, descriptive statistics, and creation of choropleth maps. The following is a brief list of projects the GeoData Center has accomplished, some of which are described in the remainder of this section: language survey mapping, place-names and historic sites, mapping of Indian communities, Sequoyah Fuels-Uranium Processing Plant, Saline Courthouse, grave and cemetery reclamation, NAHASDA formula negotiated rulemaking, information for tobacco and gaming state compacts, gaming site and health clinic site selection analysis, and a great deal more.

25.3.1 CHEROKEE LANGUAGE PRESERVATION

In 2001, the Cherokee Nation began taking strides toward language revitalization. For tribal members it was common knowledge that less and less of the Tribe was speaking Cherokee. Most significantly, younger generations had relatively few speakers. Language preservation and revitalization efforts were and are being modeling after the Maori of New Zealand and the Native Hawaiians, both of which have been successful with language immersion programs. Cherokee Nation received funding for the development of a language plan to initiate Cherokee language preservation efforts. The intent of the plan was to establish a baseline of who was speaking the language, where the language was being spoken, and the level of fluency being spoken. A language survey was conducted.

The development of the survey instrument actively involved Cherokee speakers; in taking this approach a culturally appropriate instrument was developed. This method of survey development empowered tribal members to raise an awareness of cultural nuances. The survey was administered in both English and Cherokee. After the survey was administered, interviewers found that surveys administered in Cherokee took less time that those administered in English.

The language survey was received by the GeoData Center in a database format. There were 40+ survey questions on the survey instrument, and additionally, there was a geographic component to the database that enabled analysis with the GIS. This involved a series of choropleth maps; these maps enabled members of the Language Project to visualize information from their language survey in a more profound way. They were able to determine where their most fluent speakers resided by zip code as well as determine where nonspeakers resided. A key point to mapping respondents by fluency levels aids in resource planning; for instance in locations where there is a high concentration of nonspeakers, this would be the optimal location for a beginning Cherokee class. A high concentration of speakers who are not yet master-level speakers may necessitate an intermediate-level Cherokee class. Maps that illustrated the aforementioned points were developed and provided to the Language Project (Figure 25.2).

There were further relationships with the data explored through the GIS, including the resident zip codes of speakers who spoke Cherokee as their first language. At the conclusion of the Trail of Tears, the most traditional of Tribal members located in areas that bore a resemblance to their aboriginal lands. This terrain is rugged. It is in these pockets where the most traditional are found, where the culture is still thriving, and this is where many of the master-level speakers reside. The Language Project group also rated the Cherokee Language on Fishman's Scale for language and found that the language is in danger; it is one generation away from being lost. This is evidenced by the results based on the number of speakers by age and gender. There were no speakers under the age of 40 and no speakers who were women of childbearing age. Although this is a sample, it is most likely indicative of the situation the Tribe is in presently. Additionally, as a result of there being so few speakers under the age of 40, it is difficult to recruit fluent Tribal members to earn a college degree that will enable them to teach in an immersion program.

25.3.2 MAPPING INDIAN GRAVES

Within the tribal boundary of Cherokee Nation there are Indian graves and cemeteries that are at risk of the location being lost forever. Sometimes these locations are in deeply wooded areas where only the last surviving elder of a family knows the location. Loss of land is a perennial problem tribal people face, and one of the implications of this problem of being shuffled from place to place is that graves do not "shuffle" as well. The way in which Cherokees marked their graves puts the graves at risk for loss of the location and recognition. Historically, Cherokees marked graves with a small footstone and a larger stone for the headstone, graves marked in this way have no signifier of name or date of birth. This places the graves in a risky situation. When land is conveyed into the hands of a non-Indian the significance of a Cherokee footstone and headstone may go unrecognized as a grave to the untrained eye, and the stones may simply be discarded if regarded as debris. In locating graves and cemeteries, many times the process is initiated by a Tribal member. In one case, an elderly Tribal member knew of several gravesites that he assisted the tribe in locating. Previously, the GeoData Center worked closely with the Historic Preservation officer and traveled along to perform GPS fieldwork in

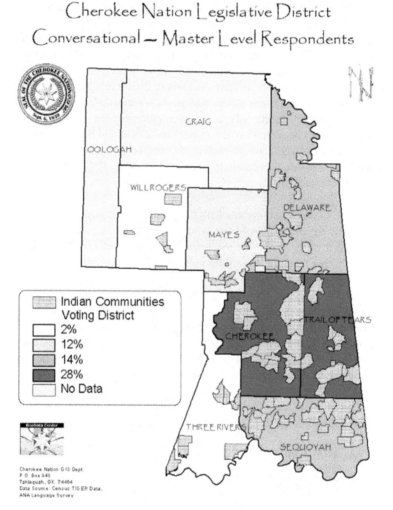

FIGURE 25.2 Language survey map illustrates the distribution of conversational to master fluency respondents by county.

collecting positions of endangered cemeteries and graves. Tribal members frequently initiated the fieldwork and shared knowledge of graves and cemeteries. This required travel often to remote areas within Cherokee Nation, which are sometimes in wooded high-relief terrain, where it can be difficult to acquire the adequate number of satellites for an accurate GPS position. These projects involved collecting the following: a point for each grave, a polygon that delineated the boundary of the cemetery, photos of the grave, if available, data pertaining to the name of the individual and the year of birth and year of death. Upon completion of data collection, maps were created, and families were supplied with maps of the gravesite and cemetery. One of the challenges to this project is the flood of Cherokee citizens able to locate graves. This is simply an issue of having the manpower to handle the increased caseload that would be required to locate all Indian graves within the Tribal jurisdiction.

25.3.3 SALINA CEMETERY RECLAMATION

The Historic Preservation Office and Cultural Resource Center received a report that the city of Salina was planning to construct a building where an Indian Cemetery was believed to be located. Local Tribal members had always maintained that there was a cemetery located at the site, but non-Indians were skeptical of this. The city's intent was to proceed with the construction of the building; the burden of proof fell upon Cherokee Nation to demonstrate that the proposed construction site housed a cemetery.

The Cultural Resource Center gathered documents, which included a map of the Benge Cemetery, depicting grave plots annotated with names. Cherokee Nation staff went out to the site along with personnel from the Kaw Nation. Kaw Nation personnel operated a Ground-Penetrating Radar (GPR), and GeoData Center staff operated a GPS unit. The GPR revealed scores of graves; the graves corresponded with the historic map of Benge Cemetery. As a result of the rediscovery of the cemetery construction was halted. This effort proved successful.

25.3.4 PLACE NAMES AND HISTORIC SITES

The documentation of traditional knowledge is important for Indian tribes. The GeoData Center worked on an initiative to map traditional Cherokee place names; information was received from Cherokee Elders, as well as historical documents. This involved gathering information relating to traditional place names and Cherokee syllabary writing of the place names. From this information a cultural map was created that depicted Cherokee place names. Historically, maps created by explorers ignore Indigenous names, and in many ways this gives the appearance of lack of presence of tribal peoples [10]. The exercise of mapping Cherokee places enables the Cherokee Nation to assert its presence in the modern world.

Another form of traditional mapping involves delineating where tribal people believe their community boundaries are. This project entailed interviewing tribal members, placing maps in front of them and asking them to sketch their community boundaries. From these maps spatial data was developed in the GIS to create an Indian Community layer. This data layer was submitted to the United States Census Bureau under a program named Census Designated Places and, subsequently, was a level of geography for which Census 2000 data was enumerated. This data layer has also become a central component of much of the mapping efforts at the Tribe and is used as an overlay with various types of socioeconomic data. In overlaying this data, patterns of data that spatially coincide with the Indian Communities offer a deeper understanding of the socioeconomics of Indian Communities.

25.3.5 INDIAN HEALTH

Several Indian clinics are operated within the tribal jurisdiction, as well as two hospitals. Information collected about patients is entered into a database, and this system allows for tables to be retrieved for health data analysis. The tribe is presently addressing several health disparities among its tribal membership. GIS is used for resource allocation as well as tracking disease prevalence and determining clinic catchment areas. The GeoData Center created a scenario of a proposed clinic site and modeled a twenty-five mile drive, overlaid on a choropleth map of the Cherokee Population. This method was used to determine gaps of coverage in service areas of the Indian health facilities. In this particular case, this analysis was used to determine the optimal location of a new clinic, which turned out to be south of the initially proposed site.

GIS is being utilized to assess the disease burden upon the tribe, in particular diabetes and cancer.

The health division of the tribe maintains a Cancer Registry of tribal patients who are diagnosed with cancer. In mapping the cancer burden, it was mapped at a zip code level using database information collected in the Cancer Registry. Information such as the number of cancer patients by health facility, cervical cancer cases, brain cancer, breast cancer, and gastrointestinal cancer were mapped; however several other cancers were mapped. This data is to be used for resource allocation; in mapping later-stage cancer in tribal members, one is able to determine which areas of the tribal jurisdiction need more cancer screenings. Future analysis will continue with analysis of cancer stages as well as screening site visits

25.3.6 METHAMPHETAMINE LABS, RISK, AND PROTECTIVE FACTORS RESEARCH

One tribal initiative has been to reduce methamphetamine abuse within the Cherokee tribal jurisdiction over the coming years. The tribe has recognized that there is a problem with substance abuse, particularly methamphetamines and the production of methamphetamines. Manufacturing of methamphetamines has been increasing over the past five years, and raids are becoming commonplace. Methamphetamine labs that manufacture the drug are scattered throughout the heart of the homeland of the majority of the Tribal members. Cherokee Nation devised a methamphetamine lab task force to address and reduce methamphetamine abuse within Cherokee Nation's Tribal jurisdiction. This research performed by Cherokee Nation GeoData Center involves collecting data relevant to Substance Abuse and Mental Health Administration (SAMSHA) Risk and Protective Factors, collaborating with Methamphetamine Task Force members to determine the appropriate factors to analyze in Indian country, performing statistical methods, and mapping the information. The goal of this research is to map areas within Cherokee Nation's Tribal jurisdiction that are at highest risk of substance abuse.

SAMSHA has a set of Risk and Protective Factors, which are to be used as indictors in the geographic analysis.

- Risk Factor: A condition that increases the likelihood of substance use or abuse or a transition to a higher level of involvement with drugs
- Protective Factor: An influence that inhibits, reduces, or buffers the probability of drug use or abuse or a transition to a higher level of involvement with drugs

There are geographic areas that are found to lend themselves to substantially more crime and social ills than others; a review of socioeconomic information of the region indicates this. In analyzing the SAMSHA indicators, it can be determined which geographic areas are more conducive to substance abuse. These areas may then be delineated according to the severity of risk (i.e., high risk, moderate risk, low risk).

In gathering and researching for data, one of the limitations is Tribal land status in Oklahoma, which differs from that of reservation tribes. The population within the 14-county jurisdiction does not contain one homogenous tribal group, but it is racially diverse, consisting of many Tribal affiliations and races. Therefore, in obtaining data from public agencies, when data is cross-tabulated by race it cannot be assumed that all American Indians in the Cherokee Nation jurisdiction are Cherokee. This makes it difficult to analyze substance abuse as it relates specifically to Cherokee citizens.

The first phase of the research project involved a review of Substance Abuse and Mental Health Administration's (SAMSHA) Risk and Protective Factors by GIS staff. This involved determining the appropriate corresponding socioeconomic data that needed to be obtained. There were seven major themes, and within each of these themes there are a series of Risk and Protective Factors or indicators that correspond with the theme.

The major themes were:

1. Family history of substance abuse
2. Family management problems
3. Family conflict
4. Parental attitudes and involvement in drug use, crime, and violence
5. Early and persistent antisocial behavior, alienation, and rebelliousness
6. Friends who engage in the problem
7. Early initiation of problem behavior

GeoData staff gathered data from the following agencies: U.S. Census, Oklahoma Department of Environmental Quality, Oklahoma State Election Board, National Archive of Criminal Justice Data, ORIGINS at the University of Oklahoma, Oklahoma Department of Education, Oklahoma Department of Human Services, Oklahoma Department of Mental Health–Center for Health Statistics, Oklahoma State Bureau of Investigation, Center for Disease Control–National Center for Health Statistics, Federal Bureau of Investigation, Oklahoma State Department of Vital Statistics, Oklahoma Substance Abuse Services, Oklahoma Department of Health.

The data obtained consisted of hundreds of fields of table data. All of this data was categorized into the aforementioned themes within a database. It was necessary to reconfigure the data the GIS system could recognize; this involved cleaning the tables and developing an appropriate database schema that would integrate into the GIS. This information was then presented to members of the Meth Task Force to be gleaned of the most relevant indicators for American Indian populations.

The second phase involved collaborating with Cherokee Nation Meth Task Force members. This component was significant in that it allowed the Tribe to assess all of the Risk and Protective Factors and determine which of the factors apply to American Indian populations. This collaboration identified indicators that applied specifically to Indian communities in northeastern Oklahoma. Particular indicators, which may point to a risk in mainstream populations, may not necessarily apply to American Indian populations. These indicators then become priority indicators on which the subsequent analyses in this research are based. These indicators become imperative, guiding the Tribe to geographic areas at the most risk for substance abuse. After the indicators were reduced further, the next step involved reducing each of the seven themes down to one value for each of the 14 counties.

The challenge of coalescing the numerous indicators into more manageable number was met by creating risk indices for each of the Risk and Protective Factor themes for each of the 14 counties in the Tribal jurisdiction. An index is created for each of the themes, based on the prioritized indicators for each theme.

There are seven Risk and Protective Factors indices:

1. Family history of substance abuse index
2. Family management problems index
3. Family conflict index
4. Parental attitudes and involvement in drug use, crime, and violence index
5. Early and persistent antisocial behavior index
6. Friends engage in problem behavior index
7. Early initiation of problem index

Each Risk and Protective Factor theme is processed through a sequence of standardization as outlined below:

A. Divide the data distribution into four equal parts; take the maximum value and subtract the minimum value. Compute the range and class using the following formulas.

   The range is the highest value minus the lowest value, i.e.:

   Maximum Value − Minimum Value = Range
   Range/4 = Class

B. Compute the four quartiles.

   The data distribution will be divided into quartiles (4 equal parts).

   Q1: 1st quartile equivalent up to 25th percentile
   Q2: 2nd quartile equivalent up to 50th percentile
   Q3: 3rd quartile equivalent up to 75th percentile
   Q4: 4th quartile equivalent up to 100th percentile

   Q1 = Minimum Value + Class
   Q2 = Q1 + Class
   Q3 = Q2 + Class
   Q4 = Q3 + Class

C. Data distribution will be ranked into the appropriate quartile range. Data that falls within the following quartile ranges will be coded with a corresponding value from the list below in #4.

D. A new attribute field will be created.

Values will be coded as follows:

Q1 = 1
Q2 = 2
Q3 = 3
Q4 = 4

E. There should be a newly quartile code fields for each of the seven Risk and Protective factor themes; this is then imported into a database with a separate table for each Index. There were a total of seven tables in the database with a field that contains the geography (i.e., county) and newly created index values.

F. Creation of final index values.

The index values or codes for all of the Risk and Protective factors are summed at the county level. These can then be divided into three classes, high risk, moderate risk, and low risk. The final product is a table that is then mapped using GIS software to graphically demonstrate the counties at the highest risk within Cherokee Nation. This end product was well received by Community Services for use in resource allocation to the highest risk areas (Figure 25.3).

25.3.7 BUSINESS AND POLICY DECISIONS

Cherokee Nation has a business sector — Cherokee Nation Enterprises (CNE) that it operates as a separate entity from the Tribal government. CNE oversees the operation of several casinos and convenience stores, as well as other entities. The GeoData Center's work with CNE has involved creating spatial analyses of social and economic variables that its target market possesses. This has been coupled with other site analysis techniques such as mapping traffic counts, existing competition, and buffers to assess the proximity of competing businesses. This type of analysis gives CNE an edge in determining new sites for business development at optimum locations.

From the policy standpoint, the tribe deals with the state government often. The Cherokee Nation is one of approximately 38 federally recognized tribes in the state of Oklahoma. Prior to entering into agreements with the state or any other federal agency the tribe makes every effort to analyze the information from various aspects to ensure any policy decisions will benefit the tribe. The GeoData Center prepared analyses for the tribe's justice department to assist it in visualizing the Indian gaming climate of the region. Tulsa has several Indian casinos located within the city, one analysis involved determining how the catchment areas for the various Indian casinos must share the customer base. This particular analysis also demonstrated how additional casinos would take away from the Cherokees' existing customer base.

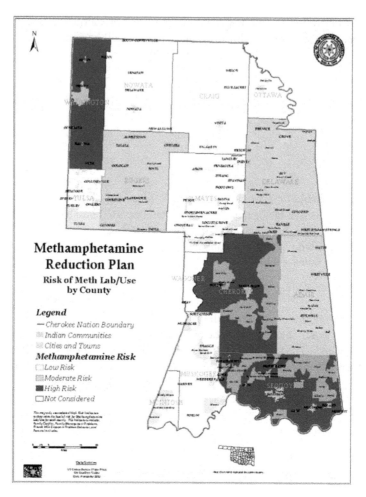

FIGURE 25.3 Map of areas at risk for substance abuse. This is the final map after summing index values by county.

25.3.8 Native American Housing and Self Determination Act (NAHASDA): Formula-Negotiated Rulemaking

The premise of the Native American Housing and Self Determination Act (NAHASDA) is to allow tribes to make decisions that are best for their unique situations in the area of housing. Cherokee Nation participated in Formula-Negotiated Rulemaking with tribes across the county with the Department of Housing and Urban Development (HUD). The rulemaking involved a series of meetings that reviewed the funding formula. The GeoData Center's scope of involvement was serving on workgroups that reviewed the variables used in the funding formula, the weights of the variables used in the funding formula, the data source of the variables, and regulatory language relevant to tribal census geographies and formula areas. These meetings provided a voice to Cherokee Nation as well as the Oklahoma caucus. The GeoData Center prepared spatial analyses of each variable used in the funding formula. This helped to make better decisions that affected tribal peoples living in the state of Oklahoma.

25.3.9 ARKANSAS RIVERBED AUTHORITY

The Arkansas Riverbed had been an ongoing unsettled land claim, between the federal government and three tribes (Choctaw, Chickasaw, and Cherokee). The Cherokees hold a 1/2 interest in the riverbed, the Choctaws hold a 3/8 interest, and Chickasaws hold 1/8. The tribes had never seen the Riverbed holdings in their entirety on one map described with the various classes of claims. The tribal leaders as well as the Cherokee National Council needed an illustrative representation of "made land" and "lost land" as a result of the meandering of the Arkansas River. This was to be used as a tool, in part, to decide whether to relinquish claim to "made land" where squatters resided.

The following layers were developed by the GeoData Center; the source data was derived from the Bureau of Land Management, United States Geological Survey, and a private contractor.

The following layers were developed:

1. Channel of the riverbed
2. Cadastral Survey Line — 1990 BLM Survey Line — most recent high water line
3. Benham Line — high water line prior to the release of the BLM
4. Government lots
5. Coal deposit/below Kerr Reservoir
6. Gas wells
7. Dams
8. Historical allotments
9. Digital elevation model

The system has been used as a tool to exercise the tribe's sovereignty and the tribe's land rights. After assessing the final map and determining the terrain of the land, the usability of the land, loss of resources such as gas and oil deposits, the three tribes involved made a decision to relinquish their claim to "made land" concentrated in the lower reach of the Arkansas river. A more precise definition of "made land" is land gained by accretion. This decision led the three tribes involved into the Arkansas Riverbed Settlement.

The Cherokee Nation originally had a large land holding in the southeast. The Nation held approximately 4.5 million acres. As a result of Indian Removal and the Dawes Act, their holdings have dwindled to nothing and have risen to 45,000 acres. The corner of northeastern Oklahoma prior to 1887 was Cherokee Nation. The Dawes Act or The General Allotment act of 1887 called for the allotment of Indian Land. The tribal land base shifted from communal to individual land holdings; this was to promote assimilation by deliberately destroying tribal relations. There would be no more ties to a communal land base, which was central to Indian culture. The premise was to assimilate Indians into white mainstream America, by way of allotting them land and having them become farmers. The implications of the Dawes Act are devastating. Many Tribal members did not have a firm hold on the economic dealings they would now enter into with their land holdings. The Dawes Act also served to dissolve the Tribes' land base. The allotted land would remain in trust for 25 years; Indians were not to be believed to survive beyond one generation. Allotment was used as "the principal tool" of the old policy of destruction of tribal life and the cause of "poverty bordering on starvation in many areas, a 30 percent illiteracy rate, a death rate twice that of the white population, and the loss of more than 90 million acres of Indian land" [11]. "The allotment policy was a failure. The Indians, for the most part, did not become self-supporting farmers or ranchers" [12].

Fast-forwarding to the twenty-first century, the Tribe has a fraction of the original land base. Many tribal members struggle to survive in today's world while preserving their culture; they face the difficulty of coexisting in two separate cultures.

The Indian Land Consolidation Act was a vehicle by which Cherokee Nation could purchase lands and put them into trust with the federal government and rebuild their tribal land base. The tribe has acquired approximately 45,000 acres to date. The GeoData Center has developed all of the Tribal land parcels using a GIS system and standards for mapping cadastral land. This allows the land to be tracked.

It is used in making decisions in acquiring more land with the goal of creating a land base that is more contiguous. Individual Restricted Land owners are Cherokee citizens who still have possession of their families' allotments assigned via the Dawes Act. Several issues face Indian landowners; one, for example, is adverse possession. Simply put, adverse possession allows a squatter to acquire land that he or she does not own. The adjacent owner knows that more times than not the Indian land owner is not knowledgeable of land law. So the squatters may fence their property, and each year they move their fence over, acquiring more of the Indian landowner's land by adverse possession. This land is original allotted land and it is passed down to the heirs. It is very common for heirs to have a lack of proficiency in the arena of land law. Adverse possession is a predatory practice on the part of the instigator. Most often the instigator is a non-Indian looking to get free land. For this reason, the GeoData Center has developed a layer of allotted lands in the GIS, but it is not allowed outside of the tribe. The problem of Indians losing their land continues to play itself out today. For this reason, GIS land information is closely guarded information.

25.4 CONCLUSIONS

The Cherokee Nation GeoData Center has developed numerous applications in the context of social justice for its citizens. Over centuries, Cherokees have lost their land and have been subjected to atrocious human rights injustices. Today Cherokee Nation is grasping technology with one hand while holding to the tradition and culture with the other hand. Cherokee citizens suffer the same social ills as other Indigenous peoples of the world, as well as shared inequities with racial minorities of the United States. Cherokee Nation has used GIS to provide compelling evidence in the areas of disease burden, language preservation, data preparation for government-to-government negotiations, and cultural preservation in terms of documenting traditional knowledge. There is a global movement in the realm of Indigenous mapping. It is being used as supporting evidence for things such as territorial claims [14]. In summary, it is being used to advocate and empower marginalized peoples struggling to have their voice heard.

REFERENCES

1. Gilbert, W.H., Jr., *The Eastern Cherokees*, Smithsonian Institution, Washington, DC, 1943,1, 77.
2. Apiti, M., Indigenous Mapping in Aotearoa, presented at the International Forum on Indigenous Mapping, Vancouver, B.C., Canada, March 11–14, 2004.
3. Ortega, F.M. and Morales, M.M., Seri Mapping for the Traditional Management Plan for the Sargento Estuary, Desemboque, Sonora, Mexico, presented at the International Forum on Indigenous Mapping, Vancouver B.C. Canada, March 11–14, 2004.
4. Johansen, B.E., Reprise/forced sterilizations. *Native Americas, Akwe:kon's Journal of Indigenous Issues*, 14, 43–47, 1998.
5. Pevar, S.L., *The Rights of Indians and Tribes*; 3rd ed., Southern Illinois University Press, Carbondale and Edwardsville, IL, 2002, p. 2.
6. Smith, C., *The Cherokee Nation History Book*, Cherokee Nation, Tahlequah, OK, 2000, p. 12.
7. Pevar, S L., *The Rights of Indians and Tribes*, 3rd ed., Southern Illinois University Press, Carbondale and Edwardsville, IL, 2002, p. 4.
8. Pevar, S.L., *The Rights of Indians and Tribes*, 3rd ed., Southern Illinois University Press, Carbondale and Edwardsville, IL, 2002, p. 5.
9. Pevar, S.L., *The Rights of Indians and Tribes*, 3rd ed., Southern Illinois University Press, Carbondale and Edwardsville, IL, 2002, p. 5.
10. Poole, P., Indigenous lands and power mapping in the Americas. *Native Americas, Akwe:kon's Journal of Indigenous Issues*, 14, 34–43, 1998.
11. Dippie, B.W., *The Vanishing American: White Attitudes and U.S. Indian Policy*, Wesleyan University Press, Middletown, CT, 1982, p. 308.
12. Prucha, F.P., *The Indians in American Society*, University of California Press, Berkeley, 1985, p. 48.

26 GIS and Participatory Diagnosis in Urban Planning: A Case Study in Geneva

Aurore Nembrini, Sandrine Billeau, Gilles Desthieux, and Florent Joerin

CONTENTS

26.1 INTRODUCTION

Territory is the living environment of all human activities, however diversified and contradictory they might be. Therefore, the strategies and policies established for territorial management are often the source of debate and conflict. This difficulty in reconciling diverging interests quickly led land planners to more and more consider participatory processes as part of their work. If collaborative decision-making is increasingly appearing to be a precondition for successful planning [1,2], then information sharing is a precondition for collaborative decision-making [3]. Furthermore, because a decision is the end result of a process, the level of participation and information sharing at the different stages of that process must also have a strong influence on the degree to which the parties involved agree on the decision that is adopted [3].

Using the above observations as its starting point, this chapter describes an experiment in which the participatory process forms part of the initial phase of a decision-making process (i.e., at the moment when the concerned actors become aware of the problems and build their motivation to act). This experiment was carried out with a group of residents in a Geneva city neighborhood. The participatory process described here takes the form of a *diagnosis,* which can be schematically considered as the collection, synthesis and prioritizing of a number of concerns and issues in the neighborhood. It was designed as a cognitive process based on the sharing of information. This information sharing was supported by the use of some 20 spatial indicators.

This chapter begins with a brief theoretical presentation of the links between the concepts of participation and the use of GIS in land planning. It then describes an experimental participatory diagnosis for urban planning in a Geneva neighborhood. The chapter focuses on the elaboration and use of cartographic indicators to support this participative diagnosis. It concludes in favor of participatory approaches that emphasize the process more than the GIS.

26.2 PUBLIC PARTICIPATION AND GIS

Recent developments in urban planning have involved trying to link the concepts of participation with information technology such as GIS, virtual reality, and the Internet [4–9]. Many of them are clearly optimistic about the power of information technology to modify participation and power relationships. By improving access to information, these technologies are seen as means for changing the flow of information and communication, and therefore, as means for bringing down social barriers and increasing the individual and collective power of citizens. Internet-related technologies in particular are sources of hope for bringing about wider participation [8]. However, alongside this optimistic vision there are more cautious and even pessimistic visions claiming that greater access to information can reinforce disparities in terms of its use and can lead to the exclusion of certain social groups [10].

The relationship between GIS and society created a debate of this kind in the 1990s. Positions were polarized, with on the one hand those highly favorable to GIS (Dobson, Openshaw, and Goodchild in [11]), and on the other those more critical of and sensitive to the social impacts of these tools ([10], Taylor in [11]). Out of this debate came, in 1998, the concept of public participatory geographical information system (PPGIS) through the Varenius Project [6] in response to increasingly strident criticisms of GIS, which was seen as a form of positivist, elitist and anti-democratic technology [12]. One of the objectives of this project was to develop alternatives to the conventional use of GIS, which was for the most part understood and dominated by technical experts alone, in order to broaden the range of users and to foster the involvement of nonexperts, such as citizens and fringe groups, in the making of decisions concerning them. Research on the use of GIS in a participatory context has proposed a variety of approaches. Most of the research has focused on technological and methodological innovations allowing the public greater interaction with GIS [5,13]. GIS in these cases is used to communicate information, which is built on facts and considered to be objective and rational, even though it conveys a specific point of view, usually that of a government or of specialists [4].

Other research, although much more rarely undertaken, has focused on a genuine dynamic for exchanging information with the public, by seeking to add local knowledge to GIS [7,9,14]. Such research, which is based on local participation and emphasizes the knowledge and perceptions of residents, is part of a bottom-up approach [9], in contradistinction to more traditional, top-down approaches.

The experiment described in this chapter did not favor either approach, but sought instead to integrate them. More exactly, it involved finding a process that would allow us to link information from government (top-down) with that coming from residents (bottom-up).

26.3 PROPOSAL FOR A PARTICIPATORY PROCESS IN SAINT-JEAN, GENEVA

26.3.1 OBJECTIVES AND MOTIVATION

The overall purpose of the Saint-Jean experiment was to test the use of urban indicators in the diagnosis process. Cartographic indicators were used to correlate knowledge and preferences. We hypothesized that this linking of information would set into motion a learning process for formulating a diagnosis. Sharing and linking information would help create an overall vision, or in more concrete terms, consultation of indicators at various levels (neighborhood or metropolitan area) or in various sectors would allow the residents involved to set their concerns within an overarching framework before assigning a priority level to the various issues. This learning process had to do not only with the situation and functioning of the neighborhood but also, or especially, with the ability to shift from a local and personal experience to one more general in nature [15]. It was felt that generating this flow of information, which was at the heart of the process, would allow opinions to be built up and to evolve.

26.3.2 CONTEXT

The Swiss law on land-use planning allows public participation on a project in the form of public consultation, when the project is almost completely defined by decision-makers. Remarks of concerned people are collected, but significant modifications to the project are difficult to bring about, because public participation occurs

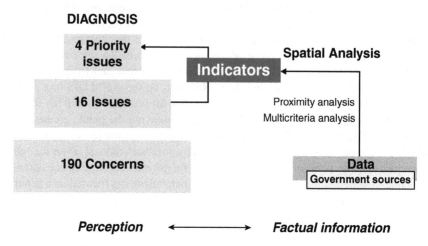

FIGURE 26.1 Indicators used in the pyramidal process.

very late in the process. Citizens also have the right to initiate a referendum on the project. This Swiss particularity is often used, but also often ends in a rejection of the project without any modification.

A detailed study of three land management conflicts in Geneva identified five decision-making processes, including or not including public participation. This study suggested that participation should be open from the very start of the process (i.e., the phase where the problem was defined), and not just at the resolution phase [3]. It is during the initial phase (problem setting), which often corresponds to the diagnosis phase in land planning, where the persons involved develop their reasons for action. Participation in this phase helps to make sure that the agreement reached through the process deals not only with the form of intervention arrived at but also with its relevance and usefulness. The Saint-Jean experiment was performed to test the hypothesis that the public can be involved also in the problem identification phases, not only in the later parts of the process, and to test specific tools such as cartographic indicators and methodologies for their use in the diagnosis process.

26.3.3 ORGANIZATION AND PLANNING OF THE SAINT-JEAN EXPERIMENT

The participatory diagnosis process was not initiated by an authority and does not belong to a general planning process. It has been proposed to the residents, during a public meeting called Forum that is held almost every month. The Saint-Jean residents have been strongly involved in many projects for almost 10 years. This process should help them coordinate their involvement in the neighborhood development.

The neighborhood diagnosis process consisted of four steps (Figure 26.1): formulation of "concerns," definition of "issues," use of "indicators" to determine the situation in the neighborhood, and formulation of the "diagnosis."

The information gathered from the neighborhood residents and from governments was gradually synthesized during the process. The many concerns of the residents (phase 1) were grouped into issues (phase 2). Then, with the help of spatial

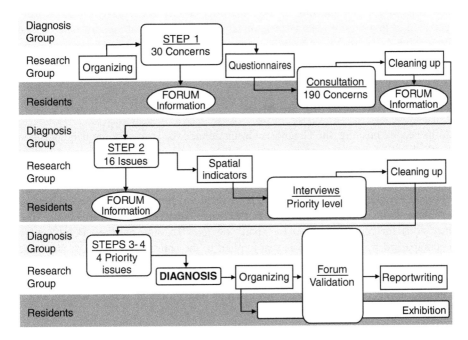

FIGURE 26.2 Information flow during the process.

indicators, priority levels were assigned to all the issues (phase 3) to determine the highest-priority issues, which would then constitute the main element of the diagnosis (phase 4). This synthesis was the main contribution by the residents to the process, which can be presented in the form of a pyramid (Figure 26.1).

The process ran from September to December 2002, and concluded with a presentation at a public Forum held to describe its stages and results at the Maison de Quartier (community center). In concrete terms, the process consisted in helping a working group identify and select the priority issues for the diagnosis. The working group, called the Diagnosis Group, was made up of a dozen residents, who volunteered after a call for participation was made at a Forum previously held in June. This group took part in all steps leading up to the formulation of the diagnosis. The research group directed the process; it also gathered and processed the information (Figure 26.2). It provided liaison between the Diagnosis Group and the residents of the neighborhood. Each step was preceded or followed by obtaining opinions from part of the population. The result of these consultations helped stimulate the reflections of the Diagnosis Group and the formulation of opinions.

26.3.4 The Four Phases in the Neighborhood Diagnosis

26.3.4.1 Phase 1: Identify Concerns

The first step involved determining the concerns about the situation in the neighborhood and the ways in which that situation was evolving, through the question: "What concerns do you have about your neighborhood?" A preliminary work session was held to ask this question of the Diagnosis Group. Some thirty concerns were gathered at that session. Next, questionnaires were developed and presented to the population for purposes of validating this first set of concerns. The persons who were consulted were also invited to add other concerns. As a result of this consultation, the 30 initial concerns created 190 new concerns about the neighborhood.

26.3.4.2 Phase 2: Define the Issues

At its second meeting, the Diagnosis Group defined the issues for the neighborhood on the basis of the identified concerns. Each issue covered several concerns dealing with the kind of development desired for the neighborhood.

The concerns that were obtained referred to concrete perceptions of the neighborhood's residents. Even though they were local and personal, they reflected general problems more difficult to perceive. For example, very precise concerns (e.g., "The bus stops for route 7 are too far apart" or "No bus stop at Rond Point") were encompassed by a general issue that applied to the entire neighborhood: "Increase and diversify public transit services." It should be noted that the simple combining of concerns greatly facilitated the incorporation of a personal experience into a more general problem. This distancing process may be considered as the emergence of a form of political competence [15].

26.3.4.3 Phase 3: Evaluate the Importance of Concerns

A series of indicators was then produced in the form of maps shown on a laptop computer during in-depth interviews with residents, in order to help clarify the various issues (see also Section 26.4 and Figure 26.5). In looking at the maps, the interviewees could compare the situation in their neighborhood with the situation in other neighborhoods, and on a smaller scale, compare their street with other streets in the neighborhood. The sharing of this information allowed the residents to give an opinion about the situation in the neighborhood and the relative importance of each issue. Opinions were obtained by way of five questions that sought to establish a priority level for each issue:

1. Is the information clear?
2. Does this indicator correspond to your perception?
3. Is the situation in Saint-Jean better than in other neighborhoods?
4. Does this indicator give relevant information about the issue?
5. Is this issue of important concern in Saint-Jean?

These questions did not prevent in-depth discussion. The use of indicators was intended only to support the discussion.

26.3.4.4 Phase 4: Formulation of the Neighborhood Diagnosis

The opinions gathered regarding the priority level to assign each issue served as the basis of the work by the Diagnosis Group to formulate the diagnosis. Issues were given priority levels as follows: high priority, important, and less important.

The Diagnosis Group accordingly identified the following four high-priority issues:

- Develop social infrastructures and improve communication among associations.
- Resolve problems in parking regulations (public and private).
- Manage motor vehicle traffic, in particular by reducing traffic and speed limits.
- Stabilize or increase availability of low-rent housing.

26.4 USE OF SPATIAL INDICATORS

26.4.1 ROLE OF INDICATORS

The linking of information is based on two major sources and kinds of information. With regard to information sources, we distinguish information produced by government from information produced by the residents, which is based primarily on their real-life experiences in the neighborhood. The information from government is collected and managed with computer tools that facilitate its use and accessibility. Information from government is usually given greater weight in urban development than is information from residents. The latter kind of information is often complementary, but at the same time it can be quite vague and difficult to obtain.

In terms of types of information, it is useful to distinguish information that describes facts from information that describes values and preferences [16]. Statistics coming from population censuses represent forms of factual information produced by government, whereas strategies, such as development plans, represent objectives and preferences. For their part, residents provide factual information when, for example, they speak about the presence of an infrastructure or its state of disrepair, and they express preferences when requesting better security for a school. But these two types of information are usually intertwined.

In our view, an ideal participatory process allows all parties involved to express these two kinds of information. The spatial indicators that were used in the process helped to achieve this objective (Figure 26.3). Indeed, the concerns expressed by the residents in the first phase of the process combine facts (e.g., traffic density) and preferences (e.g., acceptable noise levels). The indicators that are developed to provide information on these concerns used essentially factual data produced by government (e.g., noise levels in decibels). This categorization is somewhat schematic, because the selection of calculation parameters and the classification of values assigned for the cartographic indicators are also expressed as preferences. These aspects will be discussed further in the sections on spatial analysis (26.4.3) and on representations (26.4.4).

Lastly, if the process entails allowing the various interested parties to express their preferences, it should also allow preferences to be modified by discussions and interactions. Thus, the "ideal" participatory process is one that is, above all, deliberative and seeks to establish the public interest through the modification of individual opinions [15].

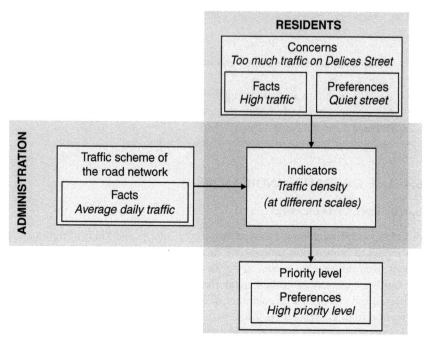

FIGURE 26.3 Role of indicators for linking information.

26.4.2 Definition of Indicators

Land management entities need relevant information that they can use directly in the decision-making process, without being overwhelmed by details [17]. This information, which is restricted to the essentials, should allow the parties involved to develop an overall idea of how the land in question should be managed. Indicators are used to meet these needs. According to Maby [18], the fundamental purpose of indicators is to present and delineate phenomena having an impact on a system such as a territory. Indicators can be defined as providing empirical and indirect interpretations of reality, but not as the reality itself [17,19]. They are arrived at through the selection or the aggregation of data. Reducing information to its essential aspects promotes better understanding of complex phenomena and allows citizens with divergent concerns to use that information.

One or more sufficiently representative and relevant indicators were proposed for each issue by the research group. Indicators definition was not discussed with the participants because of lack of time, but also because the participants were found to consider it was an expert's work.

The relation between the issue and the indicator or indicators was indirect and partial in most cases, since the issue was difficult to fully assess. Three situations may arise. In the first situation, the issue is evaluated by a single indicator considered as significant to the issue. Most indicators in our neighborhood diagnosis are in this category (Table 26.1).

TABLE 26.1
Examples of Issues Represented by a Single Indicator

Issue	Indicator
Better use of space for cyclists and pedestrians	Distribution of public spaces
Increase and diversify public transportation services	Level of public transportation service (proximity to bus stops)
Regulate parking problems (public & private)	Parking density per adult resident
Reduce traffic flow	Average no. of vehicles per day
Provide recreation activities in other areas of the neighborhood	Proximity to activities
Provide zoning for small businesses	Proximity to small businesses
Stabilize or increase available low-rent housing	Proportion of low-rent housing
Improve respect for equipment and safety in public spaces	Rate of vandalism and petty crime

TABLE 26.2
Examples of Issues Represented by a Set of Indicators

Issue	Indicators
Improve pedestrian safety	Width of sidewalks
	Safety devices near the schools
Develop social infrastructures	No. of daycare spaces per resident
	Level of proximity to public meeting places (cafes, restaurants, library, etc.)
Improve the attractiveness of public spaces	Level of proximity to green spaces
	Density of public space facilities/installations (WC)

In the second situation, the issue is represented by a set of indicators [20], which gives a profile of the issue and which allows it to be evaluated according to its different aspects (Table 26.2).

Lastly, in the third situation, issues can be represented by aggregated indicators (Table 26.3).

However, far from being mutually exclusive, these different modes of representation were complementary, and all were used for adjusting to difficulties, according to the situation and the lack or abundance of available information. Thus, the issue "Improve the attractiveness (safety, enjoyment) of bicycle/pedestrian paths" was represented by a set of indicators, one of which was also an aggregated indicator (Table 26.4).

26.4.3   COMPUTATION AND USE OF INDICATORS

Some indicators were computed directly by the representation of relevant information. However, most of the spatial analysis was carried out using a combination of

TABLE 26.3
Issues Represented by Aggregated Indicators

Issue	Aggregated Indicator	Elementary Indicators
Reduce speed of vehicle traffic	Residents exposed to noise	Population density Vehicle traffic noise level
Make the neighborhood safer	Rate of vandalism and petty crime	Average annual no. of crimes Number of residents Number of jobs

TABLE 26.4
**Issues Represented by a Set of Indicators One
of Which Is an Aggregated Indicator**

Issue	Aggregate Indicators	Elementary Indicators
Improve the attractiveness (safety, enjoyment) of bicycle/pedestrian paths	Develop bicycle trails	
	Attractiveness of public spaces for cycling and walking	Sidewalk/path surface Density of lighting Density of businesses and cafes Density of public benches Average no. of vehicles a day

spatial-thematic operations. Following is an illustration of how one indicator was computed.

26.4.3.1 Proximity Level

Several issues had to do with the offer of goods and services near residential areas in order to promote a good quality of life in the neighborhood. Ideally, these issues should be evaluated by an indicator that can measure the time and/or distance (anisotropic) required to reach services. An evaluation of the proximity level based on the distance as the crow flies (isotropic distance) can, however, be deemed representative enough with regard to pedestrian accessibility to services in the neighborhood.

The level of proximity of a set of points (originating address) to a set of destinations, i.e., services (Figure 26.4), can be evaluated by considering proximity fuzzy sets [21]. The proximity level of a destination situated at an interval between 0 and 1 is defined by a fuzzy set (μ_j) of the distance (d_j) represented as being between two thresholds: threshold $S1$, below which a destination is considered to be close (level = 1), and threshold $S2$, above which a destination is considered to be far (level = 0) (Figure 26.4). The overall proximity level for the address of origin is thus calculated by an amount that weighs in the proximity levels of destinations in terms of their relative importance, referred to as opportunity (k_j):

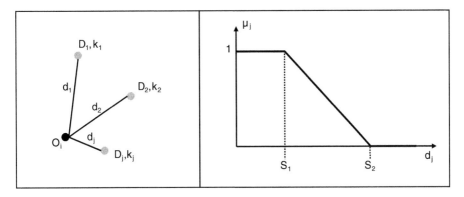

FIGURE 26.4 Proximity fuzzy sets.

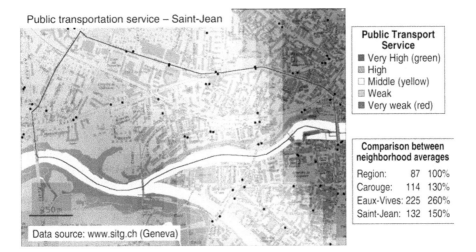

Public transportation service – Saint-Jean

Public Transport Service
■ Very High (green)
▨ High
☐ Middle (yellow)
▨ Weak
■ Very weak (red)

Comparison between neighborhood averages

Region: 87   100%
Carouge: 114   130%
Eaux-Vives: 225   260%
Saint-Jean: 132   150%

Data source: www.sitg.ch (Geneva)

FIGURE 26.5 Level of public transportation service.

$$P_{Oi} = \Sigma \; \mu j \; kj$$

The proximity level for the point of origin can be interpreted as being the average number of services (if all kj = 1), or the average quantity of the characteristics of a service (if kj ≠ 1) in proximity to an address within an interval of given thresholds.

For example, the indicator Level of Public Transportation Service (Figure 26.5) is measured in terms of proximity to bus stops, using as the opportunity level the average number of times the bus passes during one hour. Its value for a location can be interpreted as the hourly number of buses serving a nearby stop (above threshold S1). The indicators that use this method are represented by interpolating proximity values with addresses, which makes it possible to present information in a continuous manner.

26.4.4 REPRESENTATION

An indicator must measure or evaluate a situation (or its evolution) in comparison to a reference point. This comparison can be in space or time, and can be established by a legal or conventional standard. Our objective was to enable residents to evaluate the situation of their neighborhood in comparison to the neighborhood or the entire city. Spatial comparison was used to present the indicators by way of maps using two different scales, one for the neighborhood and one for the city.

The indicators are represented by way of an ordinal measurement scale in reference to average values for the entire map (e.g., average housing rental costs within a designated perimeter). A double color range is adopted as the graphic convention for all indicators. The values of an indicator at a specific point therefore have to be interpreted in comparison to the average value (indicated in yellow), with shades of green indicating a favorable situation (lowest rental costs) and shades of red indicating unfavorable conditions (most expensive rental costs). The variations among these three colors express the relative variations in the indicator's values. As a complement to this visual and quantitative comparison, each map includes an inset giving a synthesis of statistical data for purposes of comparing the neighborhood with metropolitan Geneva or with two other neighborhoods in the city.

By putting the emphasis on comparison, we are implicitly asking residents not to interpret literally the colors on the map. Indicator values in shaded orange or red do not mean a highly problematic situation for a particular zone. They mean instead that the situation is more problematic than the neighborhood average. The residents had to determine whether or not the differences in color intensity constituted a problem, and it was precisely that kind of opinion we were seeking from the interviews. The color for a zone could change when going from the neighborhood scale map to the regional scale map, because the average in the neighborhood was generally not the same as in the metropolitan area. Even though certain zones changed color with the change in scale, the relations among them were stable. This relativity of colors reinforced the message. The maps did not serve to identify zones that were satisfactory or problematic; they only enabled comparisons between locations in them.

The use of colors is inconvenient for representing absolute values when the spatial structure is of irregular size. Nevertheless a single convention was adopted for all indicators, so as to standardize the reading grid and to facilitate the consultation of maps, their readability, and their interpretation during individual interviews. Understanding a map's legend requires adjustment and a certain amount of time, which we wanted to use instead for discussing the phenomena involved, which in some cases are complex enough.

26.4.5 USE OF GIS IN THE PROCESS

Using GIS in a participatory process has both advantages and drawbacks. Technical capacities and skills can engage the interest of participants and can enhance the interaction between interveners by making modifications immediately possible. But using GIS can also weigh down interaction and discussions, because even though technological tools are constantly evolving, they have not yet shown that they can follow the rate of a debate and will probably never do so. At a more fundamental level, GIS, like other computerized technology, tends to draw attention to itself. Some interveners may feel it has the capacity to resolve the most difficult problems. At worst, the process may be seen as a means to support GIS (or the GIS technician) in solving a problem.

Therefore, before each meeting with neighborhood residents, we questioned the role we wanted to give to GIS tools. Overall, we decided to emphasize the dynamic of the actors involved and to deemphasize technology. In other words, advantages derived from technology would be considered in the process only if their impact on the dynamic between the parties was felt to be positive or acceptable. This principle led us to limit the use of GIS at interviews with local interveners, in terms of communicating information and gathering comments or responses. No effort at direct interaction concerning processing, such as modifying the thematic maps or spatial analytical or multicriteria operations, was made during the discussion phases.

26.5 DISCUSSION AND CONCLUSION

This chapter has described the use of cartographic indicators in the carrying out of a participatory diagnosis for a neighborhood in Geneva. The indicators formed part of a decision-making process centered on the diagnostic phase. We started from the hypothesis that using indicators, or in broader terms, making information available, has meaning only when it is part of a more comprehensive process, starting from the construction of reasons for involvement in that process and ending with the making of decisions [3].

Residents of the neighborhood were invited to consider the situation in their neighborhood or in the neighborhood as a whole in comparison to general issues affecting, in turn, the neighborhood and the entire city. This change of scale was used throughout the process and at different levels. First, at the thematic level, it was important to begin by identifying highly detailed individual and local concerns in small group discussions in order to gradually arrive at a consensual overall diagnosis. Next, at the spatial level, indicators emphasizing comparison were used to ask residents to compare their situation with the situation in the rest of the neighborhood or in the entire city. The indicators thus gave residents the opportunity to move from one information level to another and to see the relative importance of issues in their neighborhood in comparison to how those issues played out on different scales.

The Saint-Jean experiment opens up avenues for reflection on the role of geographic information in a participatory process. It was seen that in conveying information and representations, cartographic indicators helped stimulate dialogue among interveners and allowed these latter to refine their arguments. They also generated knowledge and opinions, by helping persons at the local level to objectify their representations or to critically assess the representations created by governmental information sources. The information thus helped anchor the discourse and created a common point of reference. In this sense, the experiment showed the importance and usefulness of disseminating territorial information in a participatory process, even though doing so is a sensitive matter from several points of view.

Indeed, maps provide a specific representation of the territory that does not necessarily correspond to the representation of local interveners. They can be difficult to understand, especially if the indicators combine several levels of information. To this cognitive consideration, the issue of the medium must be added. Maps are not usually seen as supports for discussion, and computer tools have disadvantages that can counterbalance their advantages. The use of GIS in this experiment stimulated dialogue among the residents in some situations. But in other situations, the participants were uncomfortable with the presence of this computer-based tool. In these cases, it made the dialogue more difficult.

If GIS technologies can provide important support, it is only a component of the overall process. It helps the process by allowing the participants to develop an overall idea of the problems (here, the problems facing a city neighborhood) and to understand the representations of other parties involved. Local expectations can thus be focused on a more fundamental process, during which the actual strategic issues, not the details of a development project, are discussed. The participatory process thus becomes a cognitive process that helps the parties involved grasp the complexity of territorial concerns from the point of view of sustainable development. Geographic information provides content for the learning process, and GIS tools facilitate the flow of information among the people involved. In others words, one could consider that this experiment shows that GIS can contribute to bringing down social barriers. Even if the results of this experiment seem to confirm this optimistic point of view about GIS technology, we are convinced that the most important point remains the social process in which GIS takes part.

REFERENCES

1. Couclelis, H. and Monmonnier, M., Using SUSS to resolve NIMBY: how spatial understanding support systems can help with the "Not In My Back Yard" syndrome, *Geogr. Syst.*, 2, 83, 1995.
2. Dente, B., Fareri, P., and Ligteringen, J., *The Waste and the Backyard. The Creation of the Waste Facilities: Success Stories in Six European Countries*, Kluwer Academic Publishers, Dordrecht, 1998.
3. Joerin, F. et al., Information et participation pour l'aménagement du territoire, *Revue Internationale de Géomatique* numéro spécial "SIG et développement du territoire," 11 (3–4), 309, 2001.
4. Aitken, S.C. and Michel, S.M., Who contrives the "real" in GIS? *Geogr. Inf. Plann. Crit. Theor. Cartography Geogr. Inf. Syst.*, 22 (1), 17, 1995.
5. Al-Kodmany, K., Using visualization techniques for enhancing public participation in planning and design: process, implementation, and evaluation, *Landscape Urban Plann.*, 45 (1), 37, 1999.
6. Craig, W.J., Harris, T.M., and Weiner, D., *Community Participation and Geographic Information Systems*, Taylor and Francis, London, 2002.
7. Elwood, S. and Leitner, H., GIS and community-based planning: Exploring the diversity of neighborhood perspectives and needs, *Cartography Geogr. Inf. Syst.*, 25 (2), 77, 1998.

8. Kingston, R. et al., Web-based public participation geographical information systems: an aid to local environmental decision-making, *Comput. Environ. Urban Syst.*, 24 (2), 109, 2000.
9. Talen, E., Bottom-up GIS. A new tool for individual and group expression in participatory planning, *J. Am. Plann. Assoc.*, 66 (3), 279, 2000.
10. Pickles, J., *Ground Truth: The Social Implications of Geographical Information Systems*, Guilford Press, New York, 1995.
11. Schuurman, N., Trouble in the heartland: GIS and its critics in the 1990s, *Prog. Hum. Geogr.*, 24 (4), 569, 2000.
12. Weiner, D. et al., Community Participation and Geographic Information Systems. Position paper, NSF-ESF Workshop on Access and Participatory Approaches in Using Geographic Information, Spoleto, Dec. 6–8, 2001.
13. Shiffer, M.J., Towards a collaborative planning system, *Environ. Plann. B*, 19, 709, 1992.
14. Weiner, D. and Harris, T., Community-Integrated GIS for Land Reform in South Africa, Proceeding of GISOC'99, University of Minnesota, Minneapolis, June 1999.
15. Talpin, J., Elitisme et délibération dans la pensée politique de Pierre Bourdieu, revue Sens Public no 1, Lyon, 2004.
16. Nembrini, A. and Joerin, F., Use of geographical information in public participative processes, Proceedings of 2nd Annual Public Participative GIS Conference, Portland, July 20–22, 2003.
17. OCDE, Mieux comprendre nos villes. Le rôle des indicateurs urbains, OCDE, collection Développement territorial, Paris, 1997.
18. Maby, J., Approche conceptuelle et pratique des indicateurs en géographie, in *Objets et Indicateurs Géographiques*, Maby, J., Eds., collection Actes Avignon no. 5, 2004.
19. Von Stokar, T., et al., Planification Directrice Cantonale et Développement Durable: Un Outil de Travail, Publication de l'Office Fédéral du Développement Territorial (ODT), Berne, 2001.
20. Gallopín, G.C., Indicators and their use: information for decision-making, in *Sustainability Indicators: Report of the Project on Indicators of Sustainable Development, SCOPE 58*, Moldan, B. and Billharz, S., Eds., John Wiley & Sons, Chichester, 1997, p. 13.
21. Joerin, F. et al., Une procédure multicritère pour évaluer l'accessibilité aux lieux d'activité, *Revue Internationale de Géomatique*, 11 (1), 69, 2001.

27 Visualizing Alternative Urban Futures: Using Spatial Multimedia to Enhance Community Participation and Policymaking

Laxmi Ramasubramanian and Aimée C. Quinn

CONTENTS

27.1 INTRODUCTION

This chapter describes a year-long collaboration between the Village of Oak Park, IL and the University of Illinois at Chicago (UIC) in which faculty, graduate students, village staff, citizen activists, and volunteers came together to develop neighborhood character plans for two commercial business districts in Oak Park. The unique contribution of this project is the integration of computer-mediated visualization and communication tools within traditional community organizing, planning, and decision-making processes.

In the United States, neighborhood planning without citizen participation is well near impossible, in part because citizen participation is often mandated by local laws. Additionally, many local governments consciously solicit citizen input in planning processes, because they believe that genuine citizen participation can improve the quality of the plans being made, and active citizen involvement imbues planning decisions with a certain legitimacy that ensures successful implementation. However, national trends suggest civic engagement in public life is on the decline [1]. These declining trends are also reflected in neighborhood planning activities.

The bureaucratization and routinization of citizen participation in the United States has become counterproductive. Invitations to participate in planning processes often attract well-established community stakeholders who are more likely to hold entrenched policy positions. Sometimes called the "vocal minority," these stakeholders engage in community decision-making processes in order to further a specific policy agenda, thereby avoiding a consensual approach to plan-making at all costs. On the other hand, citizens involved in local government planning are often invited to comment on finished products (plans), rather than being invited to make meaningful contributions during the plan-making process. Being asked to review a plan where many critical decisions related to density, scale, and community character have already been made (behind closed doors) reduces participatory planning to a token *public comment* period. In these situations, there is no real expectation on anyone's part that public comments influence the final outcome, one way or another. The timing and poor management of the *public comment* process can sometimes cause even the most well-intentioned citizen to take on an adversarial position vis-à-vis the plan being proposed.

Several practical considerations also limit active citizen involvement. The schedule and format of public meetings usually restricts citizen involvement to brief comments or prepared statements, inadvertently curtailing or eliminating detailed analyses and discussions. The frequency of unproductive meetings causes burnout even among dedicated community planning advocates. Effective participatory planning limits the authority of professional planners. "Public participation costs time and attention; to the extent that it introduces political and interpersonal complexities into decisions, it compromises planners' autonomy and efficiency" [2].

While remaining cognizant of these institutional and political barriers to citizen participation, in this chapter, we investigate the pros and cons of using computer-mediated communication and visualization techniques to facilitate and mediate participatory planning at the local level.

This is because we believe that one of the more significant barriers to citizen participation in local government planning is the absence of appropriate tools and processes that can help effectively manage local planning activities (e.g., zoning changes, site selection, development review) that are heavily reliant on a unique type of interactive communication that integrates qualitative and quantitative data, individual and community experience, and local histories and knowledge.

The participatory action research project described in this chapter is an innovative university–community partnership between an urban university and a progressive urban community. Over a 12-month period, faculty and graduate students from UIC worked to develop character plans to shape the redevelopment of two retail business districts in the Village of Oak Park, drawing on their professional expertise and the opinions and views of stakeholders. The plans used various digital multimedia applications to allow citizens to make meaningful contributions to the planning process and help shape the final outcomes. A suite of digital applications anchored the planning process, which was carefully designed to increase informed citizen involvement in neighborhood planning decisions and to strengthen the planning capacity among Village staff.

The chapter is organized as follows: Section 27.2 describes the history and the character of the Village of Oak Park; Section 27.3 provides a brief description of the Village's Comprehensive Plan, which guides planning in the Village, and describes the UIC team's approach to managing participation and process; Section 27.4 describes the UIC toolbox of computer-mediated planning applications developed for this initiative and its use within the UIC process; and Section 27.5 reviews the benefits and constraints associated with the use of digital technologies. The chapter concludes by identifying key technical and institutional factors that help determine the sustainability of computer-mediated neighborhood planning.

27.2 OAK PARK, ILLINOIS — AN URBAN VILLAGE

27.2.1 BACKGROUND AND HISTORY

The Village of Oak Park, incorporated in January 1902, is a thriving community of approximately 53,000 people, known for its architectural heritage and grassroots-driven politics. Within its 4.5 square miles lives a diverse mix of people with different cultures, races, ethnicities, professions, lifestyles, religions, ages, and incomes. Primarily a residential community proximate to the city of Chicago (Figure 27.1), Oak Park is the birthplace and childhood home of novelist Ernest Hemingway. Architect Frank Lloyd Wright lived in Oak Park from 1889 to 1909 and began his long career

FIGURE 27.1 Map of Oak Park.

using the Village as a testing ground for his design theories. Twenty-five buildings in the Village, the largest grouping in the United States, were designed by Wright, including his first public building, Unity Temple, a Unitarian Universalist church. There are many architecturally significant homes, ranging from Victorian to Prairie style, in the Village's three historic districts.

The Village consists of a diverse housing stock with approximately 23,700 housing units in total, with a little more than half being owner-occupied. Of the 23,700 housing units, over 10,000 are single-family dwelling units. In 2002, the average single-family home cost approximately $320,000, with condominiums being approximately $150,000. With the growth of the Chicago metropolitan region, Oak Park faces intense development pressures and consequent planning challenges that pit proponents of economic development (who argue for increasing residential density and the diversity of housing options) against those who advocate a more measured growth strategy. Anonymous comments from visitors to our project web site articulated some of these tensions:

> We are not against improving the neighborhood, but hope that the charm and diversity of Oak Park will not be replaced by a slicked-up corporate mall look. (Anonymous Comment 1)

I attended the meeting last night.... I had the feeling that the tenor of the group was anti-high density. Just to let you know that there are residents of Oak Park who have other opinions. I am a proponent of density for areas like Oak Park Avenue and the Harrison Street district. They are close to the El [public transit], and make good candidates as areas for high density housing.... (Anonymous Comment 2)

Debates about density in Oak Park eventually focus on the physical character of the neighborhood, and residents often express their concerns about how increased densities and "incompatible" activities, lifestyles, and housing options would negatively impact the character of their community. As will become clearer later in this chapter, the UIC team used computer-mediated communications tools to delineate and clarify discussions about density and community character.

27.2.2 PLANNING IN OAK PARK

Oak Park is governed by a board of trustees led by the Village president (elected representatives) who set the policy agenda. The trustees appoint a Village manager (typically someone with planning and public administration experience) to run day-to-day Village affairs. "Most Oak Parkers [are] proud of Oak Park's heritage, comfortable with its way of life, willing to admit the need for change, yet afraid of that change" [3]. Despite these apprehensions, the Village Board used policy instruments to create a racially and economically diverse community. For example, a nationally recognized diversity statement adopted by the Village underscores its commitment to continue to support its fair housing philosophy. This philosophy requires that housing opportunities are offered equally to all persons, regardless of race, economic status, gender, age, ethnicity, sexual orientation, disability, religion, political affiliation, or any of the other distinguishing characteristics that all too often divide people in society.

In its *1990 Comprehensive Plan,* an advisory document under state law [4], the Village set out general goals related to housing, transportation, and parking, public facilities, economic development, and public participation. This *Plan* is a long-range policy guide for the future physical and social development of the Village. The *1990 Comprehensive Plan* is predicated on the community's commitment to human values: a sense that the Village exists for its citizens, that the physical manifestation of the community — housing, parks, businesses, streets, etc. — are there to meet the needs of its constituents.

The *1990 Plan* focuses on six goals in five general areas. These general areas are critical elements of Oak Park's commitment to improve the quality of life for its citizens. They include:

- **Housing.** Housing is seen as an important element to preserve and enhance Oak Park's stable residential environment so persons of all ages, races, and income levels can continue to live in Oak Park in sound, affordable housing.
- **Transportation and Parking.** This aspect aims to preserve the residential character of neighborhoods and improve the health of business districts

while achieving the safe, fuel-efficient and cost-effective movement of people and goods within and through Oak Park.

- **Public Facilities and Services.** Public facilities and services should be provided in the most efficient manner, especially those public services and facilities that maintain Oak Park as a desirable community.
- **Economic Development.** This element attempts to expand the Village's tax base in order to maintain a high level of services, programs, and facilities and to encourage a broad range of convenient retail and service facilities to serve Oak Park residents and others.
- **Citizen Participation.** This is one of the most critical elements to the planning process as it seeks to maintain a high level of citizen involvement in Village affairs.

When the Village chose to proactively support the continued growth and development of retail business districts in the community (the subject of this chapter), it consciously attempted to increase public involvement in the planning process, thereby making the process open and inclusive. This openness was a key factor in the development of plans for the retail business districts and was affirmed by the Village Board of Trustees, who anticipated that citizens would develop a ten-year to a twenty-year vision for each business district and recommend strategies to achieve that vision.

There are presently twelve business districts in Oak Park, including three Tax Increment Financing (TIF) districts. Of these districts, the Oak Park-Eisenhower Avenue retail business district (henceforth referred to as Oak Park Avenue) and the Harrison Street retail business district (henceforth referred to as Harrison Street) were selected for further analysis and redesign.

The Eisenhower Expressway distinctively divides Oak Park Avenue. An elevated train stop (commonly referred to as the "El") is a median station accessible from the bridge over the expressway. The retail business district includes a range of stores including a grocery store, a laundromat, a coffee shop, an arts and crafts store, and an ale house. While generally popular in the community, the ale house has attracted a fair share of criticism related to noise and public health concerns. The street is accessible by both public and private transportation, and there is angle parking available on this street. Oak Park Avenue has a sense of being a thoroughfare (Figure 27.2).

The Harrison Street district already has a strong identity as an "Arts District." It has an interesting mix of single-family homes, multifamily apartment buildings, and commercial storefronts, many of which are occupied by various kinds of arts-related businesses. The street itself has strong nodes and is a pleasant tree-lined street with sidewalks. The street is accessible by both public and private transportation, and parallel parking is available on sections of the street. Harrison Street is one of the main entrances into the Village from the Eisenhower Expressway and typically represents the first impression of the Village (Figure 27.3).

FIGURE 27.2 Oak Park Avenue.

FIGURE 27.3 Harrison Street.

27.3 PROJECT DESCRIPTION

27.3.1 Scope and Goals

While the *1990 Comprehensive Plan* was an excellent guiding document, it did not provide the level of clarity or specificity needed to articulate the look and feel of Oak Park's many business districts. From the Village's perspective, the goals of the UIC-Oak Park planning initiative can be described thus:

- Design a process by which neighborhood business district plans can be developed, using "best practices" in public process and technical planning theory.
- Develop a set of tools for creating the process using innovative electronic media, financial models, and other state-of-the-art methodologies.
- Produce two neighborhood business district plans (for the Oak Park Ave. and Harrison St. districts) as a product of the process that will articulate the values of the community and the design character of the selected district(s), including height, density and use, parking and traffic, pedestrian access, and landscape features.

27.3.2 Principles of Civic Engagement

The Village of Oak Park enjoys a strong tradition of civic engagement. Both the 1979 and the 1990 Comprehensive Plans consciously recognized and emphasized citizen participation as one of the critical elements of Oak Park's commitment to improve the quality of life for its citizens. When the UIC-Oak Park project began in August 2002, the UIC team developed and initiated a planning process that built upon the Village's commitment to civic engagement. The participatory planning process was guided by the following principles:

- **Fairness.** Ensure that all participants have equal opportunity to express opinions, offer ideas and advice.
- **Respect.** Acknowledge and recognize the participation of individuals and groups, regardless of their views.
- **Inclusion.** Include interests and voices of those directly affected by the plans, but also those who did not participate, or whose participation did not receive meaningful attention.
- **Relevance.** Focus citizens' testimony, advice, and deliberation on issues related to the purpose and context of the project.
- **Competence.** Solicit, support, and use the skills and knowledge of participants to improve the quality of the process and the creation of the plans.

27.3.3 Working from the Bottom Up

The process of planning and design began with an urban design studio. In this studio, graduate students developed design ideas for the revitalization of the two districts

FIGURE 27.4 El, design example.

based on their analysis of data and information they gathered from readily available sources. The data collection phase was not artificially separated from the planning and design phase. As students listened to opinions and preferences of immediate neighbors, they began to design proposals that were based on precedent (best practices from other communities) and the expressed preferences of active stakeholders and neighbors who were likely to be directly affected by any type of physical change (Figure 27.4). Student ideas encompassed a wide range — some were modest proposals for a block of buildings or storefronts in one of the districts, while some proposed radical changes for the entire area such as a proposal for an entry archway linking a public park across the street in the City of Chicago with Harrison Street. It is useful to note that none of the design ideas were bounded by impact analyses at the initial stage.

The UIC team organized the varied design ideas into two scenarios — high-impact and low-impact. The impacts analyzed included economic and fiscal impacts, traffic and transportation impacts, and the visual changes that were likely in the physical environment. These scenarios were presented to the community at large. The scenarios were then reviewed by a group of stakeholders who then used the scenarios to develop a more realistic plan for each district. The final 20-year plans presented to the clients resulted from an interactive process where the physical changes proposed (e.g., increasing the percentage of retail business activity) were adjusted depending on the potential positive and negative impacts that could result. In the end, the UIC team made professional judgments about the intensity and type of development in each business district, which were informed by feedback from citizens with very diverse and sometimes conflicting agendas.

27.3.4 WORKING FROM THE INSIDE OUT

Instead of reacting to proposals for commercial improvements shaped by outside developers, the creation of neighborhood-specific character plans enabled Village residents, stakeholders, and officials to anticipate development. Plans that built upon the goals and interests of local residents provided a more attractive framework for assessing and approving future development proposals than plans based solely on developer input. The process started with the interests and expectations of the residents, owners, investors, and consumers who live, work, and shop in the two districts, and then moved outward to include Village staff, officials, and residents from other parts of Oak Park. Those who most likely faced the immediate consequences of any planned changes played a crucial, but not exclusive, role in shaping the goals and objectives of the plans.

These principles guided the UIC team in designing a process that was transparent, flexible, and adaptable to different situations. Conceptually, the process worked to identify critical issues early on in the process, listen to many different voices and identify community dynamics, clarify value statements, review differences of opinion, express conflicts, and most importantly, develop strategies to negotiate consensus. In addition, the principles and the process helped the UIC team find a balance between the temptations of some individuals and groups, who sought to turn the process into an end in itself, with the desire of others, who wanted to tailor the participatory process to fit their preconceived outcomes.

27.3.5 THE UIC PROCESS IN PRACTICE

27.3.5.1 Soliciting Participation and Communicating Project Information

The UIC team began the project by inviting the residents (renters and owners) living in or adjacent to each planning district to a kick-off meeting in September 2002. Additionally, UIC members actively contacted different groups and organizations throughout Oak Park, inviting their members to participate in the planning process. UIC team members gratefully listened to anyone who wanted to talk about the plans or the process. The UIC team used a variety of approaches to organize and communicate project information through letters to local newspapers, flyers, posters in windows of neighborhood businesses, and by making presentations in different community settings.

One of the major computer-mediated communication tools developed to organize and present information was the UIC Oak Park Project web site, which was launched in the same week that the project officially began. The web site was designed to evolve as the project evolved, with information usually added within 24 hours. Along with a web form, an e-mail account was established so citizens could e-mail directly to the UIC team, providing two methods for online communication. The web site was widely publicized, using bookmarks that were handed out at community events and placed in the library and local bookstores. The project web site became a neutral electronic space, where citizens used the feedback form to write directly to the UIC team, providing both positive and critical comments about the project as it evolved. The web site was taken offline and archived at the conclusion of the project.

27.3.5.2 Meeting Citizens

The UIC process included six types of opportunities for citizen encounters and input. Each is described briefly below. They included:

A. **Public Meetings.** Five public meetings were held at the Village Hall over the ten-month project's duration. Each public meeting typically lasted two hours. During the first hour, the UIC team shared the evolving project find-ings with citizens through short presentations and responded to questions from the audience. These presentations were televised and rebroadcast on the local government-access television channel. During the second half of the meeting, the team typically engaged attendees in a discussion or activity that was appropriate to the needs of the project, such as goal setting. Data relevant to each community meeting was available on the project web site. Synthesis and notes from each meeting were also posted on the site within forty-eight hours of the meeting.

B. **Stakeholder Meetings.** Stakeholders were identified through an open application process managed by the UIC team. Over seventy-five individ-uals volunteered to participate. The UIC team eventually selected thirty individuals to represent the community's interest, in consultation with the Village. Stakeholders included citizens representing individual interests as well as community interests. These members agreed at the outset to participate in at least three meetings and an intensive one-day planning charrette described later in the section. The stakeholder group met face-to-face three times and virtually (over e-mail) several times between March and May 2003. They served as a liaison between UIC and the community at large, and also advised UIC in the development of the plans. The stakeholder meetings were intended to create a shared sense of com-munity and provide opportunities for stakeholders to work together. The agenda for these meetings was advertised by the Village approximately 48 hours in advance of each meeting. It should be noted that the stake-holder groups held additional meetings (when UIC representatives were not present) in June and July 2003.

C. **Community Meeting and Planning Charrette.** In April 2003, UIC pre-sented the first draft plans for each district at a community meeting and planning charrette. These two distinct but interlinked events took place at the cafeteria of the Percy Julian Middle School on the evening of April 3, 2003, and then all day of April 5, 2003. The community meeting on the third gave the entire Oak Park community an opportunity to review and critique the first draft of plans for each district. Over 150 citizens attended this meeting. After presentations by the UIC team, stakeholders led discussions among small groups of citizens. Each group began the process of plan review and helped to identify the trade-offs that were needed to build community consensus. A design charrette is a time-bound and structured decision-making process that architects use to build con-sensus about the form and character of buildings. At the planning charrette, the stakeholders worked in two groups in a structured process facilitated by UIC faculty. They developed a stakeholder plan that took into account the feedback gathered from citizens and based on their analysis of the community's needs. Village staff was on hand to provide clarifications regarding existing design guidelines, parking, and zoning regulations and to answer general questions about Village priorities. In addition, the UIC team provided immediate response to impact analysis questions, so that the stakeholders were able to come up with a draft plan for each district. These plans were refined over a series of meetings held between May and June. Computer-mediated applications described in Section 27.4 were used intensively during this intensive working day to view existing conditions of the two neighborhoods, review design precedents, and run impact analyses of proposed design solutions.

D. **Design Studio Events.** At the start of this project, the Village provided UIC with a workspace at 828 S. Oak Park Avenue. The space was critical to the development of UIC's approach to the participatory planning process. Having the space allowed UIC to open the planning process to the community as well as provide the community with an opportunity to participate in the design studio. Urban planning and design, especially when it involves the physical redevelopment of a neighborhood, is often a mysterious process for residents and ordinary citizens. What they often see is the finished product — a report, a plan, a drawing, or a perspective rendering. Often these plans or drawings are presented as expert-driven solutions that have emerged from a rational planning process supported by evidence and analysis. However, the reality of planning is that it is messy, iterative, and often influenced by values and desires that may conflict with hard evidence. In the initial stages, anyone who wandered into the studio was able to get a sense of how the students were developing their ideas and solutions to address perceived needs of each of the corridors. The studio was open during prespecified hours on some weekdays, weekends, and evening hours. While resource constraints precluded full-time staffing of the studio, a phone with voicemail was installed to receive comments and questions. A computer with Internet access provided citizens with an opportunity to receive training on how to use the project web site and receive customized training on the use of the digital applications developed for this project.

E. **One-on-One Conversations.** Along the way, UIC team members had several one-on-one conversations about substantial issues concerning the project with a variety of citizens and business owners. Some of these individuals eventually became part of the stakeholder process because of these discussions. Others opened their doors to offer the UIC team space to hold meetings and helped the team in other ways. The team used both professional and personal networks to support the idea of planning as a civic project, rather than a political one. These conversations were informal, but were conducted on the phone, by e-mail, and in person. At each of the public meetings described earlier, the UIC team spent time lingering before and after the meeting, to be available to residents who wished to discuss specific issues.

F. **Meeting with Village Staff and Officials.** In addition to meetings described above, UIC team members met on a biweekly basis with Village staff. Both groups shared updates and discussed specific issues related to the project as it evolved. These meetings were critical in keeping all parties informed of the various stages of the project. All five community meetings and subsequent advertising for public participation were planned and scheduled at these biweekly meetings. These meetings allowed the UIC team to present Village staff and officials with important citizen actions and allowed the Village to communicate their priorities. The UIC team also had regular weekly meetings to insure that all members were kept abreast of current developments to avoid any confusion or conflict. UIC team members also presented project progress at three separate study sessions to the Village Board in addition to two formal presentations: the first being the presentation of the project concept and the second being the presentation of our report and findings.

27.4 INNOVATIONS IN COMPUTER-MEDIATED COMMUNICATION AND VISUALIZATION

27.4.1 WHAT IS VISUALIZATION?

Visualization, in its broadest sense, is a communicative process that relies on encoded meanings that can be transferred from creators and organizers of information to users and receivers of the same information [5]. Tufte [6] proposes that visualization is as much an art as a science, where the processes of arranging data and information in order to achieve representation, communication, and explanation are consistent, regardless of the nature of substantive content or the technologies used to display the information. Every successful visualization strategy and product is a purposeful design intended to evoke cognitive relationships in the viewer. Visualization, for the purposes of this chapter and this project, includes two- and three-dimensional representations of spatially referenced data, photographs, video, and other artifacts that emphasize nontextual communication. Interactive spatial multimedia systems are one such visualization tool. These systems integrate video, sound, and text with maps to facilitate discussions about the characteristics and attributes of built and natural environments. They have been implemented in a variety of academic and professional settings [7–14].

27.4.2 WHY USE ELECTRONIC VISUALIZATION TOOLS?

The interactive digital applications or computer-mediated communication tools developed for the UIC Oak Park project are designed to facilitate public discourse about local planning issues. Electronic visualization and communication tools enhance and complement traditional methods of citizen participation at different stages of the plan development process. Interactive applications improve access to both qualitative (images, plan drawings), and quantitative (spreadsheets, tables, charts) information that participants can access before they attend meetings or other public discussions. Additionally, the tools enable participants to interact directly with the information and products displayed as electronic movies, images, maps, tables and reports available on line. In this planning process, the technology also enhanced the quality, speed, and convenience of communication among participants who used e-mail and web postings to reach the UIC team. Citizens and staff were able to view and use maps that were annotated with images in small group settings or online. Citizens also used web-based surveys that combined both still and panoramic movie images to provide an assessment of existing physical conditions in the two business districts and comment on the suitability and appropriateness of different design/planning solutions

27.4.3 INTERACTIVE DIGITAL APPLICATIONS USED IN THE PROJECT

Different types of interactive digital applications were developed over the course of the project. The applications were developed by faculty and research assistants at the Great Cities Urban Data Visualization Lab (GCUDV), a research center within the College of Urban Planning and Public Affairs. An interactive online narrative of the project is available at: http://urban.hunter.cuny.edu/~laxmi.

27.4.3.1 Online Visual Preference Surveys

Citizens need to have conversations about existing conditions of any planning area as they plan for change. Typically, citizens have these conversations at a meeting without the support of visual information. The visual preferences survey can facilitate a more comprehensive discussion about planning issues such as character, density, and safety concerns. It can be adapted for use in group settings or it can function like a traditional online survey that provides feedback directly to the planning staff. The two surveys designed and tested through this project provided citizens a simple way to convey their ideas regarding design issues in the study areas.

27.4.3.1.1 Survey of Existing Conditions

Through the "Survey of Existing Conditions," citizens responded to photographs and panoramic views of the existing retail business district. The survey presented 18 photos for Harrison Street and 17 for Oak Park Avenue. Each of the photographs addressed planning and design issues such as the appearance of buildings, types of businesses, environment, traffic, and safety. Users were able to react to the photographs by selecting "like" or "dislike" buttons. In addition, users were provided with an opportunity to add a brief comment about their reasons for their preference.

27.4.3.1.2 Survey of Best Practices

The "Best Practices Survey" had photos and drawings from various sources showing examples of successful urban design from locations facing similar challenges to Oak Park Avenue and Harrison Street. These examples helped people understand the range of design solutions which could be considered for Oak Park. Both surveys were tested between December 2002 and March 2003 and were officially launched on March 10, 2003. The Existing Conditions survey received 84 responses, while the Best Practices survey received 45 responses.

27.4.3.1.3 Future Uses of Online Surveys

The surveys were designed to be extremely adaptable and easily modified to accommodate the needs of subsequent planning projects. For example, Village staff can maintain the current survey configuration but simply change the pictures and criteria by which people rate the pictures. Over a period of time, the Village will gather an image database. It will then be possible to create a searchable online image database of existing conditions that is available online and through an internal local area network. This database can be used to facilitate discussions during meetings of Village staff, appointed commissions such as the Planning Commission, and during meetings of the Village leadership.

27.4.4 NAVIGATIONAL AND REPRESENTATIONAL APPLICATIONS

Navigational and representational applications allow citizens to take a virtual tour of these business districts. From a design/planning perspective, this application facilitates virtual walkthroughs, akin to a conventional site visit. Using a conventional point-and-click user interface, users walking along the virtual street can pause at specific vantage points to get a sense of what lies ahead and look back to where they have just come from. They can get a 360-degree panoramic view of the area as well. This navigation tool facilitates discussion among stakeholders about existing conditions and can be further adapted to incorporate proposed changes and modifications.

27.4.4.1 Annotated Maps

Aerial views of each business district served as a base map for the project. UIC students subsequently added a walk-through of the street embedded with visual cues (images and 360-degree panoramas) and text notes. This application was designed to be available on a stand-alone computer (such as in a library or a community kiosk), where different comments could be saved and made visible (if desired) to the other users. During the project, this application was available for use by citizens in the studio. The "annotated" map could be integrated into a conventional Power-Point presentation. This application was designed to be easily integrated into the day-to-day planning work of the Village. Village staff were trained in the use of this application and felt that it could be used for making presentations to the board, at meetings with developers, and in meetings with citizen groups.

27.4.4.2 Planning Portal

Students in the technology studio designed a web-based planning portal to enhance the workings of small businesses and small business associations. Using the example of the Harrison Street Business Alliance (HSBA), the group explored how Internet technology could be used to improve communication among the members of a local business alliance, as well as their outward communication to the community of local residents and customers. The class created a simple template, which included information specific to the HSBA and its surrounding area, which could easily be altered to fit the needs of other such groups. The site was intended to be straightforward and easy to maintain, so that organizations without significant resources or expertise would be able to benefit. The site included an open public area (with business information, a calendar of events, directions, and a virtual tour), as well as a password-protected area for members only, where businesses could share information about their district and their association (through discussion boards and file sharing functions). This planning portal was not used during the project but was designed as a functioning prototype.

27.4.5 ONLINE PLANNING TOOLS

Online planning tools allow users to communicate information to decision-makers and to other citizens over the Internet without the use of any specialized software. These applications, when used in the context of a planning project allow decision-makers to gather data about specific neighborhood-level planning issues as and when feedback is needed. Used carefully, these online planning tools can complement and enrich participatory processes.

27.4.5.1 Sketch Tools

The two sketch tools designed for this project are modeled after applications developed earlier at the Urban Data Visualization Lab [15]. These applications allow users to draw on maps and submit comments accompanying their drawings to the planning team. This application provides users with a means of singling out specific intersections, blocks, or areas that require further scrutiny in the planning project and communicating the details of their concerns to the planning team. The sketch tool was made available to users over the Web. In each instance, users who went to the website were presented with a base map of the study area. They were able to identify an area of the map by choosing to draw with a line, a point, or a rectangle shape. Once they had selected an area on the map, they are then asked to type comments that corresponded to the area previously identified. They were then asked to "submit" the information, which was saved in a database. After users had submitted their comments, they were able to see "other views," which showed them areas that other users have selected and the comments that were submitted. These applications were developed during the course of the planning project and were presented to the public in April 2003. By changing the base map, the sketch tool can be used for other projects, where the Village solicits citizen input about place-specific changes.

27.4.5.2 Oak Park Community Mapping

The Oak Park Community Mapping Tool was designed to improve community decision-making by giving local stakeholders access to various kinds of data and information through an interactive community mapping web site. Through these applications, users can view maps including census data, consumer expenditure data, land use, businesses, and public amenities. The information provided through the community mapping tool is generally available to the public, but this tool would give people access to unique "local" information in a simple, accessible, and convenient format. This application was conceptualized as a studio project, but it was not fully developed or used in the UIC Oak Park project.

27.5 BENEFITS AND CONSTRAINTS ASSOCIATED WITH THE USE OF DIGITAL TECHNOLOGIES

Innovations in electronic communication and visualization offer great promise to enhance citizen participation [16]. However, they cannot and should not be seen as a complete substitute for face-to-face meetings or other forms of direct citizen involvement. When used creatively, these technologies can improve the quality and the efficiency of public discussions and debates and help build community consensus around specific planning issues.

In a pragmatic sense, the use of these interactive applications can increase participation among those citizens who are unable to attend face-to-face meetings. However, the tools collectively offer additional advantages. These applications protect the privacy of respondents and therefore allow citizens to share "unpopular" or "minority" opinions without the fear of personal attacks or criticism. For instance, the survey of existing conditions contained an image of a neighborhood ale house, which was also perceived as an unwanted land use by residents who lived in the area. The web-based survey suggested that a majority of respondents actually appreciated the ale house and frequented it regularly! The outcomes of the survey do not imply that there are no problems associated with the ale house. However, the issues that were brought up in the survey concerned noise pollution, sanitation (in the alley behind the ale house), and lack of parking — all planning issues that can be managed without having to close down the facility. The level of detail provides additional information to planners and decision-makers who are considering approvals of similar land uses in the same area.

Interactive applications make it possible for users to become proactive rather than just reactive in thinking about the future of their community. The applications developed for these two business districts can be readily adapted for use in other business districts.

The interactive tools are likely to place an additional burden on limited staff resources, since the presence of interactive technologies is likely to raise expectations among citizens who anticipate personalized and immediate responses to questions and complaints. Citizens and users need to be educated about the value and benefits associated with these technologies. Although the Oak Park community has a high level of access to technology, some people are likely to feel overwhelmed and intimidated. The adoption and use of digital technologies must be accompanied with educational support and training programs for citizens to use the new tools.

Through this project, UIC worked with a core group of staff in developing their skills to use and work with the applications developed for this project. However, additional support for training staff will be necessary to realize the complete benefits of technology adoption.

27.6 CONCLUSIONS

The Village Board, at the recommendation of the Village Planning Commission, adopted the majority of the recommendations and guidelines established through this year-long participatory planning process. The implementation of the process developed is a measurable reward showing the project's success and recognizing the sustainability and replicability of the UIC process.

A brief history and description of the Village of Oak Park and its current planning processes was presented in order to fully understand the scope and purpose of the collaborative planning project that forms the core of this chapter. The authors then described the two retail business districts that were the focus of the planning effort, elucidated the digital applications developed for this project and explained how they enhanced the quality of the deliberations among the citizens. Specifically, these applications allowed stakeholders to describe and critique the existing physical and socio-spatial characteristics of the retail area, identify community needs, and evaluate alternative planning proposals.

The unique contributions of this project are:

- The use of different computer-mediated communication and visualization techniques to facilitate specific urban design and planning activities at the micro-neighborhood scale
- The emphasis and attention placed on the participatory process itself, in which faculty and staff invested time and effort designing and implementing a genuinely participatory process that was infused at different stages with innovative electronic technologies
- The development of planning capacity (essential to sustainable development), among the citizenry and the planning staff at the Village
- The creation of a stakeholder group representing different community interests that continues to work collaboratively to champion and shape the development of two retail business districts
- The inclusion of voices not typically heard in participatory planning processes (e.g., youth, elderly, renters, self-employed people)

Some of these contributions are more directly related to the development and use of information technologies than others. However, digital technologies reinvigorated an already engaged community by creating proactive (rather than reactive) participatory planning processes. In addition, some of the techniques that were used to revitalize the participatory process were made possible because of the team's decision to use computer-mediated communication and visualization technologies. For instance, the use of a safe/neutral electronic space (the project web site), which allowed citizens to post anonymous comments, made it possible for citizens to share their real thoughts, for example about high density. The team quickly learned that not all Oak Parkers were anti-high density, contrary to anecdotal evidence, but instead learned that all citizens were very concerned about having an open planning process. Likewise, the team was able to post documents and plans (work in progress) on the project web site to show how ideas about the planning and design of the retail business districts evolved over time. In some instances, the technology was invisible to the end user. For example, the team used GIS as a back-end data processing and analysis tool. Traditional GIS-generated artifacts such as high-quality maps and analyses of cadastre data were integrated into easy-to-use document formats such as PDF files and PowerPoint. Overall, the project demonstrates how computer-mediated communication and visualization technologies can be used to enhance traditional participatory planning. It requires extra time and a lot of patience, but an engaged citizenry aids the planning process.

ACKNOWLEDGMENTS

Sincere thanks are due to our UIC faculty colleagues, Professors Saurav Dev Bhatta, Kazuya Kawamura, Rachel Weber, and Tingwei Zhang. The contributions and support provided by Professor Charles Hoch, the project director, deserves special mention. We also thank the Oak Park Village Board of Trustees and the staff, including the Village Planner Craig Failor for helping to facilitate the partnership. This research was conducted while Dr. Ramasubramanian served as Research Assistant Professor and Associate Director of Great Cities Urban Data Visualization Program and Lab at the University of Illinois at Chicago (UIC)'s College of Urban Planning and Public Affairs, where graduate research assistants Nina Martin, Nidhi Vaid, and Ramki Srinivasan were actively involved in the design and development of the interactive applications described in this chapter. The contributions of UIC graduate research assistants Helen Edwards, Howard Fink, Xin Li, Jennifer McNeil, William Neuendorf, Amanda Perkins, Rachel Scheu, and Praveen Shangunathan are also gratefully acknowledged.

REFERENCES

1. Putnam, R., *Bowling Alone: The Collapse and Revival of American Community,* Simon & Schuster, New York, 2000, ch. 15.
2. Carp, J., Wit, style, and substance: how planners shape public participation, *J. Plann. Educ. Res.,* 23, 242, 2004.
3. Goodwin, C., *The Oak Park Strategy: Community Control of Racial Change,* University of Chicago Press, Chicago, 1979, ch. 3, 36.
4. Illinois Compiled Statutes, ch. 60, article 105–35.
5. Shannon, C.E., A mathematical theory of communication, *Bell System Tech. J.,* 27, 379, 1948, http://cm.bell-labs.com/cm/ms/what/shannonday/paper.html.
6. Tufte, E.R., *Visual Explanations: Images and Quantities, Evidence and Narrative,* Graphics Press, Cheshire, CT, 1997, p. 9.
7. Câmara, A., Gomes, A.L., Fonseca, A., and Lucena e Vale, M.J., Hypersign — a navigation system for geographic information, in *Proceedings of the Second European Geographical Information Systems Conference,* Brussels, 1991, 175.
8. Fonseca, A., Gouveia, C., Ferreira, F.C., Raper, J., and Camara, A., Adding video and sound into GIS, in *EGIS 93 Conference Proceedings,* Genoa, 1993, p. 176.
9. Jones, R.M., Edmonds, E.A., and Branki, N.E., An analysis of media integration for spatial planning environments, *Environ. Plann. B Plann. Design,* 21, 121, 1993.
10. Laurini, R. and Milleret-Raffort, F., Principles of geomatic hypermaps, in *Proceedings of the 4th International Symposium on Spatial Data Handling,* Zurich, 1990, p. 642.
11. Shiffer, M.J., Towards a collaborative planning system. *Environ. Plann. B Plann. Design* 19, 709, 1992.
12. Shiffer, M.J., Interactive multimedia planning support: moving from stand-alone systems to the World Wide Web. *Environ. Plann. B Plann. Design* 22, 649, 1995.
13. Shiffer, M.J., Spatial multimedia for planning support, in *Planning Support Systems: Integrating Geographic Information Systems, Models and Visualization Tools,* Brail, R.K., and Klosterman, R.E., Eds., ESRI Press, Redlands, CA, 2001, p. 361.

14. Ramasubramanian, L. and McNeil, S., Visualizing urban futures: a review and critical assessment of visualization applications for transportation planning and research, in *Proceedings of the City Future Conference,* Chicago, 2004. http://www.uic.edu/cuppa/cityfutures.
15. Al-Kodmany, K., Extending geographic information systems (GIS) to meet neighborhood planning needs: recent developments in the work of the University of Illinois at Chicago, *URISA J.,* 12, 3, 2000. Available online at http://www.urisa.org.
16. Ramasubramanian, L., Building communities: GIS and participatory decision-making, *J. Urban Technol.,* 3, 67, 1995.

Part III-E

Learning from Practice: GIS as a Tool in Planning Sustainable Development

SDI and Public Administration

28 SITAD: Building a Local Spatial Data Infrastructure in Italy

Piergiorgio Cipriano

CONTENTS

28.1 INTRODUCTION

The need for harmonized geographic information (GI) for urban and territorial planning, environmental evaluation and monitoring, and disaster management, is strictly related to the availability of services to search, retrieve, and access data.

This chapter illustrates the importance of spatial data infrastructures (SDIs) in order to search, retrieve, and access GI within a community of users/producers, through the use of metadata catalogs and web application (services) to find and visualize data.

As a practical example, the text focuses on an ongoing project of a regional SDI in Piemonte (Italy) as part of an e-government program of the Regione Piemonte authority.

In the final part of the chapter the "lessons learned" within this experience are proposed.

28.2 THE NEED FOR SPATIAL DATA INFRASTRUCTURES

"The term spatial data infrastructure (SDI) is often used to denote the relevant base collection of technologies, policies, and institutional arrangements that facilitate the availability of and access to spatial data" [1]. SDIs provide services to discover, evaluate, and access spatial data for users and providers within all levels of government, the commercial sector, the nonprofit sector, academia, and for citizens in general. Public administration departments strongly need to easily find useful information to manage many activities. Actually, public sector information almost always has a "spatial dimension," and many data collected can be easily referenced to spatial context.

Since production and maintenance of spatial data are very expensive activities, the use (and reuse) and the distribution of spatial data have been encouraged within European, national, and local initiatives; current developments in geographic information (GI) technologies (GIS software, web services, databases, open standards) allow spatial data to be produced and distributed through web browsers, GIS desktop clients, PDAs, portables, and other devices.

In order to promote e-government services (administration-to-administration, administration-to-business) many initiatives currently are undertaken worldwide at national and international level on spatial data infrastructures: more than 120 countries in the world are developing national SDIs, and many of them are actively working as part of transnational programs.

In Europe, INSPIRE (Infrastructure for Spatial Information in Europe, http://www.ec-gis.org/inspire/) represents the main important initiative undertaken on geographic information by the European Commission. Its goal is "an open, cooperative infrastructure for accessing and distributing information products and services online" [2].

INSPIRE, according to its common principles, envisages a distributed network of databases, linked by common standards and protocols to ensure compatibility and interoperability of data and services. In fact, by ensuring that electronic data content and services residing at national and regional organizations are implemented according to common standards, they become easily accessible and can be combined seamlessly across administrative borders, thus creating what can be called the technical part of a spatial data infrastructure (SDI).

The current state of the art in information technology makes it possible to realize SDIs based on distributed databases. In a number of Member States, SDIs are being implemented. "The fact that there are still difficulties in seamlessly combining data or services from different Member States resides in the differences in how a location on the Earth is defined, how a geographic phenomenon is represented, how data is documented, and how information and services are delivered" [2].

In July 2004 the INSPIRE Proposal for a Directive was adopted by the Commission. This represents a major step for the use of geographical information in Europe as a contribution to environmental policy and sustainable development.

28.3 SDIs AT REGIONAL SCALE: AN EARLY EXPERIENCE IN PIEMONTE (ITALY)

Some European countries having the least-developed national SDIs (due to the weakest coordination at the national level) have, on the other hand, excellent examples of regional SDIs, thanks to good coordinating mechanisms at that level [3].

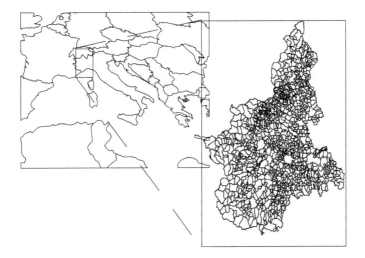

FIGURE 28.1 Administrative fragmentation in Piemonte region.

This issue is much more emphasized in cases of high fragmentation of local authorities. The problem of small and medium authorities is a typical setting in European countries characterized by different levels of local government; in Italy, for instance, 20 regions, 103 provinces, and more than 8100 municipalities are responsible for local government on different themes and functions.

The Piemonte region is distinguished by an enormous number of municipalities (1206); 15% of the total is concentrated in Piemonte, while areas and population represent, respectively, just 7% and 8%. The 1206 municipalities, 41 mountain communities, 32 municipalities, unions, and 8 provinces, represent a highly fragmented puzzle of local authorities operating in the Piemonte region (Figure 28.1).

The high fragmentation of the public sector represents one of the factors that led to the idea of a regional SDI, also driven by the following aspects:

- The great involvement of local public authorities in activities regarding spatial information. Regione Piemonte, Provincia di Torino, and Città di Torino are three main examples of public sector organizations in Piemonte collecting, managing, distributing, and using spatial data at regional, provincial, and municipal levels.
- RuparPiemonte. Many of the regional authorities are already connected within the regional Public Administration Network (RuparPiemonte) and encouraged to use web-based services and applications to manage their own information.
- The presence of CSI-Piemonte (http://www.csi.it), a regional consortium of 51 local public administration authorities founded in 1977 by law. CSI-Piemonte is involved in several e-government projects and coordinates many activities among associated bodies on Information and Communication Technology (ICT), data-exchange and data-sharing services, and geographic information systems.

The project for a regional SDI in Piemonte is called SITAD (Sistema Informativo Territoriale Ambientale Diffuso), and it points toward a local infrastructure to facilitate the coordination of public sector departments to collect, manage, distribute, and reuse spatial data concerning environment, urban planning, natural resources, pollution, and other themes.

Actually, Regione Piemonte administration already began to collect, describe, and diffuse geographic information and environmental data in the early 1990s.

Many services are already available on the web in both an "Internet" version and RuparPiemonte version (access is restricted to regional public authorities only). Some examples are:

- Repertorio Cartografico (http://www.regione.piemonte.it/repertorio/) represents the collection of geographic data and static maps of Regione Piemonte and contains the list of available (and downloadable) data and maps, with related metadata.
- Motore di Ricerca Spaziale (http://gis.csi.it/motore/servlet/login) is an alternative search service to discover data, maps, and webGIS applications defining subject(s) and/or geographic extent. The search engine is based on a webGIS application, used also to visualize data.
- MosaicaturaPRG www.regione.piemonte.it/sit/argomenti/pianifica/urbanistica/siurb/prg.htm. is a webGIS application to visualize geographic data derived from municipal master plans mosaic (currently more than 1100 municipalities out of 1206).

Provincia di Torino administration and Torino Municipality (respectively, the main province and the capital of the region) are also involved in the web distribution of geographic information through:

- Provincia di Torino — Web Cartografico: web service to visualize and download geographic data (base, infrastructures, master plans, roads, and other data) for the whole provincial area (http://www.provincia.torino.it/web_cartografico/).
- Città di Torino — SIT on line: webGIS service available in intranet version (municipal employees only) and Internet version; geographic data are efficiently focused on the 1:1000 scale base map, maintained up-to-date very 3 months (http://sit.comune.torino.it).

Other examples could be taken into consideration, but we can simply assume the three major ones as representative cases in the regional panorama.

Table 28.1 summarizes the present situation in terms of available Internet services concerning GI diffusion at the three levels of local government in Piemonte.

The SITAD project aims to build up a regional SDI, and it has been realized by the Regione Piemonte GIS department, as part of the regional Administration-to-Administration (AtoA) e-government program, due to the long experience developed in GIS activities by Regione Piemonte itself and Provincia di Torino and Città di Torino. During the first year (2003), the project was mainly focused on the collection of use cases of such stakeholders.

TABLE 28.1
GI Diffusion and Web Services
in Piemonte Region

	Metadata Collection and Web Diffusion	Search-and-Retrieve Services	GI: WebGIS Application	GI Download Services	Static Maps: View-Only Services	Static Maps: Download Services
Region	*	*	*	*	*	*
Provinces			*	*	*	
Municip.			*			

The basic idea of the regional infrastructure is more ambitious compared to INSPIRE plans, because it includes not only spatial data but also other multimedia information. Compared to INSPIRE, greater emphasis has been given to the real use of the data, and for this reason several services and web applications were specifically developed. All INSPIRE components are supported (catalogs, metadata, standards and interoperability, core data, etc.), so it will be possible to use the current regional SDI as the building block of INSPIRE in Italy [4].

28.4 METADATA CATALOG AND SERVICES: PUBLISH, SEARCH, RETRIEVE, AND ACCESS GEOGRAPHIC INFORMATION

In the new regional SDI services for metadata collection, data search-and-retrieve, data download, and static maps visualization, are available to every public sector authority; other users (private sector, citizens) can use the catalog for searching and accessing information:

- Catalog search-and-retrieve services. Searches by thesaurus keywords, subject, data provider, temporal coverage, and geographic extent.
- Catalog consultation services. As a result of the search operation, a list of entities (geo-data, tables, images and static maps, documents, and webGIS services) is presented to the user; every user can view metadata elements, and, depending on his/her group profile, access data.

- Access services. The visualization and download operation is made available by the data provider in relation to the type of user (group profile); i.e., public authorities' employees are allowed to access some information, not available to citizens (apart from metadata).

Homogeneity and integration between spatial data does not strictly mean "same data and same structures"; the main key point is the description of data collected and maintained by public sector organizations, within the same metadata structure. This issue drove the first phase of the project analysis to the consideration of a unique metadata catalog for the whole region, open to every public sector organization within the regional area.

The activities of the first year (2003) were mainly focused on the development of web services to compile and publish metadata related to spatial information and multimedia (regarding environment and spatial planning), and services to search-and-retrieve resources and access geographic data (visualization, download).

Following the development phase of such services, the second year was focused on the involvement of public sector stakeholders at regional, provincial, and municipal levels, in order to increase the knowledge on the "metadata issue."

Through an Intranet regional network among Public Administration, the metadata catalog is accessible by registered users for "describing" the information they manage (live GIS data, static maps, databases, other documents, and information services).

The web wizard application allows users to easily compile metadata and indicate where, how, and who can access data. The metadata catalog has been structured according to ISO19115 DTD schema and designed to be filled up through a web application or by harvesting remote metadata repository (e.g., from provincial or municipal level) via XML.

The metadata catalog management by multi-users at local levels can also be a concrete answer to the need for a greater involvement of small and medium-size authorities, often excluded by ITC programs and e-government initiatives.

Since different organizations (e.g., departments of Regione Piemonte, Provincia di Torino) underlined the opportunity of having "their own" portal to access and query the catalog, a multilayout interface has been produced. For this reason we developed a unique catalog gateway (a regional "geoportal"), accessible from different web sites (public sector organizations' web sites) with different layouts and different search-and-retrieve criteria.

Data access is offered online, via web services. Spatial data (live GIS data) are accessed via online mapping services (web map services) and served dynamically to clients. The architectural schema of spatial servers has been designed to be platform and proprietary independent, in a web-mapping approach based on OpenGIS Consortium (OGC) specifications; with standards-based interoperable Web mapping, each map server implements a common interface, a messaging protocol such as the WMS interface for accepting requests and returning responses [5].

Information available in different formats (such as text, static images, videos, an tables) and described in the metadata catalog, is made accessible via http, ftp protocols, through visualization and/or download services. Accessibility is intended to be customized on the user's profile (users enter the search engine with or without browser certificates, and can access information according to their own user profile).

28.5 THE "METADATA ISSUE"

Metadata represent the most important component in a SDI catalog search-and-retrieve service and directly affect the use of geographic information. At the same time, metadata are probably the most boring activity in the collection and management process of geographic information.

The problems increase if we want to consider also nonspatial information, assuming every type of data that could not be directly considered spatial or spatially referenced. Texts, images (such as photos, drawings, and sketches), videos, and audio files are important information for environmental planning, evaluation, and monitoring. Such information could be easily geo-referenced, but which metadata elements are needed to describe such information?

In a SDI initiative this information could be thought of as digital data, often available (accessible) via web. This consideration highlights the need for a "light" standard to describe many data formats, and possibly to make it possible for "metadata unskilled" people. As a minimum set of elements we can consider the Dublin Core Multimedia Initiative specifications (http://www.dublincore.org).

The Dublin Core (DC) initiative aims to simplify metadata catalog search-and-retrieve services on the web and can also be efficiently used for geographic information. DC is a "light" metadata profile and, for this reason, will never completely substitute a specific geographic metadata (i.e., ISO19115:2003, or FGDC-STD-001-1998, ...) but can be seen as a subset of a "technical," detailed metadata.

DC elements are: title, creator, subject, description, publisher, contributor, date, type, format, identifier, source, language, relation, coverage (temporal, spatial), and rights; every element can be defined by a set of 10 attributes derived from the ISO11179 standard. At the European level MEGRIN and MADAME projects have already evaluated the use of the DC standard as a main schema for metadata searching (discovery level).

As a useful and interesting example of DC metadata, we can consider the INSPIRE position papers at http://www.ec-gis.org/inspire/; the papers produced are described (first page of the document) using DC elements.

Within the regional SDI in Piemonte, DC standard represents the "entry" metadata level, common to every type of information contained in the metadata catalog.

The 15 DC elements are "core" references of a second and fully detailed level, such as ISO19115 for geographic data. At the same time, the use of DC elements facilitates the data entry operation by unskilled metadata users. Within the regional Public Administration Network a web-based service is made available to authorized users for metadata entry and management.

The main task of the SITAD project concerning metadata is related to applications for metadata entry and management; on the basis of the use cases list, different software/platform solutions were compared, looking simultaneously at commercial, freeware, and ad hoc solutions. According to the subsidiarity principle, a web metadata entry service allows local authorities (municipalities, provinces) to declare and describe information (geographic or not) they are responsible for and manage. They also are allowed to define accessibility criteria to external users ("who-access-what"). Live GIS data (SDE layers, shapefile, CAD drawings, raster images) can be automatically described in the catalog through a Java web module (developed on the basis of off-the-shelf developing solution). Layers described, maintained "at the level where this can be done most effectively" according to INSPIRE, can be visualized through a multimap service viewer developed on a ESRI ArcIMS® cluster (Environmental Systems Research Institute, Inc., Redlands, CA). The viewer has been treated as an "empty" box, logically connected to remote spatial servers (mainly on ESRI ArcIMS platform), where map services linked to spatial data (dB or file systems) run. Figure 28.2 represents the presentation logic, the business logic, and the data logic of the infrastructure developed in 2003. From different portals, users interact through WAI-compliant interfaces (i1, i2) and access the catalog (MTD) to write/publish metadata or query (application A1). Query results can be evaluated as "metadata report" (m) and/or accessed through download/download services. Live GIS data map generator application (A2) serves maps derived from map services running on distributed map servers (S1, S2, S3, ... Sn); therefore, data can be maintained at the level where map servers are available, at the regional, provincial or municipal level.

At the moment, GIS data are served using commercial GIS server solution adopted by Regione Piemonte, on a Linux cluster of up to 14 spatial servers. In order to reach OpenGIS recommendations we are going to completely apply OGC-WMS 1.1.1 specifications [6]. This will allow different GIS server solutions to share geographic information, in a unique visualization tool (web browsers, GIS desktop clients, PDAs), independently from either data structures and formats, or the software used. Web map services developed in such configurations can be restricted to public administration users through the Regional Public Administration Intranet (RUPAR-Piemonte), or available to private sector and citizens via Internet portals.

28.6 BUSINESS AND SOCIAL BENEFITS

As an AtoA e-government project, the regional SDI represents the logical connection of several independent GIS projects undertaken during the last decade and, at the same time, of some AtoB (Administration-to-Business) and AtoC (Administration-to-Citizens) e-government projects involving GIS services and web mapping at the regional level. In this scenario, SITAD activities will cut down the total cost of data management and the cost of the increasing number of map services running at regional scale, on the one hand through the reuse of map services (and data) already available, while on the other hand adding value to spatial information existing at municipal level with the central metadata catalog and the web metadata entry application.

The cost to develop the regional geoportal during the period 2003–2005 (corresponding to 0.5 €/inhabitant) is totally funded by Regione Piemonte through the e-government program. Benefits are expected at the economic level with the diminishing of costs sustained at the moment by pubic administration organizations (thus, indirectly, by citizens) for activities related to data production, collection, and management. At the social level, on the other hand, a deeper integration between spatial

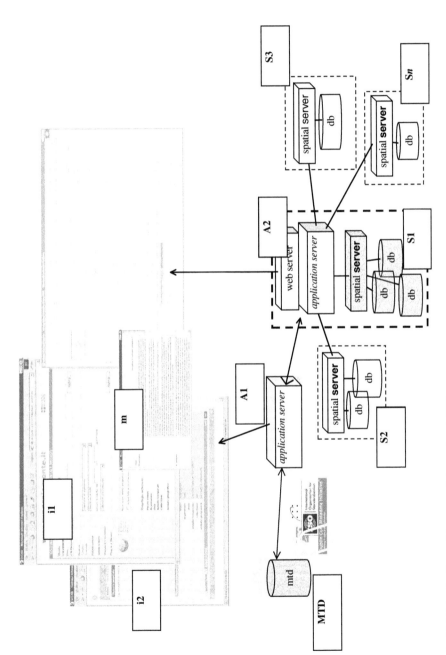

FIGURE 28.2 The architecture of the regional SDI.

information managed by different organizations will facilitate better decisions on relevant issues like natural environment protection and monitoring, urban planning, housing, infrastructures, cultural heritage, wildlife preservation, and other themes (especially after the 2006 Winter Olympic Games hosted in Turin province, that will have a deep impact on land management).

According to future European policies (derived from INSPIRE outcomes) Regione Piemonte has already been working on a very important aspect: pricing policy to be adopted within the regional SDI. More than 30 spatial data sets (including a list of more than 50 spatial layers covering the whole region) can be downloaded free of charge; at the same time, some official digital maps are also free of charge and downloadable as graphics files, while the cost for the paper version is about 10–20€ each.

The policy reflects the fact that the maintenance of spatial data is an obligation of the regional authorities for their own needs of governance. The costs are therefore absorbed by the public authorities and not charged to the citizens, whereas the maps require some extra manual work and are mainly used by citizens or professional users ready to pay for them [4].

28.7 LESSONS LEARNED AND FURTHER DEVELOPMENTS

The project briefly described in this chapter is a "state-of-the-art" picture of new developments and implementations to existing services to build up a SDI in a local context. During the first phase of the project the focus has been on technology issues (metadata database design, web application, standards for interoperability, and so on).

As a three-year project, it was decided to redefine 2004 and 2005 activities in order to concentrate efforts on two organizational aspects: stakeholders' involvement and end user and data providers' requirements have been representing the main critical issues in the development of the infrastructure. During 2004 the SITAD project has been focusing on two main sets of activities. The first one concerns the integration between services developed in 2003 (the search-and-retrieve metadata catalog and the multimap service webGIS viewer) and other existing data-exchange services built up by Regione Piemonte in the past. The second set of activities concerns the big challenge of "nontechnological" aspects of building a SDI.

The work has been organized as a list of interconnected activities: these activities are finalized to disseminate the idea of the regional SDI and to practically involve public sector stakeholders. This part of the project has been regarded as a "Community Demonstration Project" (CDP) model, to prove how spatial data dissemination could be very helpful for many items and activities managed by public sector organizations (http://www.fgdc.gov/nsdi/docs/cdp.html).

The aim of CDPs is to demonstrate practically how sharing geographic data and maps helps problem solving at different levels of administrations to extend standards and data to the private sector, according to EU directives on access to public sector information [7].

CDPs are going to be set as a mix of workshops, lessons, regional guidelines collection, and distribution and experimental activities, open to public organizations such as provincial and municipal GIS departments, urban planning departments, environment departments, and other interested organizations.

As a future task (2006 onwards), CDPs will be structured as a sort of "e-learning" program, based on the mix of "live" events (workshops, meetings, etc.) and web interactive demonstrations.

REFERENCES

1. Nebert, D., The SDI Cookbook, GSDI Eds., 2001, 8, www.gsdi.org/gsdicookbookindex. asp.
2. Smits, P. et al., INSPIRE Architecture and Standard Position Paper, JRC, 2002, http:// inspire.jrc.it/reports/position_papers/inspire_ast_pp_v4_3_en.pdf.
3. GINIE, Spatial Data Infrastructures: Recommendations for Action, 2002, http:// wwwlmu.jrc.it/ginie/doc/PG_SDI_en.pdf.
4. Annoni, A., Lessons from the Italian NSDI, INSPIRE, 2004, 15, http://inspire.jrc.it/ reports/AANSDI_Italy_FinalApproved_v12en.pdf.
5. Kolodziej, K., OpenGIS Web Map Server Cookbook, OGC, 2003, 11, http://www. ogcnetwork.org/docs/03-050r1.pdf.
6. de La Beaujardiere, J., Web Map Service Implementation Specification, OGC, 2001, http://www.opengis.org/docs/01-068r2.pdf.
7. European Parliament, Directive 2003/98/CE of the European Parliament and the Council of 17 November 2003 on the re-use of public sector information, 2003, http:// europa.eu.int/.

29 Local GIS: Implementing the Urban Spatial Enabled Information System

Walter Oostdam

CONTENTS

29.1 INTRODUCTION

This chapter presents an implementation case study undertaken at the municipality of the city of 's-Hertogenbosch in The Netherlands. It describes how GIS is entering the mainstream IT of the organization by implementing the Urban Spatial Enabled Information System (USEIS).

The implementation of such a system is a step-by-step process that is very complex, because it effects different aspects of the whole organization, including the IT infrastructure, the different software, processes, and workflows used in the organization, organizational aspects, and of course, the people who will eventually use the system. In the organization of the municipality hundreds of different products are produced, using their own processes and workflows. Some of them are relatively easy to automate; others are very difficult because of their complexity. Therefore, the implementation of the system is cut into smaller pieces, in the form of projects, where each project focuses on the implementation of a single department or on a specific product, process, or a typical technical aspect of a component of the system. In this case the USEIS is not a software system that can be bought from a vendor. It is a collection of several large software systems (a document management system, a large GIS viewing system, called Geonet, and a drawing management system), automated or semiautomated connectors (developed in house or by externals), procedures, and workflows, the latter two automated or manual. Although the ambition is to automate as many aspects as possible, it is not always possible, either because they are too complex or because it is not feasible to automate them. If the latter is the case, the reason is often that the costs do not balance with the end result. The system provides means to search for and retrieve information stored in the GIS layers of Geonet, the document management system, and the drawing management system. These means are offered in two flavors: either geographical, using a map, or administrative, using the metadata search capabilities of the document management system.

Using Internet technology ensures that the information is retrievable from any computer at any time at any location, and the optical harmonization gives the user the feeling that he/she is working with one system. Therefore, the keyword for this USEIS is integration: integrating all the different components, making them "talk" to each other and providing an interface to the end user that gives him/her the illusion of working with one system, while in the background he/she is switching from one system component to another, using the strengths and possibilities of each component. The role of GIS in this system is very important, since providing the ability to retrieve information using a map or map interface is handled by the GIS component. Another important aspect of successfully implementing the USEIS is the momentum, the proper time where the important parts of the system and the right organizational circumstances are available. This will be explained in more detail later.

As stated earlier, the implementation of the USEIS is cut into different parts. One of those pieces is a project called "GIS-Bestemmingen" (GIS for zoning and development plans), an intranet-based system for viewing and retrieving information about zoning and development plans. This project is also part of the intranet-based general GIS viewing system of the municipality of the city of 's-Hertogenbosch, called Geonet, which is one of the three major components of the USEIS.

This chapter therefore focuses on the part of the USEIS that deals with retrieving and viewing information about zoning and development plans in the framework of the GIS-Bestemmingen project. However, only relevant information about this project will be explained, while more extensive information about this project can be found elsewhere [1].

29.2 THE USEIS

In this section the USEIS is described: the principle of the system, its major components, and the role of the Internet capabilities of each component.

The USEIS is an Internet-technology-based integration between three major components (software systems) that are by themselves full-grown information systems with their own IT infrastructure, software, implementation path, user interface, and application managers, and which operate independently. It allows for searching and retrieving information and documents in two ways, either geographically or administratively.

29.2.1 THE MAJOR COMPONENTS OF THE USEIS

The three major components are:

- A large, intranet-based, organization-wide GIS viewing system, called Geonet, based upon ArcIMS® technology (ArcIms and ArcViewIMS are registered trademarks of Environmental Systems Research Institute, Inc., Redlands, CA). It unlocks and brings together in one environment geographical information that is produced at the different departments in the organization.
- A drawing management system (DrMS), based upon ProjectWise® technology (Bentley Publisher and ProjectWise are trademarks of Bentley Systems, Inc., Exton, PA). All CAD drawings are stored in this system.
- An organization-wide document management system (DMS), based upon Panagon® technology (Panagon is a trademark from Filenet Corporation USA, Costa Mesa, CA). Almost all documents, other than drawings, are stored in this system.

The first two are already fully operational in the organization, and the last is in the process of being implemented, which means that some departments are using it already, while other departments are in stages of preparing themselves for implementation, and yet other departments are scheduled for implementation.

In Schema 29.1, the main relationships between these components are shown.

It should be noted that user input is forwarded to the GIS and the DMS only. This complies with the principle of two possible ways of searching and retrieving information. Also, the drawing management system (DrMS) is only accessed indirectly from the document management system (DMS). The output of the DrMS is directly forwarded to the intranet client. The DrMS acts as a slave to the DMS. This behavior will be explained in more detail later. Finally, one of the arrows between

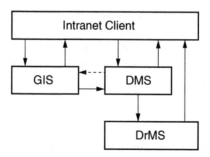

SCHEMA 29.1 Relationship between the main components in the USEIS.

the GIS and the DMS is dashed, which means that in the framework of the GIS-Bestemmingen case study this connection is not (yet) established.

29.2.2 INTERNET CAPABILITY

The USEIS is a system based upon Internet technology. Using that technology ensures that information can be retrieved on any computer at any place at any time. It also plays an important role in connecting the major components. Therefore, it is a prerequisite that the major components provide Internet-based interfaces. Another prerequisite for the major components is that these interfaces allow for customization using standard Internet scripting and programming languages like Java and ASP. The last prerequisite regarding Internet capabilities of the components is that they allow for being addressed by the other components using what can be called a "URL command." A URL command differs from a regular URL. It is a URL that is extended with parameter values that can be interpreted by the page that is being addressed. Very common examples are so called ASP pages. A URL command allows for passing through parameters, variables, etc. It thus allows for customization and steering the behavior of the component. All of the three major components fulfill these prerequisites. As stated above, the three major components are by themselves independently working systems. Their Internet capabilities act as the glue (connectors) between them and as the stucco (the uniform interface). Without it, the USEIS would remain a utopia.

29.3 THE GIS-BESTEMMINGEN PROJECT AS A PILOT FOR THE IMPLEMENTATION OF THE USEIS

This section will describe in detail how the USEIS is implemented for a part of the information of the municipality of 's-Hertogenbosch, namely the information directly related to zoning and development plans.

29.3.1 RELATIONSHIP BETWEEN GIS-BESTEMMINGEN AND "GEONET"

The project GIS-Bestemmingen is a long-term project. It started before the implementation of Geonet and was based upon technology available at that time, which was ESRI's ArcViewIMS® software, while Geonet is based upon ESRI's ArcIMS software which was, at the beginning of the implementation of this system, brand new. It seemed logical for the GIS-Bestemmingen project to switch to the newer technology, but at that moment the proceedings of this project were at such a state that that was not feasible. One reason was the difference in presenting the information in the systems and the way the search capabilities were implemented in both systems because of the different purposes they serve. However, from the beginning of the implementation of Geonet, the decision was made to use as much common information as possible (i.e., topographical backgrounds and address coordinates). Thus, from a data point of view, some integration was already made. Besides that, a link between Geonet and GIS-Bestemmingen was implemented in Geonet, allowing users to switch from Geonet to GIS-Bestemmingen. However, it soon became apparent that keeping two systems in the air that have so much in common, but are based on different technologies, was not desirable in the long term. Instead, there should be one overall system. Therefore, as part of the outcome of a report on the use of spatial information in the entire organization, one of the assignments on the task list in that report is to set up a transition path for GIS-Bestemmingen toward Geonet. Although it means that some firm adaptations regarding the search methods in Geonet are necessary, eventually GIS-Bestemmingen will cease to exist as an independent system and will dissolve in Geonet. These adaptations are foreseen in the immediate future. The fact that in the case study ArcViewIMS technology is used instead of ArcIMS technology does not influence the outcome, since the concept is based upon the use of URL commands, which are supported by both technologies. In fact, any other web-based GIS which supports hyperlinks attached to objects in a map layer could have been used for this pilot.

29.3.2 INFORMATION IN GIS-BESTEMMINGEN

When talking about information that can be retrieved and viewed by GIS-Bestemmingen, a distinction must be made between information stored in map layers and geo-databases within the GIS component and information stored in documents within the DMS and DrMS. The reason is that searching using the geographical interface is done by searching in a map, and searching using the administrative interface is done by searching in the metadata of the DMS. To relate these two kinds of information, a relationship must exist. This can be established using a traditional database method. All that is necessary is defining a key field and storing a common value or identifier that exists at both sides. In the case of zoning and development plans, this identifier is called the plan number. Each plan has a unique number. This number is stored with the shape of the boundary of the plan in the map layer, and in the metadata of a related document in the DMS, and also in the metadata of related drawings — if present — in the DrMS. Using this plan number it is, for example, possible to launch from within the GIS a query at the DMS that sounds like "Give me a list of all the documents with area number equal to 928."

29.3.2.1 Information Stored at the Document Side

Information about all documents that are part of a zoning and development plan produced by or sent to the municipality will eventually be stored and handled at the document side. Different kinds of documents require different kinds of storage and handling. At this side the administrative search and retrieval (viewing) capabilities of the USEIS are established.

29.3.2.2 Type of Documents

The process of establishing a new zoning and development plan is very complex at this time. This is one of the reasons for the revision of the Spatial Planning Act. During this process a lot of documents are produced, which all have their own value at a certain stage in that process. But the most important documents are those at the end of the process, when the new zoning and development plan becomes effective as a legal plan. These documents will be reviewed hundreds of times during their life span, until they are followed up by a newer plan. This is the reason why the choice was made to concentrate during the pilot of GIS-Bestemmingen only on those documents that have a legal state. The documents that are part of a zoning and development plan at a legal state which are important are:

- At least one drawing that is a map showing the different zones. Sometimes there are more drawings, depending on the size of the area they cover or on the location of the area in the city. Plans that are located in the historic city center require additional drawings (e.g., showing the allowed directions of ridges of houses).
- A document with regulations that exactly describe what spatial activities in a specific zone are allowed.
- A document that is an elucidation to the regulations.
- A letter containing the official approval from a higher authority, which can be the province or in some cases the state.

29.3.2.3 Format of Documents

Zoning and development plans mostly have a long life span. The oldest plan that has a legal state at the city of 's-Hertogenbosch dates from 1935. It is obvious that this plan no longer fulfils the needs of modern spatial planners. It will soon be replaced by a new plan, but until that plan has reached a legal state, the old one is in place. The production of drawings of zoning and development plans has evolved from hand drawing to CAD drawing and soon will evolve to object-oriented creation in a GIS. The production of text documents has evolved from typewriting to being produced using a word processor on the computer. In general, a move from analogue production to digital production has taken place. Although digitally produced, the printed and plotted documents are used for approval of a new zoning and development plan by placing a stamp and a signature on them. At this moment only the analogue version has an official legal state. The revision of the Spatial Planning Act will provide means for the approval of digital documents. For viewing purposes, digital documents, either drawings or text documents, are preferred over scanned documents, since measuring in vector drawings is exact, unlike measuring in a scan, and since searching for text in a digital text document is far easier than in a scanned

TABLE 29.1
Possible Document Formats for Zoning and Development Plans

	Drawings	Regulations	Elucidation	Approval
Analogue (scanned)	X	X	X	X
Digitally produced	X	X	X	

document. Therefore, the decision was made to use the digital "masters" instead of a scan of the signed paper "copy," if available. This results in a mix of analogue and digital documents that need to be enabled for retrieving and viewing. In Table 29.1 an overview of the possible formats is given for each document type. Note that the approval is only available as a scanned text, since the approval is a printed letter sent by an external organization.

29.3.2.4 Storage of Documents

Although perhaps expected, not all documents are physically stored in the document management system. Drawings are stored somewhere else. This distinction between text documents and drawings is essential for the architecture of the USEIS. Digital drawings (vector drawings) are stored in the drawing management system, scanned drawings (raster drawings) are directly stored on a hard disk on a server, and a very small part of the paper drawings will never be reviewable by means of the USEIS, because they cannot be scanned. Examples of the latter are old paper drawings that are too large for today's scanning devices or which risk serious damage by a scanning device because of age. The reason for not storing digital drawings in the document management system is that such a system may be well suited for handling and viewing documents, but not for handling the engineering content that is included in digital drawings (*.dgn files). The drawing management system, which is from the same vendor as from the *.dgn files, is optimized for handling this kind of documents. The reason why the scanned drawings are not stored in the document management system is that that system does not provide an Internet-based viewer that is capable of publishing very large raster files at high speed. Besides that, one of the demands from the users is that measuring distances and areas must be one of the functionalities provided by the viewer. To meet the mentioned requirements, Bentley Publisher® software from Bentley Systems Inc. was chosen. The use of this software has additional advantages. It is also capable of publishing *.dgn files and printing them or parts of them at any desired scale on any available printer. It is also optimized for use in conjunction with the drawing management system.

29.3.2.5 Handling of Documents

For the sake of user friendliness, the starting point of the USEIS uses only the search engine of the document management system when searching for information using the administrative method. This includes the digital drawings stored in the drawing management system and the scanned drawings stored on a hard disk on a server.

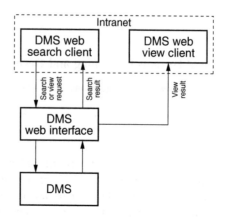

SCHEMA 29.2 Search and retrieval of text documents.

For text documents the handling is a straightforward process, since they are handled by the document management system itself. It provides Internet-based search and publishing capabilities "right out of the box." From the DMS web search client, the user defines a query, which is submitted to the DMS via the DMS's web interface. The results are displayed back in the web search client. If the user desires to view a document that is part of the query result, a request for displaying that document is passed via the web interface to the DMS, which displays the document in the DMS web view client. Schema 29.2 shows the above-described search and retrieval process for text documents.

Moreover, in some way the document management system must be aware of the existence of drawings in the drawing management system, otherwise no search on drawings is possible. Normally, metadata about a document and the document itself are stored in the document management system. In the case of drawings, metadata about the drawing are entered in the document management system as usual, but instead of physically storing the drawing into it, an html file is stored. In this html file a reference (in html language called "href") to a URL command is included. Also in the header of the html file a setting is set, which switches immediately after opening the html file to that reference. This is the essential part of the connection between the DMS, the DrMS, and the scanned drawings stored on a hard disk at a server. When the user decides he wants to view a drawing, the request is passed via the web interface of the DMS to the DMS. The DMS retrieves the html file and passes it to the drawing web view client. As soon as the html file is opened in this viewer, the URL command is immediately invoked, and the retrieval process of a drawing is started. Although retrieving scanned (raster) drawings and digital (vector) drawings uses the same principle as described above, the further handling of these documents differs, since the scans are stored on a hard disk, and the digital drawings are stored in the drawing management system.

The URL command for digital drawings contains code to invoke the web interface of the drawing management system, together with the unique' identifier of the drawing. Based upon the identifier, the drawing management system retrieves the drawing and passes it via the web interface of the drawing management system through a gateway to the publishing engine.

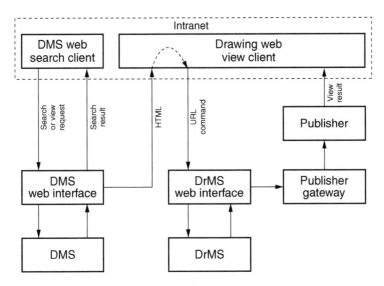

SCHEMA 29.3 Search and retrieval of digital vector drawings.

This engine "renders" the drawing in a format (in this case cgm) suitable for the drawing web view client and passes it to the client. The result is that the drawing is shown in the drawing web view client and can be reviewed and printed at any scale using the tools provided by it. In Schema 29.3 the process of searching for and retrieving of digital drawings is shown. Using this technique, the drawing management system acts as a slave of the document management system.

The URL command for scanned drawings contains code to invoke the Publisher engine, together with the name of the scanned drawing on the hard disk. This engine retrieves the drawing in I*.tiff format and passes it to the drawing web view client. The result is that the drawing is shown in the drawing web view client and can be reviewed and printed at any scale using the tools provided by the client. In Schema 29.4 the process of searching for and retrieving of scanned drawings is shown. Using this technique, the retrieval of scanned drawings can be regarded as acting like a slave of the document management system, since the user has only indirect access to the drawings.

29.3.2.6 Information Stored at the GIS Component

The most important map layer stored in the GIS component by now is the map with shapes of the boundaries (contours) of all current legally valid zoning and development plans. In some cases the granting of a building permit is also depending on additional legislation and regulations. As far as these rules have geographical reference, they are stored in the GIS component also and are made visible in a map layer that is part of GIS-Bestemmingen. Examples are archaeological protection zones, soil pollution zones, monumental protection zones, buffer zones of underground infrastructure like oil and gas pipes, zones with restrictions to garage exits, restriction zones for low fly zones of the air force, noise protection zones, stench protection zones around farms, and so on. The implementation of the USEIS in the pilot project

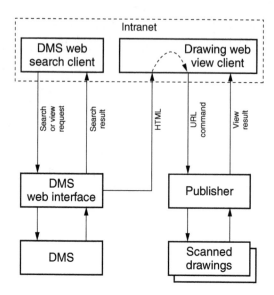

SCHEMA 29.4 Search and retrieval of scanned raster drawings.

GIS-Bestemmingen will focus on the information that is directly related to the map layer of zoning and development plans. However, the applied technique can be used for all other information stored in the GIS-component, which has relationships with documents. As described before, the relationship between the GIS side and document side is based upon a common attribute value called plan number. This value is used in the URL command to address the document side from within the GIS side. The GIS web map client is designed to show first the current legal zoning and developing plan after a spatial search. This search can be based upon an address, a zip code, cadastral parcel number or by using the Info button and subsequently clicking on a place in the map. The relating X,Y coordinate of the spatial search is passed via the GIS web map interface to the GIS. As a result, next to the map window, essential information about the current legal zoning and development plan at that location is shown in the GIS web map client. Together with that, a hyperlink is shown with the text "Plandocumenten bekijken" ("Show documents related to this plan"). In Figure 29.1 an image of the web map client is shown. If the user desires to follow that link, a URL command, which contains the plan number, is invoked to address the DMS web search client directly with the instruction to search for documents registered in the DMS with that same plan number. It therefore bypasses the interactive possibilities of the DMS web search client, which is presented when performing an administrative search on the DMS. The result of following the hyperlink is a list of documents shown in the DMS web search client that comply with the parsed plan number. From this point, the retrieval of documents related to the chosen zoning and development plan follows the same principles as described for the administrative search in the previous subparagraph. In Schema 29.5, the process of spatial searching for documents is shown. Note that there is only a connection from the GIS side to

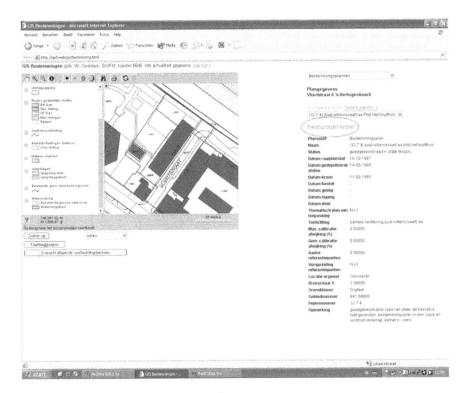

FIGURE 29.1 Example of the web map client.

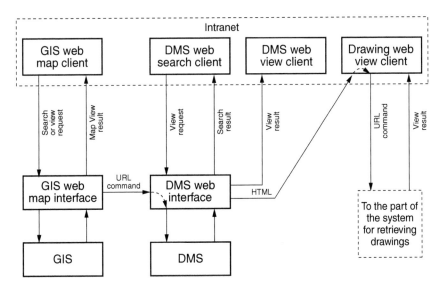

SCHEMA 29.5 Spatial search for documents.

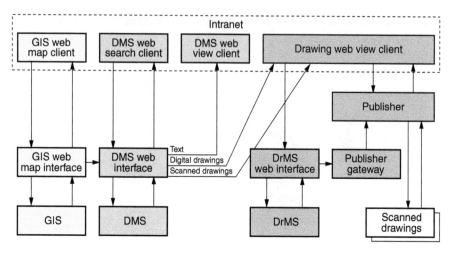

SCHEMA 29.6 Putting it all together: the total schema for search and retrieval.

the document side and not the other way around. This is intended, since the system at this stage is designed to provide a spatial and an administrative way to search for documents. The other way around, from the document side to the GIS side, is technically possible, but does not fall in the framework of the pilot project.

Schema 29.6 brings all the previous discussed handling of information together in one schema. It also shows the relationship between the intranet client environment and the three major components, namely the GIS, the DMS, and the DrMS. Note that components that belong to each other are grouped. A special role is designated for the Publisher and the Publisher gateway. This software, which is bundled, is specially designed for publishing drawings on the Internet or intranet, either vector-based or raster-based. It is therefore positioned somewhat between the major components and the intranet-based clients.

29.4   REQUIRED CUSTOMIZATION

One of the biggest advantages of this system is the small amount of customization that is needed to connect the major components and form an integrated system. Within the scope of this pilot project only two minor adaptations were necessary. The first was adapting the search template of zoning and development plans in the DMS. This template is, in fact, an ASP page. The code in this page needed to be adjusted in such a way that it was able to process parameters that are provided within the call (URL command) for this ASP page. This was a two-day job for the webmaster. The second customization was providing a solution for generating a proper html file that contains the URL command for addressing the drawing management system from within the document management system. The syntax of the code in this file is always the same, except for only two numeric values, which are different for each drawing that needs to be registered in the document management system.

These two values are together the unique identifier of a drawing in the drawing management system. The organization-wide agreement about registering and storing of documents in the document management system is that the end user decides when he or she undertakes this action. This liberty makes the need for an automated registering of drawings in the document management system superfluous. Instead, the solution was to develop a small executable that asks the end user for the two values, which are displayed in the interface of the drawing management application, for a name for the html file and the place where it should be stored. After this file creation, the end user can store the file in the document management system and subsequently fill in the metadata. Programming this routine was a half-day job. The fact that only these small adaptations were necessary proves the power and simplicity of the deployed technology. Besides that, it is also very cheap compared to the costs of the major systems and the Publisher.

29.5 USER RIGHTS IN THE USEIS

Two of the three major components of the USEIS, namely the document management system and the drawing management system are protected by a user login and password. This protection extends to the use of these systems using Internet technology. Every employee has a user account at the document management system. After logging in, the system knows which documents the employee has access to and also what type of access (read only, write, etc.). When performing a search in the document management system, the system returns only a list of documents to which the user has access. The same counts for the drawing management system, except that not all employees have an account at this system, only those who are directly involved in producing and approving drawings. However, the organization-wide agreement is that all employees must be able to review all approved drawings that are stored in the drawing management system. To comply with this agreement, a guest account has been created in the system, which has read-only access to all approved drawings. Using this account's user ID (guest) and password (guest) ensures admission to the drawing management system when at a certain moment this information is requested.

Another difference with the document management system is that at the drawing management system the type of access rights depends on the state of a drawing in its workflow. In contrast with the document management system, the drawing management system also uses workflows for drawings. The last state of all drawing is the state called "approved."

Working with workflows could lead to the situation that a search in the document management system results in showing a drawing in the returned list, but when the user decides to view the drawing and enters his user ID and password for the drawing management system, access to view the drawing is denied. In that case, the drawing is in a state for which the user has no access rights. Schema 29.7 shows the process of retrieving a drawing for viewing using a search in the document management system.

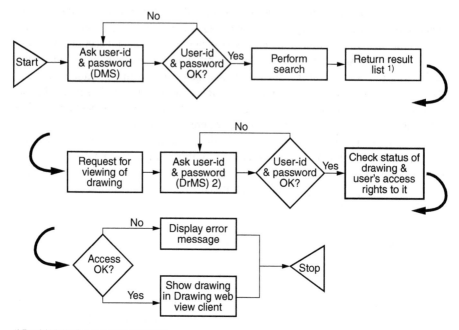

1) Result is depending on user's access rights
2) User account or guest account

SCHEMA 29.7 Handling digital rights when retrieving a drawing for viewing.

29.6 EXTENDING THE PILOT GIS-BESTEMMINGEN TO THE USEIS

The results of the pilot were positive and, as such, highly appreciated by the employees directly involved in consulting zoning plans. However, the aim of the USEIS is to involve all of the documents that have a geographical reference and to establish a bidirectional connection between documents and their geographical counterparts. The case study provides only a connection from the GIS to the DMS and not vice versa. In order to fully implement the USEIS the following issues need to be addressed:

- Instead of using ArcViewIMS, ArcIMS must be used to establish the connection with the DMS.
- The connection needs to work for any kind of document with a geographical relationship instead of only working for zoning plans.
- The connection from the DMS toward the GIS needs to be established also.
- The process of geocoding documents needs to be automated as much as possible.

The switch from ArcViewIMS to ArcIMS will not cause severe problems, since both applications are capable of dealing with URL commands. It is more a matter of rewriting programming code.

To address the issue of being able to make a connection to any kind of geo-graphical-related document and an object in a map layer, a general and uniform identifier needs to be established, since it is not feasible to program a "connection rule" for each possible connection. Without going into detail, this can be established by assigning an object ID to each object in a map layer and assigning a document ID to each document in the DMS and registering in a database table the relationship between both IDs, taking into account that these relationships could be one to one, one to many or many to one. After establishing this database table, which describes the relationship between documents and their geographical counterparts (objects), it is possible to make the connection from the DMS toward the GIS also.

Automating the process of geocoding documents is very important. The less the end users and the administrators are bothered with entering information, the greater is the chance that the correct connection information is stored in the system. The more rules regarding this information are known and are automated, the higher the success of the system will be. Describing and implementing these rules is a major point of attention when fully implementing the USEIS. All the above-described issues need to be addressed before the USEIS can be completely established.

29.7 REASONS FOR ESTABLISHING THE USEIS

In this section the factors are described why now, at the time of writing and not sooner nor later, the implementation of the USEIS, in this case by means of the pilot project GIS-Bestemmingen, is feasible. These factors are related to management support, organizational changes, the availability of the technical prerequisites, and external catalysts.

29.7.1 MANAGEMENT SUPPORT

From a management point of view the awareness of the importance of geographical information within the organization by the management has led to the assignment of a small project team to investigate and write a report about the role and value of spatial information for the entire organization for the next five years. Part of that report should be a list of activities and tasks that should be carried out to accomplish the goals defined in it. The added value of the spatial component of information in the fields of analyzing, middle- and long-term planning and also in the field of the daily practice of keeping the law and the fulfillment to local and national regulations by citizens and companies located in the city boundaries was made evident in this report. One of the most important outcomes of this report was a recommendation to make spatial information an integral part of the entire information flow. One of the activities in the list was to establish a connection between the document man-agement system and the GIS viewing system. Another one was to establish a con-nection between the drawing management system and the document management system. This report was approved by the general management in the spring of 2003. Therefore the implementation of the USEIS has the support from the management, since it establishes connections between the three systems mentioned in the report.

29.7.2 ORGANIZATIONAL CHANGES

A short while ago the organization has moved to a new building. All of the different departments that were located in different buildings scattered around the municipality are brought together in this new building. By itself, this is not the real challenge, but this new building is accompanied with a new way of working at an office, called flexible workplaces. The employees no longer have their own rooms, computers, or filing cabinets. Instead, they roam about the building and pick a workplace that suits their needs at a given moment, depending on the work they plan to do. Also, it is the aim of the management to reduce the size of the physical (paper) archives and replace them with digital archives. The goal is to reduce the use of paper documents as much as possible, since it is one of the prerequisites for working with flexible workplaces. An employee cannot carry his own filing cabinet around inside the building. These two factors have lead to the decision to introduce an organization-wide used document management system. One of the first steps of implementing the document management system is the definition of the metadata that should be stored with a document and, above all, what kind of documents are to be stored. As the first departments started with this implementation, it soon became apparent that the end users wanted to have access to drawings and GIS information that are related to documents stored in the document management system. Therefore, an important issue during the implementation of that system is defining what documents have a relationship with a geographical object, either real or virtual. Otherwise, no connection with the GIS viewing system can be made. This is also valid for the connection between the document management system and the drawing management system. It is obvious that these definitions should be made as early as possible, preferably before a department is going to implement the document management systems. Adding relational information afterwards is an almost impossible task. The need to define these relationships acts as a momentum. Since a large number of departments are in the process of implementing the document management system, now is the moment to ensure that necessary relational data are defined. With an example implementation in the framework of the project GIS-Bestemmingen, the need for this data can be demonstrated easily and acts as a motivator for the teams in charge of the implementation of the document management system for their departments.

29.7.3 TECHNICAL PREREQUISITES

The model of the USEIS was created by the author at the beginning of 2002. At that time, it was based upon the (Internet) technology provided by the providers of the three main software components. However, it was a pure theoretical model, since an Internet-based viewing application was operational at the organization only from the GIS-component. With the start of the implementation of the document management system in early 2003 at a few departments which acted as pilot projects, the Internet-based viewing and retrieving software of the document management system became available. Recently the Internet-based viewing and retrieving capabilities for drawings in the framework of the project GIS-Bestemmingen were installed. At the same time a small but very crucial ASP page was developed for the project mentioned, which searches for all the documents in the document management system that belong to a certain development or zoning plan. This resulted in the connection between the GIS, the drawing management system and the document management system. With these two latest developments, the implementation of the USEIS accelerated. From a technical point of view the proof was given that the theoretical concept works.

29.7.4 EXTERNAL CATALYSTS

There are two major external factors that act as catalysts for the implementation of the USEIS in the framework of the GIS-Bestemmingen project.

The Public Counter 2000, in Dutch Overheidsloket 2000, is an initiative of the Dutch government that stimulates the municipalities to arrange their information in such a way that a citizen who needs a certain service can obtain all the necessary information relating to that service at one counter, therefore making both providing and obtaining this service in a much more efficient and customer friendly manner. This initiative was taken because very often a citizen needed to go from one clerk to another, visiting different counters at often different buildings before he had the right information. To obtain the goal of this initiative, it is necessary that previously scattered information is related to each other and integrated. The objective of the USEIS is to fulfill that requirement.

Secondly, the upcoming revision on the Spatial Planning Act plays an important role as a catalyst in the implementation of the USEIS. It explains why the emphasis was put on the GIS-Bestemmingen project to act as a pilot project. The Spatial Planning Act is designed to balance the spatial needs of an ever-growing population (housing, commerce, and industry) and, at the same time, to maintain an attractive space to live in with respect for nature, wildlife, countrysides, recreation, tourism, and cultural heritage. However, this law was written in 1965, at a time when zoning and development plans, which are legally subject to and part of this law, were hand drawn at drawing boards and the belonging regulations were typed out on a typewriter. The speed of the changes in demands of spatial use, inherent to the current speed of life and the growing importance of the availability of and need for digital information, has led to the design of a new version of the Spatial Planning Act [2].

One of the outcomes of this law will be that in the near future instead of the analogue zoning and development plans the digital version will become the law, including digital approvals and signing. The move to the digital world requires that viewing of zoning and development plans using digital methods will become necessary and imperative. Making use of Internet technology to achieve this is the most logical approach, because in that way retrieval of this information is possible at any place at any time by anyone.

29.8 CONCLUSION

Providing an integrated system like the USEIS means a huge improvement for the organization. Employees will have almost all information directly available on any place at any time. This avoids the need for time-consuming digging in analogue archives to find the appropriate documents and also lowers the risk of omitting important relevant information about a certain issue. The spatial enabling of this information provides added value, since it reveals relationships between documents that cannot be revealed using the document management system only. The USEIS therefore helps to achieve the goals set in the national initiative of the Public Counter 2000. The USEIS combines the power of the three major components in one front office search and retrieval system, yet leaving each component intact as fully operational, independent systems used in the back office. The pursued approach in the pilot GIS-Bestemmingen also means that the USEIS is dependent on these major components. This implies the risk that if one of those systems fails, a part of the USEIS is also out of order.

The pilot GIS-Bestemmingen shows that, with relatively simple adjustments, the USEIS can be built. Basically, it consists of a set of fixed attributes stored at the GIS side and the document side, two intermediate database tables and a bunch of clever URL commands, which fully use the Internet possibilities provided by each component. From the user perspective, serious attention should be paid to finding a user-friendly way to store geographical information with address-based documents. Also, synchronization between user logins and passwords is a topic that contributes to the user-friendliness of the system. It is obvious that the implementation of the USEIS is a gradual process. With every department, where the document management system is being rolled out, a piece of the puzzle is added. Fortunately this allows for reflection and evaluation that can lead to improvement and perfection. Prerequisite for this is that at an early stage of the implementation process, awareness of a possible geographical reference of a document type is present and taken into account. Adding spatial references afterwards to documents always involves extra efforts in terms of manpower and money.

Although the USEIS is intended for use within the organization of the municipality, the fact that it is based upon Internet technology allows for the information stored in it to be accessed by the citizens of the city 's-Hertogenbosch also by publishing it or parts of it on the municipality's web site. This provides the citizens with an opportunity for more involvement in all the activities that are taking place in their residence, because they can access relevant information directly from the Internet at any time without the need for going to the office of the municipality. This potentially reduces the amount of requests posed at the information desk of the municipality and therefore could contribute to a more efficient civil service.

REFERENCES

1. Oostdam, W., Publishing Zoning/Development Plans and relevant relating Information on the Intranet, CORP, TU Wien, February 2001 http://213.47.127.15/corp/archiv/papers/2002/CORP2002_Oostdam.pdf.
2. Ministry of Spatial Planning, Revision of the Spatial Planning Act, Ministry of Spatial Planning of The Netherlands, 2002, http://www2.minvrom.nl/pagina.html?id=7352#.

Index

Index

Milton Keynes UK
Ingram Content Group UK Ltd.
UKHW021927071024
449327UK00022B/1717